T0205655

# GREEN CHEMISTRY

Fundamentals and Applications

# GREEN CHEMISTRY

## Fundamentals and Applications

*Edited by*

**Suresh C. Ameta and Rakshit Ameta**

Apple Academic Press Inc.
3333 Mistwell Crescent
Oakville, ON L6L 0A2
Canada

Apple Academic Press Inc.
9 Spinnaker Way
Waretown, NJ 08758
USA



First issued in paperback 2021

*Exclusive worldwide distribution by CRC Press, a member of Taylor & Francis Group*


ISBN 13: 978-1-77463-269-7 (pbk)
ISBN 13: 978-1-926895-43-7 (hbk)



**Library of Congress Control Number: 2013945482**

---

**Library and Archives Canada Cataloguing in Publication**

---

Green chemistry: fundamentals and applications/edited by Suresh C. Ameta and Rakshit Ameta.

Includes bibliographical references and index.
ISBN 978-1-926895-43-7
1. Environmental chemistry. 2. Environmental chemistry–Industrial applications.
I. Ameta, Suresh C., author, editor of compilation II. Ameta, Rakshit author, writer of introduction, editor of compilation

TP155.2.E58G74 2013          660          C2013-904995-9

---

Apple Academic Press also publishes its books in a variety of electronic formats. Some content that appears in print may not be available in electronic format. For information about Apple Academic Press products, visit our website at **www.appleacademicpress.com** and the CRC Press website at **www.crcpress.com**

# ABOUT THE EDITORS

**Suresh C. Ameta**

Prof. Suresh C. Ameta has served as Professor and Head of the Department of Chemistry at North Gujarat University Patan (1994) and M. L. Sukhadia University, Udaipur 2002-2005), and Head of the Department of Polymer Science (2005-2008). He also served as Dean, P.G. Studies for a period of four years (2004-2008). Now, he is serving as Director of Pacific College of Basic & Applied Sciences and Dean, Faculty of Science at PAHER University, Udaipur. Prof. Ameta has occupied the coveted position of President of the Indian Chemical Society, Kotkata, and is now lifelong Vice President (2002-continue).

He was awarded a number of prizes during his career such as the national prize twice for writing chemistry books in Hindi, the Prof. M. N. Desai Award, the Prof. W. U. Malik Award, the National Teacher Award, the Prof. G. V. Bakore Award, and the Life Time Achievement Award. The Indian Chemical Society has instituted a national award in his honor.

He has successfully guided 70 PhD students. Prof. Ameta has about 400 research publications to his credit in journals of national and international repute. He is the author of about 40 undergraduate- and post-graduate-level books. He has completed 5 major research projects from different funding agencies. Prof. Ameta has delivered lectures and chaired sessions in various international and national conferences. He is also reviewer of number of international journals. Prof. Ameta has an experience of over 43 years of teaching and research.

**Rakshit Ameta**

Dr. Rakshit Ameta has enjoyed a first-class career and was awarded a gold medal for standing at first position in the M. L. Sukhadia University, Udaipur, India. He was also awarded the Fateh Singh Award from the Maharana Mewar Foundation, Udaipur for his meritorious performance. He has served the M. L. Sukhadia University, Udaipur, and the University of Kota, Kota. Presently, he is serving at PAHER University, Udaipur, India, as an Associate Professor of Chemistry. Seven PhD students are working under his supervision on various aspects of green chemistry.

He has about 60 research publications in journals of national and international repute. He has a patent also to his credit. Dr. Rakshit has organized two national conferences as Organizing Secretary at University of Kota in 2011 and PAHER University in 2012. Dr. Rakshit was elected as council member of the Indian Chemical Society, Kolkata (2011-2013) and the Indian Council of Chemists, Agra (2012-2014). He has contributed to many books. He has not only written five degree-level books but has contributed chapters in books published by Nova Publishers, USA; Taylor & Francis, UK; and Trans-Tech Publications, Switzerland.

# CONTENTS

# LIST OF CONTRIBUTORS

**Aarti Ameta**
Department of Chemistry, Guru Nanak Girls' P.G. College, Udaipur, India, E-mail: aarti_ameta@yahoo.com

**Chetna Ameta**
Department of Chemistry, M. L. Sukhadia University, Udaipur, India, E-mail: Chetna.ameta@yahoo.com

**Garima Ameta**
Department of Chemistry, Guru Nanak Girls' P.G. College, Udaipur, India, E-mail: garima_ameta@yahoo.co.in

**K. L. Ameta**
Department of Chemistry, Modi Institute of Technology and Science, Lakshmangarh, India, E-mail: klameta77@hotmail.com

**Noopur Ameta**
Department of Chemistry, M. L. Sukhadia University, Udaipur, India, E-mail: noopurameta@yahoo.com

**Rajat Ameta**
Zyfine Cadila, Ahmedabad, India, E-mail: ameta_rajat@rediffmail.com

**Rakshit Ameta**
Department of Chemistry, Pacific College of Basic & Applied Sciences, PAHER University, Udaipur, India, E-mail: rakshit_ameta@yahoo.in

**Rameshwar Ameta**
Department of Chemistry, S. M. B. Govt. P.G. College, Nathdwara, India, E-mail: ameta_ra@yahoo.com

**Rohit Ameta**
HASETRI, JK Tyre, Kankroli, India, E-mail: rohit16ameta@yahoo.com

**Suresh C. Ameta**
Department of Chemistry, Pacific College of Basic & Applied Sciences, PAHER University, Udaipur, India, E-mail: ameta_sc@yahoo.com.

**Surbhi Benjamin**
Department of Chemistry, Pacific College of Basic & Applied Sciences, PAHER University, Udaipur, India, E-mail: surbhi.singh1@yahoo.com

**Indu Bhati**
Department of Chemistry, M. L. Sukhadia University, Udaipur, India, E-mail: indu_0311@yahoo.co.in

**Shipra Bhardwaj**
Department of Chemistry, Govt. P.G. College, Kota, India, E-mail: sidsidsmart@yahoo.co.in

**N. P. S. Chauhan**
Department of Polymer Science, M. L. Sukhadia University, Udaipur, India, E-mail: narendrapalsingh14@gmail.com

**Neelu Chouhan**
Department of Pure & Applied Chemistry, University of Kota, Kota, India, E-mail: niloochauhan@hotmail.com

**Abhilasha Jain**
Department of Chemistry, St. Xavier's College, Mumbai, India, E-mail: jainabhilasha5@gmail.com

**Nirmala Jangid**
Department of Chemistry, M. L. Sukhadia University, Udaipur, India, E-mail: nirmalajangid.111@gmail.com

**Yuvraj Jhala**
Department of Chemistry, B. N. P.G. College, Udaipur, India, E-mail: yuivrajjhala@gmail.com

**Sangeeta Kalal**
Department of Chemistry, M. L. Sukhadia University, Udaipur, India, E-mail: sangeeta.vardar@yahoo.in

**Anil Kumar**
Department of Chemistry, M. P. Govt. P.G. College, Chittorgarh, India, E-mail: anilchohadia@yahoo.co.in

**Neelam Kunwar**
Department of Chemistry, Pacific College of Basic & Applied Sciences, PAHER University Udaipur, India, E-mail: neelamkunwar13@yahoo.com

**Shikha Panchal**
Department of Chemistry, Pacific College of Basic & Applied Sciences, PAHER University, Udaipur, India, E-mail: shikha_dpr99@yahoo.co.in

**Arpit Pathak**
Department of Chemistry, M. L. Sukhadia University, Udaipur, India, E-mail: arpitpathak2009@gmail.com

**P. B. Punjabi**
Department of Chemistry, M. L. Sukhadia University, Udaipur, India, E-mail: pb_punjabi@yahoo.com

**Ajay Sharma**
Department of Chemistry, Govt. P.G. College, Sirohi, India, E-mail: Ajay395@gmail.com

**Bhoopendra K. Sharma**
Department of Chemistry, G. G. Govt. P.G. College, Banswara, India, E-mail: bhoopendrasharma@ymail.com

**Hari Shankar Sharma**
Department of Chemistry, Govt. P.G. College, Bundi, India

**Sanyogita Sharma**
Department of Chemistry, Pacific Institute of Technology, Udaipur, India, E-mail: sanyogitasharma22@gmail.com

**Shewta Sharma**
Department of Chemistry, M. L. Sukhadia University, Udaipur, India

**Vikas Sharma**
Jain Mandir Road, Kota Jn., Kota, India, E-mail: vikascy@yahoo.com

**Anuradha Soni**
Department of Chemistry, B. N. P.G. College, Udaipur, India, E-mail: dranuradhasoni@gmail.com

**Dipti Soni**
Department of Chemistry, Pacific College of Basic & Applied Sciences, PAHER University, Udaipur, India, E-mail: soni_mbm@rediffmail.com

**Paras Tak**
Department of Chemistry, Pacific College of Basic & Applied Sciences, PAHER University Udaipur, India, E-mail: parastak2011@gmail.com

**Jitendra Vardia**
Amoli Organics Pvt. Ltd., Vadodara, India, E-mail: jitendravardia@yahoo.com

**Ritu Vyas**
Department of Chemistry, Pacific Institute of Technology, Udaipur, India, E-mail: ritu24vyas@gmail.com

**Yasmin**
Department of Chemistry, Techno India NJR Engineering College, Udaipur, India, E-mail: ali_yasmin2002@rediffmail.com

# LIST OF ABBREVIATIONS

| | |
|---|---|
| 2,4-D | 2,4-dichlorophenoxyacetic acid |
| 2-MeTHF | 2-Methyl tetrahydrofuran |
| ABS | Acrylonitrile-butadiene-styrene |
| ACCN | 1,1′-Azobis(cyclohexanecarbonitrile) |
| AIBN | Azo-bis-isobutyronitrile |
| AL | Alkaline liquid |
| AMBI | 5-amino-6-methyl-2-benzimidazolone |
| AOPs | Advanced oxidation processes |
| Ap–CVD | Atmospheric chemical vapor decomposition |
| BDO | 1,4-Butanediol |
| bmimBr | 1-Butyl-3-methylimidazolium bromide |
| BVMOs | Baeyer–Villiger monooxygenases |
| Bzf | 3-Furfuryl-8-methoxy-3,4-dihydro-2*H*-1,3- benzoxazine |
| Bzs | 3-Octadecyl-8-methoxy-3,4-dihydro-2*H*-1,3-benzoxazine |
| C2C | Cradle-to-cradle |
| CaLB | Candida antarctica lipase B |
| CAN | Ceric ammonium nitrate |
| CCL | Candida cylindracea lipase |
| CHP | Combined heat and power |
| CNHs | Carbon nanohorns |
| CNTs | Carbon nanotubes |
| $CO_2$ | Carbon dioxide |
| CPC | Cetylpyridinium chloride |
| CPME | Cyclopentyl methyl ether |
| CRs | Carbon dioxide reactors |
| CVD | Chemical vapor deposition |
| DBS | Dodocylbenzene sulphonic acid, sodium salt |
| DCP | 2,4-dichlorophenol |
| DMF | N, N-dimethylformamide |
| DMPU | N, N-Dimethylpropyleneurea |

| | |
|---|---|
| DMS | Dimethyl sulphide |
| DPPH | 2,2-Diphenyl-1-picrylhydrazyl |
| Ee | Enantiomeric excess |
| emimCl | 1-Ethyl-3-methylimidazolium chloride |
| EPA | Environmental protection agency |
| EPS | Exopolysaccharide |
| FA | Fly ash |
| FAME | Fatty acid methyl esters |
| FBS | Fluorous biphasic system |
| $FeO_x$ | Ferrioxalate complex |
| F-SPE | Fluorous solid-phase extractions |
| Hb | Hemoglobin |
| HMF | 5-Hydroxymethylfurfural |
| HOMO | Highest occupied molecular orbital |
| IPA | Isopropanol |
| LAS | Linear alkyl sulfonate |
| LCA | Life cycle assessment |
| LDA | Lithium diisopropylamide |
| LOI | Limiting oxygen index |
| LTMP | 2,2,6,6-Tetramethylpiperidide |
| LUMO | Lowest unoccupied molecular orbital |
| MA | Maleic anhydride |
| MAOS | Microwave assisted organic synthesis |
| MCL | Medium-chain length |
| MM | Micro fibrillated materials |
| MPG | Monopropylene glycol |
| M-S-H | Magnesium silicate hydrates |
| MSNs | Mesoporous silica nanoparticles |
| MW | Microwave |
| NBTPT | 1-benzyl-2,4,6-triphenylpyridinium tetrafluoroborate |
| NCW | Near critical water |
| NGOs | Non government organizations |
| NTA | Nitrilotriacetate |
| NTO | 5-Nitro-1,2,4-triazol-3-one |
| nZVI | Nanoscale zero-valent iron |
| $P_a$ | Acoustic pressure |

| | |
|---|---|
| PAHs | Polycyclic aromatic hydrocarbons |
| PCBs | Polychlorinated biphenyls |
| PE | Polythene |
| PEG | Polyethylene glycol |
| PHA | Polyhydroxy alkanoates |
| PHAs | Polyhydroxyalkanoates |
| PHB | Poly-3-hydroxybutyrate |
| PHB | Polyhydroxy butyrate |
| PHE | Phenanthrene |
| PHH | Polyhydroxyhexanoate |
| PHV | Polyhydroxyvalerate |
| PICs | Products of incomplete combustion |
| PLA | Polylactic acid |
| p-NTS | p-nitrotoluene-ortho-sulphonic acid |
| PPA | Polyphosphoric acid |
| PPCP | Pharmaceuticals and personal care products |
| PPD | p-Phenylenediamine |
| PPG | Procaine penicillin-G |
| PS | Polystyrene |
| PTC | Phase transfer catalyst |
| PTT | Polytrimethylene terephthalate |
| PVC | Poly vinyl chloride |
| RB-5 | Reactive black-5 |
| RME | Reaction mass efficiency |
| ROL | Rhizopus oryzae lipase |
| SA | Sand aggregates |
| SAIER | Strongly acidic ion exchange resin |
| $ScCO_2$ | Supercritical carbon dioxide |
| SCFs | Supercritical fluids |
| $ScH_2O$ | Supercritical water |
| SCL | Short-chain length |
| SCMs | Supplementary cementitious materials |
| SCW | Supercritical water |
| SCWO | Supercritical water oxidation |
| SFC | Supercritical fluid chromatography |
| SLS | SODIUM lauryl sulphate |

| | |
|---|---|
| SOFC | Solid oxide fuel cells |
| SP | Super plasticizer |
| SPB | Sodium perbonate |
| SSF | Simultaneous saccharification and fermentation |
| SSRIs | Selective serotonin reuptake inhibitors |
| TBH | Tebuthiuron |
| TBPH | Tetrabutyl phosphonium hydroxide |
| TBTO | Tributyltin oxide |
| TCE | Trichloroethylene |
| THF | Tetrahydrofuran |
| TMAA | Tetramethyladipic acid |
| TMG | 1,1,3,3-Tetramethylguanidinium |
| TNT | Trinitrotoluene |
| TOA | Trioctylamine |
| TX-100 | Triton X-100 |
| UASB | Upflow anaerobic sludge blanket |
| VOCs | Volatile organic compounds |
| WWTP | Waste water treatment plants |
| Xan | Xanthan |

# PREFACE

The role of chemistry in the advancement of human civilization is very significant, but this achievement has come at the cost of human health and the global environment. Many chemicals find their way up to the food chain and get circulated in the ecosystem. This book is a step to promote a strategy towards sustainable development and with an aim to create a 'Greener World.'

The utilization of green chemistry principles in different industries can therefore be viewed as both an obligation and a significant opportunity to enhance our positive impact on the global community. Green chemistry can be defined as the invention, design and application of chemical products and processes to reduce or eliminate the generation of hazardous substances. Green chemistry is not a new branch of chemistry; rather it is a thought process on existing and new tools and knowledge of chemistry in a way that it continues to contribute to the society while protecting the environment also. Nowadays, green chemistry has been established as a pathway/route for a safer world, and its crucial role in the sustenance of the environment has been reviewed in this book.

This book addresses different topics under the domain of green chemistry, such as introduction, green reactants, green catalysts, green products, green solvents, different AOPs, and the use of green processes on the industrial scale. The book also deals with the current and future impact of green chemistry and its education in order to maximize atom economy.

The authors have aimed to enlighten readers regarding the green routes that are both environment friendly and beneficial on the industrial scale as well. The basic need is to put the theories into practice, and readers are requested to read, understand, and also give suggestions, if any.

**— Suresh C. Ameta and Rakshit Ameta**

# PREFACE

[illegible]

—Suresh C. Ameta and R[illegible]

CHAPTER 1

# INTRODUCTION

RAKSHIT AMETA

## CONTENTS

The entire world is in the cancerous grip of rapidly increasing environmental pollution in its various facets like water pollution, air pollution, soil pollution, and so on. This problem is further supported by global warming and energy crisis. Environmental pollution makes life miserable on this beautiful planet, "The Earth". This is all because of use of gray chemistry to fulfill different demands of materials with varied applications, such as metallurgy, synthesis of pharmaceuticals and other chemicals, use of volatile organic solvents, polymers, dry cleaning, agriculture, use of detergents, and so on, which creates different kind of pollutions. Men cannot survive without using many of these toxic chemicals to make their life more comfortable even at the cost of their health. Therefore, there is a pressing demand all over the globe to either reduce the use of more toxic materials or to replace them with less toxic or less harmful alternates.

This can be achieved by transforming from gray chemistry to green chemistry. The term green chemistry was used first by Paul T. Anastas in the beginning of last decade of 20th century. The field of green chemistry has been excellently presented by several workers (Tundo and Anastas, 2000; Anastas et al., 2000; Ameta, 2002; Lancaster, 2002; Ameta et al., 2004; Matlack, 2008; Ameta et al., 2012).The green chemistry is in no way different from gray chemistry except the approach toward a chemical process, may be manufacture, design, and applications. Green chemistry is totally different from environmental chemistry. In environmental chemistry, one takes care of the kind of pollution, extent of pollution, and methods to combat against this pollution whereas green chemistry takes care of all these factors in advance. It is something like diagnosis of any disease and its treatment is environmental chemistry while prevention from that disease is green chemistry. There is a well-known proverb that "Precaution is better than Cure". Prevention is a green chemical pathway while cure is an environmental pathway.

Apart from the existing forms of pollutions, we are going to face and even we are facing these emerging faces of pollution today also that is, polymer and detergent pollutions. Almost all materials like metal, wood, textile, and so on, are slowly being replaced by one or other kind of polymers, which has resulted into accumulation of this polymeric material into the dumping yards. The disposal of this dumped material or its recycling is a burning problem of the day.

Soap is biodegradable but as it does not work in hard water, therefore, it is rapidly being substituted by detergent. We have almost forgotten the use of washing soap leaving aside bathing soap. These detergents are not biodegradable and, therefore, remain for years together in nearby water resources; thus, making this water unfit for its use as portable water. It is utmost necessary to find out either some substitute for these detergents or to develop methods for degrading the accumulated detergents in water.

Pharmaceutical industries are facing a problem, which is like a double headed arrow. The drugs should be toxic to bacteria or fungi but it should not be harmful to human beings, animals, plants, and so on. If an effort is being made to increase the efficacy of a drug, insecticides, weedicides, and so on, it may also increase its toxicity, which is undesirable. To keep the toxicity low and increasing the efficiency is a challenging task for a chemist because one is working to achieve two totally opposing objectives. However, green chemistry may provide some feasible solutions to this problem.

According to Anastas, green chemistry utilizes a set of 12 principles that either reduces or eliminates the use or generation of any hazardous substance in designing, manufacture and application of chemicals. This is as approach, which is based on reducing the amount of waste generated at source rather than treating this waste after it has been formed. We as the chemists are normally blamed for creating pollution, but the green chemical approach not only solve the problem of pollution but it will also provide the methods to synthesize or utilize substances in an eco-friendly manner. Green chemical approach is holistic in nature and encompasses almost all the major branches of chemical science, such as organic or inorganic synthesis, catalysis, drug discovery, material science, polymer, nanochemistry, supramolecular chemistry, treatment of waste water, and so on.

Green chemistry is also known synonymously as:

1. Clean Chemistry
2. Atom Economy
3. Benign by Design Chemistry
4. Eco-Friendly Chemistry
5. Environmentally Benign Chemistry
6. Sustainable Chemistry and also
7. E-Chemistry

To achieve green chemical pathway at laboratory as well as industrial level still exists as a challenge for chemists. Collaborative efforts are urgently needed from government, industries, academics, and non government organizations (NGOs) to face this challenge.

In a presidential address to Indian Chemical Society, Ameta (2002) has very rightly given the slogan:

Green Chemistry: Green Earth

Clean Chemistry: Clean Earth

There may be a confusion that green chemical pathway is almost benign, but it is not a perfectly true statement because there cannot be any chemical, which is perfectly benign and therefore, green chemistry diverts use of chemicals from malign to benign manner. Common salt is necessary for life, but it may develop hypertension, if taken in excess; so it is the case with carbohydrates (sugar), which are required for providing energy for daily routine life but if given in excess, it may be harmful to humans. Therefore, shifting from less benign (more malign) to more benign (less malign) process may be considered as a green chemical approach. Someone has well said that "A matter may act as poison if given in large amount, and a poison if given in very small amount may act as a nectar". This is the basic concept of homoeopathy, which deals with very small concentrations of toxic chemicals and surprisingly enough, it may cure many dreadful diseases. The efficiency of these homoeopathic medicines increases on dilutions.

The green chemical approach is governed by 12 principles given by Anastas (Anastas and Warner, 1998).These principles are important in combating against environmental pollution and for the betterment of human health. These principles are:

(i) ***Prevention***: It is better to prevent waste rather than treating or cleaning up waste after it is produced.

(ii) ***Atom economy***: Syntheses should be so designed wherever possible, to maximize the incorporation of all materials used in the process into their final products.

(iii) ***Less hazardous chemical syntheses***: Wherever practicable, synthetic methods should be designed so as to use and generate substances possessing little or no toxicity to human health and the environment.

(iv) ***Designing safer chemicals:*** Chemical products should be designed to affect their desired function while minimizing their toxicity.

(v) ***Safer solvents and auxiliaries***: The use of auxiliary substances (e.g., solvents, separating agents, etc.) should be made unnecessary, wherever possible and innocuous, when used.

(vi) ***Design for energy efficiency***: Energy requirements of chemical processes should be recognized for their environmental and economic impacts and should be minimized. If possible, synthetic methods should be conducted at ambient temperature and pressure.

(vii) ***Use of renewable feed stocks***: A raw material or feedstock should be renewable rather than depleting, whenever technically and economically practicable or feasible.

(viii) ***Unnecessary derivatization***: Blocking group, protection or deprotection, and temporary modification of physical or chemical processes should be avoided whenever possible.

(ix) ***Catalysis***: Catalytic reagents (as selective as possible) are superior to stoichiometric reagents.

(x) ***Design for degradation***: Chemical products should be designed so that at the end of their function, they break down into innocuous or harmless degradation products and do not persist in the environment.

(xi) ***Real-time analysis for pollution prevention***: Analytical methodologies need to be further developed to allow for real-time, in-process monitoring and control prior to the formation of hazardous substances.

(x) ***Inherently safer chemistry for accident prevention***: Substances and the form of a substance used in chemical process should be chosen to minimize the potential for chemical accidents, including releases of chemicals, explosions and fires.

Atom economy is an important concept in philosophy of green chemistry (Trost, 1995; Sheldon, 2000). It is important to utilize maximum number of atoms of the reactant to minimize the generation of waste products. It is defined as atom economy in green chemistry, meaning by one has to be economic in use of atoms. The atom economy is defined as:

$$\%\,\text{Atom Economy} = \frac{\text{Molecular weight of desired product}}{\text{Molecular weight of all reactants}} \times 100\%$$

Addition reactions and rearrangements are normally follow atom economy but the chemistry is not complete with only these reactions, hence, some substitution and elimination reactions are also required; thus, generation of wastes is bound to be there but the efforts of the chemists should be to produce minimum byproducts.

"Gray" process can be made "Green" by making a judicious selection of green substrate, green solvents, green reagents, green catalysts, green conditions, and so on, to synthesize a green product.

Principles of green chemistry are beautifully condensed by Tang et al. (2008) as PRODUCTIVELY:

**Principles of Green Chemistry**:

**P** – Prevent wastes

**R** – Renewable materials

**O** – Omit derivatization steps

**D** – Degradable chemical products

**C** – Catalytic reagents

**T** – Temperature and pressure ambient

**I** – In-process monitoring

**V** – Very few auxiliary substances

**E** – E-factor, maximize feed in product

**L** – Low toxicity of chemical products

**Y** – Yes, it's safe

Efforts are being made to fulfill all the 12 conditions to make a chemical process perfectly green, but it is not always practicable to satisfy all the requirement of 12 principles of green chemistry. Therefore, a chemical process is better defined as greener than the other chemical processes, which fulfills more conditions and further researches may make it still greener and this process will go on.

## KEYWORDS

- **Environment**
- **Eco-friendly chemistry**
- **Green chemistry**
- **Pollution**

## REFERENCES


1. Ameta, S. C. (2002). *Journal of the Indian Chemical Society, 79,* 305–307.
2. Ameta, S. C., Mehta, S., Sancheti, A., & Vardia, J. (2004). *Journal of the Indian Chemical Society, 81,* 1127–1140.
3. Ameta, R., Ameta, C., Tak, P., Benjamin, S., Ameta, R., & Ameta, S. C. (2012). *Journal of the India n Chemical Society, 89,* 992–1018.
4. Anastas, P. T., & Warner, J. C. (1998). *Green chemistry, theory and practice*. New York: Oxford University Press.
5. Anastas, P.T., Heine, L. G., & Williamson, T. C. (2000). *Green chemical synthesis and processes*. Washington, DC: American Chemical Society.
6. Lancaster, M. (2002). *Green chemistry: An introductory text*. London: Royal Society of Chemistry.
7. Matlack, A. (2010). *Introduction to green chemistry* (2nd ed.) London: CRC Press.
8. Sheldon, R. A. (2000). *Pure and Applied Chemistry, 72,* 1233–1246.
9. Tang, S. Y., Bourne, R. A., Smith, R. L., & Poliakoff M. (2008). *Green Chemistry, 10,* 268–269.
10. Trost, B. M. (1995). *Angewandte Chemie International Edition (English), 34, 259–281.*
11. Tundo, P., & Anastas, P. T. (2000). *Green chemistry: Challenging perspectives.* New York: Oxford University Press.

CHAPTER 2

# BENIGN STARTING MATERIALS

SANYOGITA SHARMA, NEELAM KUNWAR, SANGEETA KALAL, and P. B. PUNJABI

## CONTENTS

## 2.1 INTRODUCTION

Our environment is composed of atmosphere, earth, water, and space. Under normal circumstances, it remains clean and therefore, enjoyable. However, with increasing world population and with limited natural resources, the composition and complex nature of our environment has changed. Our world is beautiful, but the increasing use and improper disposal of the effluents from various industries is creating pollution of the environment.

The time is approaching for natural gas and petroleum production to peak, plateau, and then decline. Prices are also increased substantially in the last few decades contributing to the almost uncertainty. These trends and the uncertain future inevitably influence industrial nations. As our fossil raw materials are irreversibly decreasing supported by pollution pressure on our environment, the progressive changeover of chemical industry to renewable feedstocks for their raw materials has become an inevitable necessity (Okkerse and Bekkum, 1999).

It will have to proceed increasingly to the renewable raw material basis before natural gas, oil, and all other sources are exhausted. The over reliance of chemical industry on fossil raw material has its limits as these are depleting at a rapid pace and are not renewable. Now there is a question, when will fossil fuels be exhausted? Or when will fossil raw materials become so expensive that biofeedstocks become economically competitive alternative? Experts opine the end of cheap oil for 2040 at the latest (Umbach, 1996; Klass, 1998). This is a development that one can witness by the fact that chemical industries are now combating the increasing costs of natural oil and gas (Campbell and Laherrere, 1998).

The future is quite promising. Scientists and technologists following the trends on sustainability and natural resources are persuading industries either to use alternative resources or to develop some new approaches toward more efficient chemical processes. Thus, there is a pressing demand for the transition to a more bio-based production system, but this is hampered by different obstacles. Fossil based raw materials are relatively more economic at present and the process technology for their conversion into organic chemicals is well developed. It is basically different form that is required for transforming carbohydrates into products with industrial applications.

## 2.2 MATERIAL AND METHODS

### 2.2.1 SUSTAINABILITY

The most commonly used definition for sustainable development comes from a report by the World Commission on Environment and Development that is, "To meet the needs of the present without compromising the ability of future generations to meet their own needs." Green chemistry and green engineering are striving hard to develop new methodologies for sustainable development. Their proposals focus on:

(i) **Renewable feedstocks and raw materials:** (Dewulf and Lagenhove, 2006; Benaglia, 2009) Green chemistry needs to change starting materials into renewable feedstocks. The most desired property of basic starting material is its lower toxicity and environmental impact. Health and safety protection of workers involved and the environment is on top priority. Green chemistry is just proposing change of direction from fossil-based feedstocks into biological raw materials. There are many problems in using these materials, but in the last few years, there are some encouraging new results for large scale production and use of alternative renewable materials. The terrestrial biomass is quite complex containing high molecular weight products like sugars, hydroxy and amino acids, lipids and biopolymers (cellulose, hemicellulose, chitin, starch, and lignin), and proteins. The most important class of biomolecules produced is carbohydrates (~75% of the annually renewable biomass approximately 200 billion tons). A minor fraction (4%) of this is used by man, while the rest decays or recycles along natural pathways. The bulk of this annually renewable feedstocks (carbohydrate biomass) is polysaccharides, and their non food utilization is limited to textile, paper, and coating industries.

(ii) Oleochemistry: Oleochemicals like fats and oils (from plants and animals) as raw materials is becoming a new source of chemical feedstocks (Hill, 2000; Gutsche et al., 2008). A series of new raw materials exist in the market with variety of applications like cosmetics, polymers, lubricating oils, and so on.

(iii) Photochemistry: Green chemistry also puts a lot of emphasis on some photochemical reactions in chemical processes (Albini and Fagnoni, 2004; Ravelli et al., 2009). Light (in ultraviolet and visible region) can catalyze many reactions. Photochemistry has a great potential and quite a few interesting research findings, which were introduced in the last few years including some applications Ultraviolet as well as visible light from Sun is considered as renewable energy sources and thus, photochemistry can contribute to some of the green synthetic chemistry applications.

(iv) Biocatalysis and biotransformations: Biocatalysis is particularly a green technology with many applications, which are considered benign for the environment and energy efficient (Ran et al, 2008; Whittall and Sutton, 2009; Cheng and Gross, 2011; Tao and Kazlauskas, 2011). Enzymes have been used for many synthetic chemical routes with great advantage in the food and pharmaceutical industries. Biocatalysis is in the interface of fermentation techniques (food and alcoholic drink industries) with some other industrial processes, where enzymes are used for higher yields and low energy consumption. Biotransformations can be achieved through biocatalysis and these are considered good green techniques for a series of chemical industries and a variety of chemical products. An ecofriendly and economic biocatalytic route with lipase extracted from *Aspergillus niger* as an efficient biocatalyst has been suggested for the glycerol carbonate synthesis (Tudorache et al., 2012).

(v) **Capture or sequestration of carbon dioxide:** Green chemistry is involved in carbon dioxide reduction in chemical industries. Climate change and phenomenon of greenhouse effects due to $CO_2$ emission is considered a very important environmental problem by green chemists. Any effort to reduce $CO_2$ emissions during the industrial process is an important goal from green chemistry point of view. Also, any design in chemical processes, which sequester or capture or can use $CO_2$ is worthy from the aims of green chemistry (Holtz, 2003; Hester, 2009; Allen and Brent, 2010; Leimkuhler, 2010).

(vi) **Waste biomass as chemical feedstock, biomaterials and biofuel:** The advances in last decade were to use biomass for the production of various materials and these were quite impressive. It was known for decades that biomass from agricultural processes goes almost as waste. Biomass is considered a solution of very important problem of sustainability with increasing fossil fuel prices. In recent years, many new technologies showed the use of biomass as biofuel, raw material for the production of biomaterials, polymers and various other applications (Ravindranath and Hall, 1995; Ragauskas et al., 2006; Soetaert and Vendamme, 2009). Selective hydrogenation of alternative oils is a useful tool for the production of biofuels. Highly selective hydrogenation of non food oils like flax over non toxic heterogeneous catalyst can be used to make them suitable for biodiesel formulation (Zaccheria et al., 2009). A renewable gasoline was prepared directly from aqueous phase hydrodeoxygenation of aqueous sugar solution in a two-bed reactor (Li et al., 2011). Catalytic upgrading of bio-oil using 1-octene and 1-butanol over sulfonic acid catalysts was done by Zhang et al. (2011). It is an atom economic route for upgrading bio-oil to oxygenated fuels by simultaneous acid catalyzed reactions with olefins and alcohols. A route to liquid hydrocarbon fuels has been suggested by Case et al., (2012) by pyrolyzing mixtures of levulinic acid and formic acid salts. This one step process is operated at atmospheric pressure without catalyst or hydrogen addition.

(vii) **Biodegradation of biomass to biogas and biodiesel:** Biomass is well known for its use for biofuel, especially from organic wastes in landfills. Biomass can be used for the production of biodiesel through some chemical and physical processes. Akbar et al. (2009) prepared Na doped $SiO_2$ solid catalyst by sol-gel method for the production of biodiesel from jatropha oil. The biodiesel was produced by transesterification of jatropha oil over solid catalyst, $Na/SiO_2$, form fatty acid methyl ester with very high yield under mild conditions of operation. Biodiesel was also produced by palladium catalyzed decarboxylation of higher aliphatic esters (Han et al., 2010). It is an effective and highly selective decarboxylation approach to convert higher aliphatic esters into diesel like paraffins. The methodology of this process provides a new protocol to utilization of biomass-based resources, especially to the second generation biodiesel production. Biomass offered an opportunity for the production of 19% of energy on a global scale. Now, it is estimated that 4% of all fuel products in cars is produced from biomass. Dicyclohexylguanidine group covalently attached on silica gel is an efficient basic heterogeneous catalyst for the production of biodiesel in a continuous flow reactor (Balbino et al., 2011). Crude glycerol obtained from biodiesel waste was found suitable for the production of intracellular non-reducing sugar trehalose and relatively pure propionic acid simultaneously (Ruhal et al., 2011).

### 2.2.2 SUSTAINABLE MATERIALS

Materials produced and used in modern society are quite diverse and evolving. There are approximately 75,000 chemicals used commercially. As not a single formulation has a unique sustainability, it is useful to provide an operational definition. A sustainable material is that, which fits within the constraints of a sustainable material system. In order to be sustainable, a material must be appropriate for the system and vice versa.

Two strategies have been identified to support a sustainable materials economy. (i) Dematerialization, which involves developing ways to use less material to provide the same service in order to satisfy human needs and (ii) Detoxification of materials used in products and industrial processes.

Chemistry plays a pivotal role in production of food, materials supply for clothing and shelter, preventing disease, and providing health care products. Organic chemicals are some of the important starting materials for a large number of major chemical industries. The production of organic chemicals as raw materials or reagents for other applications is a major sector of manufacturing polymers, pharmaceuticals, pesticides, paints, artificial fibers, food additives, and so on. Organic synthesis on a large scale as compared to the laboratory scale, involves the use of energy and basic chemical ingredients from the petrochemical sector, catalysts and after the end of the reaction, separation, purification, storage, packaging, distribution, and so on. During these processes, there are many problems of health and safety for workers in addition to the environmental problems caused by their use and disposal as waste.

For a long time, the most important goal of a chemist was to prepare a compound in suitable amounts and desirable high purity from available starting materials. In a world with a continuously increasing population and limited resources, the idea of a sustainable development is of major importance for the future in the 21st century.

In the last two decades, much more attention has been paid to the effect of chemical production on the environment. It is clear that it is much better, less difficult and less expensive to develop processes and compounds that are sustainable from scratch than to change an existing chemical process or to remove a toxic chemical from the environment to reduce its potential hazard and pollution created by it. In order to do so, chemists, biochemists, engineers, and pharmacists working together in drug development or constructing new materials must think always about sustainability when they transform their ideas into any products and processes.

They must learn to judge the suitability of a chemical transformation or the use of a chemical compound within a limit of different parameters. It is not only the yield of the reaction, which counts but also which starting materials are required or used? Whether one can make these from renewable resources? Do these generate toxic by-products and how these by-products can be avoided? How much waste material is generated by this process and whether it is an energy efficient process? Asking such questions at the beginning of any chemical compound, process and technology development will lead to a proper, more efficient, and sustainable use of chemistry.

Chemistry is the science of material and its transformation, which plays a key role in the process and acts as the bridge between physics, material sciences, and life sciences. Only those chemical processes, which have reached (after careful optimization) maximum efficiency, will lead to more sustainable compounds and process. The awareness, creativity, and progressive attitude of a scientist is necessary to bring these reactions and chemical processes to maximum efficiency. The term "Green Chemistry" has been coined for all such efforts achieving this goal. Therefore, attempts have been made by them to design synthesis and manufacturing processes in such a way that the waste products are minimized so that they have no or negligible effect on the environment and their disposal is also convenient.

It is therefore necessary that the starting materials, solvent, and catalyst should be carefully chosen for carrying out reactions for example, use of benzene as solvent must be avoided at any cost since it is carcinogenic in nature. If possible, it is better to carry out reactions in the aqueous phase.

### 2.2.3 CHOICE OF STARTING MATERIALS

It is very important to make a proper selection of the appropriate starting materials. Till now, most of the organic synthesis makes use of petrochemicals and other hazardous or toxic chemicals, which affect the workers handling these starting materials. Petrochemicals are non renewable and these also require considerable amounts of energy and therefore, it is important to reduce the use of such petrochemicals by using alternative starting materials of agricultural or biological origin.

Feedstock selection largely dictates the reactions and conditions that will be employed in a chemical synthesis and it should come from renewable sources rather than depletable resources, as far as if possible ideal feedstock must be:

- Renewable
- Poses no hazards
- Converted to the desired product using few steps
- 100% yield
- 100% atom economy

#### *CHEMICAL SUBSTITUTES AND REPLACEMENTS*

Methylene chloride, benzene, and xylenes are among the top 20 starting materials produced in 1990 (Relsch, 1991). These are still used because economic losses are associated in phasing out these chemicals, but the chemical industries are introducing alternates at a rapid pace. N-Methyl-2-pyrrolidone has been commercialized as a promising replacement for methylene chloride.

Many a times, simple substitutes cannot replace these chemicals, which are unique precursors for the synthesis of secondary derivatives and materials. Formaldehyde and vinyl chloride have unique properties and tremendous industrial value, but these are also hazardous to humans. Such chemicals do not bioaccumulate, but have well known harmful threshold levels for humans. These are controlled by some regulations for their storage, transport, handling, and work place exposure. Therefore, some of the major initiatives have focused to replace them by utilizing some other renewable and environmentally benign starting materials, which are obtained from agricultural, animal and microbial resources.

## *RENEWABLE FEEDSTOCK FROM AGRICULTURE (BIOMASS)*

Some major benefits of using biomass are

- It provides renewable feedstock
- It does not contributes to net $CO_2$ to the atmosphere
- It conserves fossil fuel leading to a secure domestic supply
- It provides a platform for making use of chemical products, which is otherwise considered waste.

(i) **Chemicals from fermentation processes (Glucose fermentation):** Glucose can be obtained from various carbohydrates like starch, cellulose, sucrose, and lactose. On a large scale, glucose is produced from starch by enzymatic hydrolysis. Corn is the main source of glucose. Another important source of producing glucose is woody biomass. Improvement in processes for harvesting and processing wood cellulose could result in an alternate source of glucose, which is relatively much less expensive than corn.

Mesoporous silica nanoparticles (MSNs) with different pore size were synthesized and used as hosts to physically adsorb or chemically link cellulose for its conversion to glucose. The results show that chemically linked cellulose onto large pore MSNs exhibits a glucose yield of more than 80% with excellent stability (Chang et al. 2011).

(a) **Lactic acid:** Lactic acid (2-hydroxypropionic acid) can be produced either by chemical synthesis or by fermentation of different carbohydrates such as glucose (obtained from starch), maltose (produced by specific enzymatic starch conversion), sucrose (obtained from syrups, juices, and molasses), lactose (produced from whey), and so on (Ravindranath and Hall, 1995). Lactic acid is produced on industrial scale today mainly through the fermentation of glucose. An important step in the lactic acid production is the recovery from fermentation broth. The traditional process for the recovery of lactic acid is still far from ideal. Lactic acid exists in two optically active isomeric forms, L (+) and D (–). It is used in the food, chemical, pharmaceutical, and cosmetic industries. It is a bifunctional compound bearing a hydroxyl group and an acid group and is utilized for number of chemical conversions to useful products.

Nowadays, there is an increasing demand for biodegradable polymers that can replace conventional plastic materials and can also be used as new materials like controlled drug delivery devices or artificial prostheses. Thus, polylactic acid polymers could be an environment friendly substitute to such plastics derived from petrochemical materials.

It esterifies with itself to give two primary esterification products: the linear lactic acid lactate (2-lactyloxypropanoic acid) and cyclic lactide (3,6-dimethyl-1,4-dioxane-2,5-dione). This lactide is an important compound, because it is a monomer for the production of poly (lactic acid) or polylactide, and other copolymers. Effective conversion of D-glucose into lactic acid has been described using microwave irradiation in solventless condition with alumina potassium hydroxide.

Esters of lactic acid and different alcohols (particularly methanol, ethanol, and butanol) are non toxic and biodegradable. These are high boiling liquids and have excellent solvent properties and therefore, replace toxic and halogenated solvents for a large range of industrial uses. Lactate esters are also used as plasticizers in cellulose and vinyl resins and they enhance the detergent properties of ionic surfactants. Direct hydrogenation of lactic acid or lactates produces propylene glycol and it can be an alternative green route to the petroleum based process. Propylene glycol (1,2-propanediol) is a commodity chemical, which can be used as a solvent for the production of unsaturated polyester resins, drugs, cosmetics, and foods (Corma et al., 2007). Dehydration of lactic acid gives acrylic acid. This acid, its amide and ester derivatives are the primary building blocks in the manufacture of acrylate polymers, which find numerous applications in surface coatings, textiles, adhesives, paper treatment, leather, fibers, detergents, and so on.

Advances in fermentation and especially in separation technology have reduced the potential production cost of lactic acid. The production of lactic acid from waste sugarcane bagasse derived cellulose was reported by Mukund et al. (2007). It deals with the simultaneous saccharification and fermentation (SSF) of sugarcane bagasse cellulose to lactic acid using *Penicillium janthinellum* mutant EUI and cellobiose utilizing *Lactobacillus delbrueckii* mutant Uc3. Salt assisted organic acid catalyzed depolymerization of cellulose was carried out by Stein et al. (2010). Dicarboxylic acids combined with inorganic salts (NaCl or $CaCl_2$) afford the depolymerization of crystalline cellulose under mild conditions in water. The mechanical force and layered catalysts can efficiently depolymerize cellulosic materials (up to 84% conversion) as observerd by Hick et al. (2010). The most effective mechanocatalyst was aluminosilicates based on the kaolinite structure.

Catalytic upgrading of lactic acid to fuels and chemicals by dehydration and C–C coupling reactions was observed by Serrano-Ruiz and Dumesic (2009). They described a single reactor catalytic process to convert aqueous solution of lactic acid into a spontaneously separating organic phase that can serve as a source of valuable chemicals (propanoic acid and $C_4$–$C_7$ ketones) and can be used to produce high-energy density fuels. Simultaneous saccharification and fermentation (SSF) of cellulosic substrate to D-lactic acid using EUI cellulases and Lactobacillus lactic mutant RM2-24 was reported (Singhvi et al., 2010). The SSF was carried out in screw-cap flasks at 42°C with shaking at 150rpm.

(b) **Succinic acid:** Succinic acid is also produced in this way. Succinic acid reacts with alcohols in the presence of acid catalysts to form dialkyl succinates (Fumagalli, 1997). The esters of succinic acid with low molecular weight alcohols (methyl and ethyl succinates) find applications as solvents and synthetic intermediates for very many important compounds.

Direct hydrogenation of succinic acid, succinic anhydride, and succinates leads to the formation of the products like 1,4-butanediol (BDO), tetrahydrofuran (THF), and γ-butyrolactone (GBL) 1,4-Butanediol is a compound of quite common interest as a starting material for the production of some important polymers such as polyesters, polyurethanes, and polyethers (Weissermel, 2003). γ-Butyrolactone was synthesized highly selectively from biomass derived 1,4-butanediol by vapor-phase dehydrocyclization over novel copper-silica nanocomposite catalyst (Hwang et al., 2011). A wide variety of aldehydes and ketones can undergo the Stobbe condensation with succinic ester giving a variety of compounds. These products find applications in technical and medical field, biological activities (Moussa et al., 1982; Baghos et al., 1993).

(c) **3-Hydroxypropionic acid:** 1,3-Propanediol is a starting material for the production of polyesters. It is used together with terephthalic acid to produce polytrimethylene terephthalate (PTT), which is in turn used in the manufacture of some fibers and resins. 3-Hydroxypropionic acid on dehydration gives acrylic acid. Acrylic acid and its derivatives (esters, salts, or amides) are important compounds and used as monomers in the manufacture of some polymers and copolymers. These polymers have numerous applications such as surface coatings, absorbents, textiles, paper making, sealants, adhesives, and so on.

3-Hydroxypropanoic acid is oxidized to give malonic acid (propanedioic acid). This acid and its esters are utilized to yield a large number of condensation products. They are important intermediates in syntheses of vitamins $B_1$ and $B_6$, barbiturates, non-steroidal anti-inflammatory agents, other numerous pharmaceuticals, agrochemicals, and flavor and fragrance compounds.

(ii) **Chemical transformations of monosaccharides:** Thermal dehydration of pentoses and hexoses in acid media gives three important basic non petroleum chemicals. (i) Furfural (2-furancarboxaldehyde) obtained from dehydration of pentoses. (ii) 5-(Hydroxymethyl)furfural (HMF) from dehydration of hexoses, and (iii) Levulinic acid from hydration of HMF (Figure 2.1).

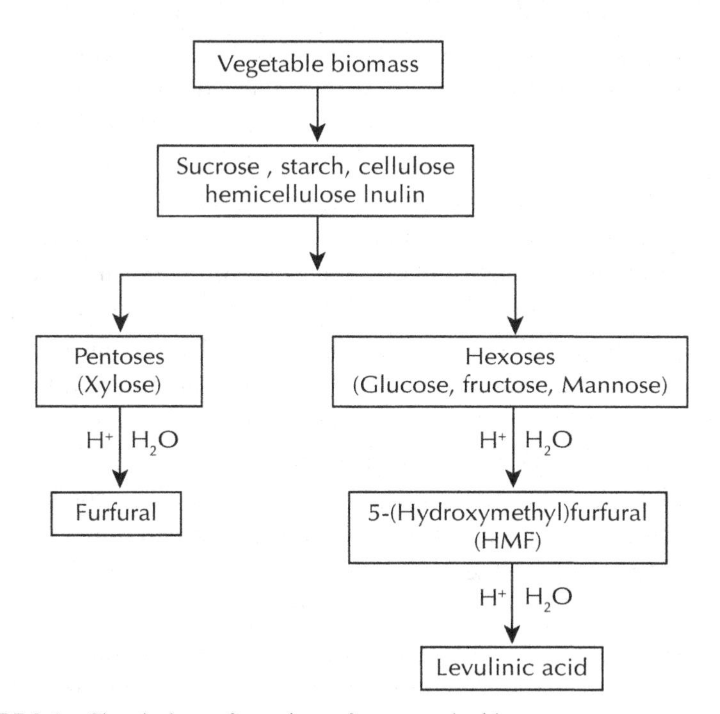


**FIGURE 2.1** Chemical transformations of monosaccharides.

5-(Hydroxymethyl) furfural possesses a high-potential industrial demand, and therefore, it has been judicious called a sleeping giant (Bicker et al., 2005) and one of the new petrochemicals readily accessible from renewable resources (Lichtenthaler et al, 1991). Some surfactants were synthesized by Gassama et al. using furfural derived 2[5H]-furanone and fatty amines (2010). Furfural obtained from biomass has been transformed into surfactants belonging to a new betaine family with two hydrophobic moieties.

5-(Hydroxymethyl) furfural and furfural were produced by dehydration of biomass derived mono and polysaccharides by Chheda et al. (2007). They present a biphasic system for acid catalyzed dehydration of various biomass derived carbohydrates to form furan derivatives, which have potential to be sustainable substitutes for petroleum-based building blocks used in production of fuels, polymers, and drugs. A simple procedure for the conversion of HMF to (5-alkyl and 5-aryl aminomethyl-furan-2-yl) methanol has been developed by Cukalovic and Stevens (2010). In this process, reactions were conducted without the use of a catalyst and under very mild conditions. Marcotullio and Jong (2010) used chloride ions to enhance furfural formation from D-xylose in dilute aqueous acidic solution. The simple addition of NaCl to an aqueous acidic solution significantly improved the yield and selectivity of furfural. Xylitol hydrogenolysis occurs efficiently on Ru/C with $Ca(OH)_2$ involving kinetically relevant dehydrogenation of xylitol to xylose and its subsequent retro-aldol reaction (Sun and Liu 2011). Wang et al. observed the direct conversion of glucose based carbohydrates into 5-hydroxymethylfurfural (HMF) via an easily prepared Sn-mont catalyst with high efficiency and good stability (2012a). The conversion of carbohydrates and lignocellulosic biomass into 5-hydroxymethylfurfural was observed by Yang et al. by using $AlCl_3.6H_2O$ catalyst in a biphasic solvent system (2012). Low cost and non toxic $AlCl_3.6H_2O$ in a biphasic medium of water/THF with NaCl additive converts raw biomass directly to high yields of furfural and modest yields of HMF (Figure 2.2).

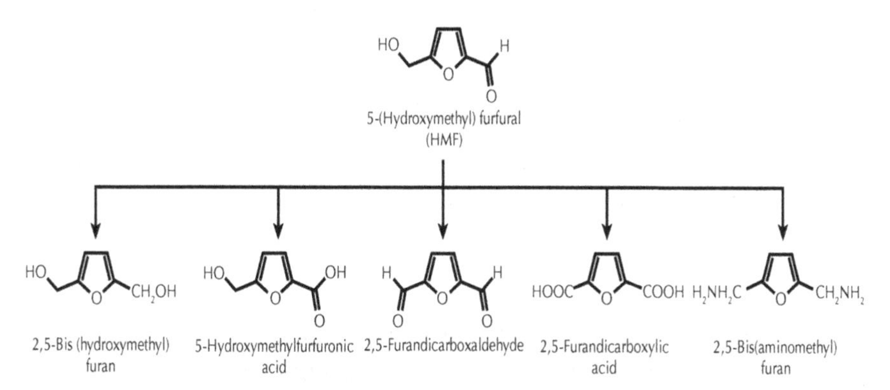


**FIGURE 2.2** Conversion of 5-(Hydroxymethyl) furfural.

These intermediates produced from HMF could replace some petrochemical-based monomers. 2,5-Furandicarboxylic acid may replace terephthalic, isophthalic, and adipic acids in the manufacture of polyamides, polyesters, and polyurethanes (Gandini and Belgacem, 1997; Moreau, et al.,2004). 2,5-Furancarboxaldehyde is a starting material for the preparation of Schiff bases, and 2,5-bis(aminomethyl)furan is able to replace hexamethylenediamine in the preparation of polyamides while 2,5-bis(hydroxymethyl)furan is used in the manufacture of polyurethane foams (Pentz, 1970) and the fully saturated 2,5-bis-(hydroxymethyl)tetrahydrofuran can be used like alkanediol in the preparation of polyesters.

(iii) **Chemical transformation of disaccharides**

(a) **Sucrose**

Sucrose is the main carbohydrate feedstock of low molecular weight chemicals. It is present in honey, sugar, fruits, berries, and vegetables. Sucrose can be functionalized and converted into different interesting additives. The hydrolysis of sucrose allows its conversion into inverted hexoses, that is, glucose and fructose. These are widely used in the food industry. Sucrose can also find applications in some polymers, such as polyurethanes. The use of sucrose for the preparation of phenolic or alkyd resins as well as polyesters, polycarbonates, and polyurethanes has been reviewed by Kollonitsch (1970).

## A. CONVENTIONAL ROUTE

1. **Formation of formic acid**

When methanol and carbon monoxide are treated in the presence of a strong base, methyl formate is obtained.

$$CH_3OH + CO \rightarrow HCO_2CH_3$$

This reaction is performed in the liquid phase at elevated pressure in industries. Typical reaction conditions are 80°C and 40 atm and base is sodium methoxide. Hydrolysis of the methyl formate produces formic acid.

$$HCO_2CH_3 + H_2O \rightarrow HCO_2H + CH_3OH$$

Efficient hydrolysis of methyl formate requires a large excess of water. Some routes proceed indirectly by first treating the methyl formate with ammonia to give formamide, which is then hydrolyzed with sulfuric acid.

$$HCO_2CH_3 + NH_3 \rightarrow H\overset{O}{\overset{\|}{C}}NH_2 + CH_3OH$$

$$2\ H\overset{O}{\overset{\|}{C}}\ NH_2 + 2\ H_2O + H_2SO_4 \rightarrow 2\ HCO_2H + (NH_4)_2SO_4$$

This approach suffers from the need to dispose off the ammonium sulfate, which is formed as a byproduct. This problem has led some manufacturers to develop some other energy efficient means to separate formic acid as the large excess amount of water is used in direct hydrolysis. In one of these processes, the formic acid is removed from the water via liquid–liquid extraction with an organic base. This method is now being replaced by a green chemical route.

2. **Green formation of formic acid**

A new method to transform carbohydrate-based biomass to formic acid has been reported by **Wölfel** et al., (2011). This process involves oxidation with molecular oxygen in aqueous solution using a Keggin-type $H_5PV_2Mo_{10}O_{40}$ polyoxometalate as a catalyst. Several water soluble carbohydrates were fully and selectively converted to formic acid and $CO_2$ under very mild conditions. The complex biomass mixtures, such as wood saw dust was transformed to formic acid, giving 19 wt % yield (11% based on the carbon atoms in the feedstock) under non optimized conditions.

Vegetable oils and animal fats can give two types of reactions. These are:

(i) Reactions of carboxyl group (Figure 2.3) and (ii) Reactions of fatty chains (Figure 2.4)

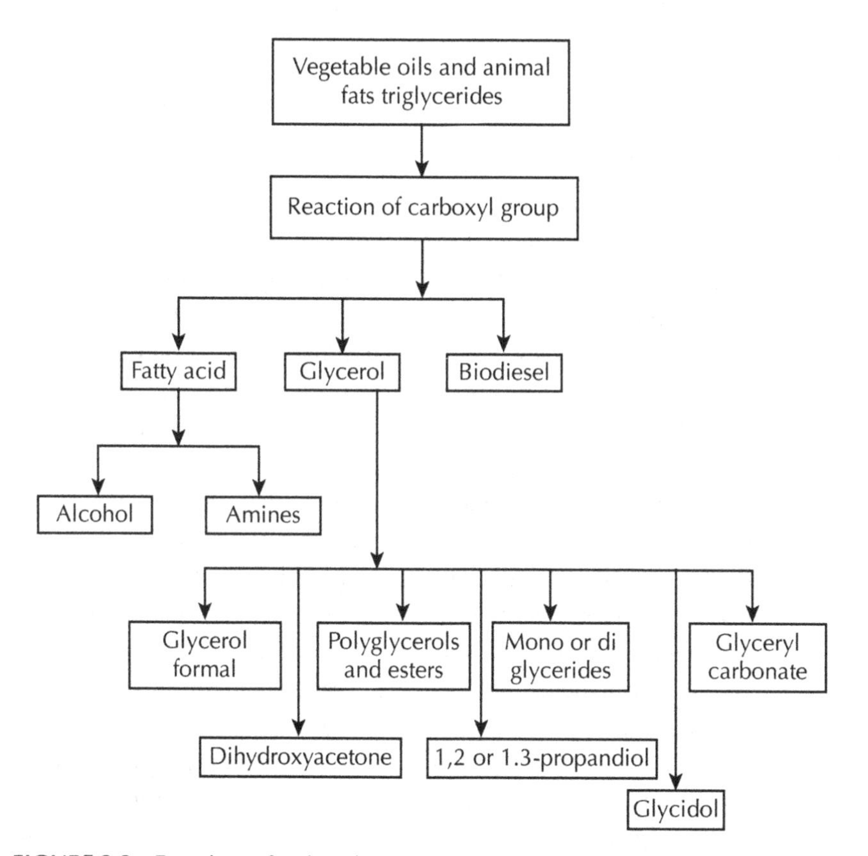


**FIGURE 2.3** Reactions of carboxyl group.

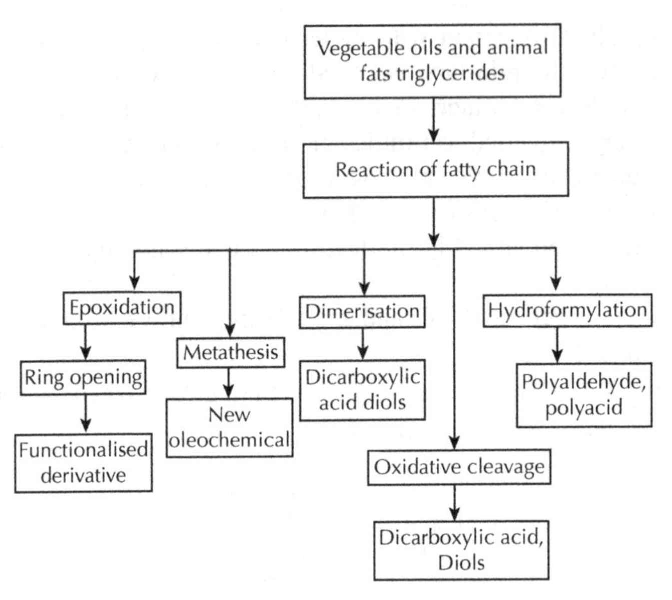


**FIGURE 2.4** Reactions of fatty chains.

Fats and oils are obtained from vegetable and animal fats, mainly formed by mixed triglycerides having fatty acid moieties. They are chemically not very different from some petroleum fractions in the sense that they contain a large paraffinic or olefinic chain. A large proportion of vegetable oils, such as coconut, palm, and palm kernel oils, come from countries with tropical climates. Soybean, rapeseed, and sunflower oils come from moderate climates. Animal fat is obtained from the meat industry, with beef tallow being the most abundant fat, and fish oil coming from the fishing industry. Biodiesel, lubricants, surfactants, surface coatings, polymers, pharmaceuticals, cosmetics, and so on, can be produced from these animal fats and vegetable oils.

Biodiesel is a diesel fuel made from renewable resources like oils derived from farm crops, such as soybeans or even from recycled vegetable oils which was left over from making fries at fast food restaurants. It is synthesized by removing glycerin from soybean or other vegetable oil. The by-product glycerin is also useful for making soap.

$$\text{Vegetable oil} + \text{Methanol} \xrightarrow{\text{KOH}} \text{Biodiesel} + \text{Glycerine}$$

Advantages

- It is a renewable source of energy, unlike fossil fuel derived petroleum diesel,
- Biodiesel on burning neither emits sulfur nor increases an overall amount of $CO_2$ in the atmosphere (the $CO_2$ released from biodiesel is balanced by the $CO_2$ taken up by plants).

Propylene glycol is synthesized by the utilization of waste glycerol from biodiesel production. This is a conversion involving dehydration of glycerol to acetol, followed by hydrogenation to yield propylene glycol.

$$\text{Glycerol} \xrightarrow[-H_2O]{} \text{Acetol} \xrightarrow[H_2]{} \text{Propylene glycol}$$

(iv) **Cellulose, hemicelluloses, and lignin**

(a) Preparation of cellulose from plant sources

(i) Separation from matrix of lignocellulose (hemicellulose and lignin) is required.

(ii) Harsh chemical processing is required.

(iii) Cellulose product may require bleaching.

b. Chemical modification of cellulose

(i) Cellulose may be modified, as it contains number of -OH groups, where other groups can be bonded to impart a variety of properties.

(ii) Rayons are made by treating cellulose with base and carbon disulfide and then extruding the product through fine holes to make thread.

(iii) Similar process is followed to make cellophane extruding it through a long narrow slot.

(c) Cellulose acetate

An ester is obtained, when most of the -OH groups present on cellulose are replaced by acetate groups by treating cellulose with acetic anhydride.

(d) Cellulose nitrate

Cellulose nitrate is obtained, when most of the -OH groups on cellulose are replaced by $^{-}ONO_2$ groups by treating cellulose with a mixture of nitric acid ($HNO_3$) and sulfuric acid ($H_2SO_4$).

i. It is used as an explosive.
ii. In the early days of moving pictures, transparent film was used for movie films, which resulted in some disastrous fires, giving highly toxic fumes of $NO_2$ gas.

(e) Synthesis of propargyl cellulose

A fast and simple reaction on exposure to microwave irradiation permits the synthesis of propargylcellulose in an aqueous alkaline medium and it is considered green. The influence of various reaction parameters such as amount of propargyl bromide, reaction time or microwave activation on the reaction efficiency has been reported (Faugeras et. al., 2012).

(f) Conversion of cellulose derived 5-(chloromethyl) furfural into δ-aminolevulinic acid

5-(Chloromethyl)furfural was prepared by the conversion of cellulose and it was further converted into δ-aminolevulinic acid in three simple chemical steps involving conversion to (i) 5-(azidomethyl)furfural, (ii) photooxidation, and (iii) catalytic hydrogenation in reasonably good yield 68%. δ-Aminolevulinic acid is a natural product with important agrochemical and pharmaceutical applications (Mascal and Dutta, 2011).

Biomass —Step 1, 80 – 90%→ Cl, O, CHO —Step 2, 68%→ $H_2N$, O, OH, O

ALA

(g) Various other feedstocks from cellulose wastes

Rumen bacteria act on cellulose wastes and then a treatment followed with lime in large fermenters; in absence of oxygen produce calcium acetate, calcium propionate, and calcium butyrate. These esters can be acidified to produce corresponding acids like acetic acid, propionic acid, and butyric acid. These carboxylic acids can be hydrogenated to give corresponding alcohols like ethanol, propanol, and butanol. Organic acids on treatment at 450°C produce ketones like acetone, methyl ethyl ketone, and diethyl ketone.

Sugar alcohol can be synthesized by hydrolytic hydrogenation of cellulose over supported metal catalysts (Kobayashi et al. 2011). Pt/carbon black BP2000 is an effective and durable catalyst for the conversion of cellulose to sugar alcohols, which promotes both the hydrolysis of glycosidic bonds and the hydrogenation of glucose (Kobayashi et al., 2011).

Lignin is a chemically complex biopolymer that is associated with cellulose in plants. It is difficult to use lignin because of its inconsistent and widely variable molecular structure. It was felt that catalytic transformation of readily available, widely distributed and renewable non food lignocelluloses to some value added chemicals is necessary. But it still remains as a great challenge as the bulk of these resources and/or the energy available from such biomass are in the form of lignocellulose, which are tight, covalent, hydrogen bond-linked matrix of carbohydrate polymers (cellulose and hemicellulose) and phenolic polymers (lignin). This results in its insolubility in common solvents.

It was first found that the ionic liquid 1-butyl-3-methylimidazolium chloride (bmimCl) was able to dissolve cellulose with 10 wt% of solubility at 100°C. It was reported that cellulose and original biomass can be efficiently hydrolyzed to reducing sugars in bmimCl with the help of a mineral acid (Li and Zhao 2007, Li et al. 2008). The catalytic conversion of lignocellulose and/or its constituents to chemicals such as 5-hydroxymethylfurfural (HMF) using ionic liquid as solvent has also been studied. It has been reported that cellulose and lignocelluloses such as corn stove could be also be converted to 5-hydroxymethylfurfural in 1-ethyl-3-methylimidazolium chloride (emimCl), while N,N-dimethylacetamide and LiCl were applied as co-solvents in the presence of Cr (II) salt and a mineral acid catalyst (Zakrzewska et al, 2011; Binder and Raines, 2009). Production of HMF and furfural from other lignocellulosic biomass such as corn stalk, rice straw and pine wood in bmimCl and 1-butyl-3-methylimidazolium bromide (bmimBr) solvents under microwave irradiation has also been studied (Zhang and Zhao, 2010).

Therefore, transformation of these lignocelluloses to useful chemicals has been recognized as an effective approach for improving upon the problem of increasing energy crisis and climatic change. An efficient catalytic transformation process for agricultural residual lignocelluloses in cooperative ionic liquid pairs was also achieved by Long et al. (2012). The promotion of the dissolution equilibrium, combined with fast, *in situ* acid catalyzed degradation of cellulose and hemicellulose, resulted in significantly greater conversion of the biomass to some important biochemicals and selective delignification. Acid treatment of carbohydrates was reported by Hu and Li (2011) in methanol rich medium, which stabilizes reactive intermediates and suppresses polymer formation and greatly promotes methyl levulinate production. A catalytic process was presented by Braden et al. (2011) for conversion of lignocellulosic biomass to liquid alkenes. Its techno economic analyses show that the economics is comparable to that of cellulosic ethanol.

(v) **Synthesis of benzoxazines:** Two completely bio-based benzoxazine monomers, 3-furfuryl-8-methoxy-3,4-dihydro-2*H*-1,3- benzoxazine (Bzf) and 3-octadecyl-8-methoxy-3,4-dihydro-2*H*-1,3-benzoxazine (Bzs), have been successfully synthesized by Wang et al. (2012b) *via* a solventless method. The results reveal that these monomers undergo homogeneous copolymerization when the Bzf–Bzs molar ratio exceeds 1:2. The furan moiety of Bzf was found to have beneficial effects on this copolymerization, which includes improvement of their cross-linking density and enhancement of thermal properties of the copolymerized resins. These effects may be due to the electrophilic aromatic substitution of the furan ring structure. These findings have significant importance in designing some new fully bio-based polybenzoxazines with desirable properties.

(vi) **Biofuel waste streams (Amino acids):** Use of biomass as a feedstock provides an alternative with a closed carbon cycle. Amino acids are quite abundant in biomass and these amino acids are highly functionalized. These properties make some efficient routes possible toward bulk chemicals saving energy as well as reagents. According to the European guidelines, 10% of all transportation fuels should be originated from biomass by 2020 (EPC, 2009). Combining with the estimated worldwide biofuel production, (IEA, 2011) this will lead to the production of about 100 million tons of protein per year, corresponding on an average of 5% (5 million tons) of each amino acid including phenylalanine. 13,000 tons of phenylalanine was produced for food applications (Demain, 2000). Substantial amounts of phenylalanine could be available as a feedstock for the production of bulk chemicals without competing with the food and feed market in years to come.

Cinnamic acid could also be obtained from plant residue streams, so that a strategy can be made for a potential route to biostyrene and bioacrylate compounds, which are important monomers for the plastic industry (Spekreijse, 2012).

(vii) **Production of γ-valerolactone by levulinic acid:** Hydrogenation of levulinic acid to γ-valerolactone has been reported by Galletti et al. (2012). It can be easily obtained in high yields, with very mild reaction conditions by hydrogenation of an aqueous solution of levulinic acid, where a commercial ruthenium supported catalyst in combination with a heterogeneous acid co-catalyst, such as the ion exchange resins Amberlyst A70 or A15, niobium phosphate, or oxide was used. All the hydrogenations were carried out at 70–50°C and at low hydrogen pressure (3–0.5 MPa). The combined effect of acid and hydrogenating heterogeneous components was also verified for the hydrogenation of aliphatic ketones to the corresponding alcohols. Acid treatment of carbohydrates was reported by Hu and Li (2011) in methanol rich medium, which stabilizes reactive intermediates and suppresses polymer formation and greatly promotes methyl levulinate production.

The efficient catalytic hydrogenation of levulinic acid is a key step in biomass conversion. It was shown that the levulinic acid could be reduced efficiently to γ-valerolactone in the presence of a catalyst *in situ* generated from $Ru(acac)_3$ and different sulfonated phosphine $R_nP(C_6H_4\text{-m-}SO_3Na)_{3-n}$ (n = 1,2; R = Me, Pr. iPr, nBu, Cp) ligands in a solvent and promoter free reaction (Tukacs et al., 2012). A highly efficient and recyclable heterogeneous Cu-$ZrO_2$ nanocomposite catalyst was developed for selective hydrogenation of levulinic acid to γ-valerolactone with complete conversion and 90–100% selectivity (Hengne and Rode, 2012).

(viii) **Biodiesel from vegetable oils versus algae:** Biodiesel is one of the most prominent renewable alternative fuels, which can be derived from a variety of sources including vegetable oils, animal fats and used cooking oils, as well as alternative sources such as algae. Issues such as land use change, food *versus* fuel, feedstock availability, and production potential have greatly influenced the search for the best feedstocks, but an issue that will ultimately determine or decide the usability of any biodiesel fuel properties as cold flow and oxidative stability have been problematic issues for biodiesel.

The fatty acid profile of any biodiesel fuel largely depends on the feedstock, which significantly influences these properties. The comparison has also been done between biodiesel derived from vegetable oils and biodiesel obtained from algae particularly in respect of fuel properties. The fatty acid profiles of many algal oils possess high amounts of saturated and polyunsaturated fatty acids. Thus, biodiesel fuels derived from algae in many cases is likely to possess poor fuel properties, that is, both poor cold flow and low oxidative stability. This observation shows that potential of production of biodiesel only is not sufficient to decide suitability of any feedstock. It is also important to know how the fuel properties of a biodiesel can be improved through modification of its fatty ester content? Algal oils are probably best produced under tightly controlled conditions as the fatty acid profile of algal oils is quite susceptible to changes under these conditions. Algal oils yielding biodiesel with the least problematic properties have been determined by reported fatty acid profiles (Knothe, 2011).

(ix) **Esterification of glycerol to methyl glycerate:** The oxidation of glycerol in methanol has been carried out by gold nanoparticles on different oxide supports ($TiO_2$, $Al_2O_3$, and ZnO) using molecular oxygen as the oxidant in a batch set-up. The main oxidation products of glycerol are methyl glycerate and dimethyl mesoxalate, which indicates that C–C bond scission occurs to a limited extent as compared to glycerol oxidations in water. Highest selectivity to methyl glycerate was observed in the case of $Au/TiO_2$ as the catalyst. The use of a base is not essential for the glycerol oxidation reaction to occur, although for $TiO_2$ and $Al_2O_3$, higher initial activities were found in the presence of sodium methoxide. Au/ZnO gives comparable activity and selectivity both; in the presence and absence of a base. Oxidation with reaction intermediates indicates that oxidation of methyl glycerate to higher oxygenate does not occur to a significant extent in methanol. An alternative pathway for the formation of dimethyl mesoxalate involving dihydroxyacetone has been proposed (Purushothaman et al., 2012).

(x) **Production of hydrogen:** Standards on the quality of hydrocarbon fuels do not compromise particularly on sulphur and aromatic contents. This is one of the major forces behind increasing hydrogen demands by petroleum refineries. The fuel standards are often based on keeping a control on environmental pollution. However, most of the commercial hydrogen production processes are based on non renewable resources, which are associated with high carbon footprints. With increasing demands of hydrogen as the fuel, the carbon footprint associated with hydrogen production will increase accordingly. Incentives for hydrogen production by green technologies will be encouraging for smooth functions of industrial processes from high to low carbon footprint. It will facilitate entry of green reforming technologies into the hydrogen market. James et al. (2011) reviewed the potential of some emerging reforming technologies for hydrogen production from renewable resources.

(a) Triacetic acid lactone

Triacetic acid lactone has been established to be a biorenewable molecule with much potential as a platform chemical for the production of commercially valuable bifunctional chemical intermediates and end products, such as sorbic acid (Chia et al., 2012).

(b) Hydrogenolysis of γ-valerolactone

γ-Valerolactone is one of the most significant cellulose-derived compounds, which is directly converted into 1,4-pentanediol. This conversion was carried out by chemoselective hydrogenolysis catalyzed by a simple but quite versatile copper–zirconia catalyst. Depending on the reaction conditions, 2-methyltetrahydrofuran could also be obtained in excellent yields.

(xi) **Valorization of corn cob residues to porous carbonaceous materials:** A strategy to utilize corn cobs to generate porous carbonaceous materials has been reported, which are used to generate biodiesel from waste oils. The focus has been increased on reducing organic wastes in industry and for providing and utilizing renewable chemicals and fuels. Valorization of wastes is attracting considerable attention nowadays, providing an alternative to the disposal of a range of waste materials in landfill sites. Particularly, the valorization of food wastes is considered to be quite promising.

Arancon et al. (2011) utilized corn cobs, a common food waste to generate microporous carbonaceous material. This material was then subsequently sulfonated to give a solid acid catalyst, which exhibited excellent activity in the simultaneous esterification or trans-esterification of waste oils.

(a) Trapping of radical in neat conditions

Shapiro (2010) reported that radicals generated during oxidation of aldehyde to carboxylic acids can be efficiently trapped under friendly environmental conditions, either in neat conditions or in water.

(xii) **Hydrogen production by sustainable process:** Hydrogen production by water splitting still remains a golden dream for scientists, because it is a route to non fossil fuel as well as a potential source of clean energy. It is quite essential to use a non polluting energy source like sunlight to drive the reaction. The production of hydrogen by the photocatalytic cleavage of water can overcome all the hurdles. This area is briefly reviewed by Bowker, focusing mainly on two routes (i) the use of sacrificial agents such as alcohols to act as oxygen scavengers and liberate hydrogen and (ii) by direct water splitting to produce both hydrogen and oxygen. The factors which are important in determining the characteristics of effective photocatalytic water splitting systems have been discussed.

(xiii) **Use of carbon dioxide:** A large amount of carbon dioxide is available in atmosphere and oceans, but its utilization as feedstock for the chemical industry is often prevented by its thermodynamic stability. Only a limited processes based on carbon dioxide as a raw material have been realized on a technical scale so far for example production of urea, methanol, or salicylic acid.

A few catalytic reactions of a lactone platform chemical have been discussed, where $CO_2$ has been used as a feedstock. Numerous other reactions can also be carried out starting from this molecule leading to versatile interesting products like acids, alcohols, or diols, aldehydes, amino acids, amines and so on in high yields. Furthermore, esters, silanes, or even some polymers are obtained using the $\delta$-lactone as a building block. Thus, in presence of efficient catalytic systems leading to high selectivities, a new approach for the utilization of $CO_2$ as a reasonable feedstock for chemical reactions has been described by Behr and Henze (2011). Copolymerization of carbon dioxide was also reported by Darensbourg and Wilson (2012). Carbon dioxide and epoxides (oxiranes) produce polycarbonates to afford high selectivity for copolymer *versus* cyclic carbonate formation.

(xiv) **Synthesis of polyoxygenated compounds:** Air, light, water, and *spirulina* are utilized in a green method to transform readily accessible furan substrates into a diverse range of synthetically useful polyoxygenated motifs commonly found in natural products **(Noutsias et al., 2012) (Figure 2.5)**.

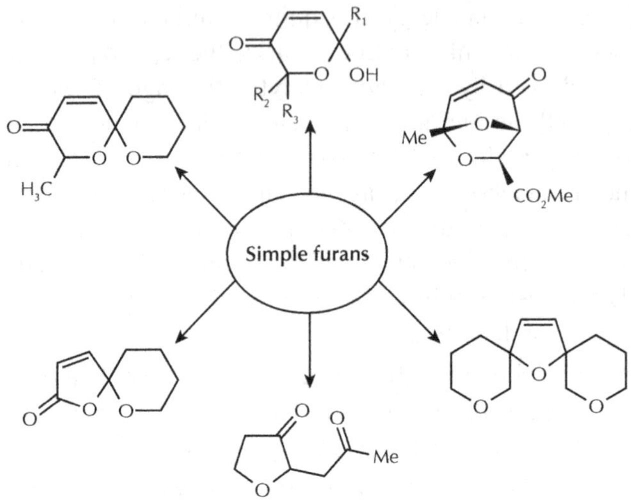

**FIGURE 2.5** Synthesis of polyoxygenated compounds.

(xv) **Waste materials as resources for catalytic applications**
The use of high-volume waste materials in catalysis or for synthesis of a catalyst has been studied by Balakrishnan et al. (2011). Waste materials derived from both; industrial and biological sources have attracted attention of chemists. The materials include red mud, aluminum dross, fly ash, blast furnace slag, rice husk, and various kinds of shell.

(xvi) **Transforming collagen wastes into doped nanocarbons**
Leather industry produces huge quantities of bio-waste that can be utilized as raw material for the bulk synthesis of carbonaceous materials. The synthesis of multifunctional carbon nanostructures from pristine collagen wastes by a simple high temperature treatment was reported Meiyazhagan et al. (2012). It was observed that the nanocarbons derived from this biowaste have a partially graphitized structure with onion-like morphology. These are naturally doped with nitrogen and oxygen, resulting in some multifunctional properties. This route from bio-waste raw material provides a cost effective alternative to existing chemical vapor deposition (CVD) methods for the synthesis of functional nano-

carbon materials. It also presents a sustainable approach to tailor nanocarbons for various applications.

(xvii) **Transforming animal fats into biodiesel using charcoal and $CO_2$:** A simple methodology for producing biodiesel has been reported, confirming that the main driving force of biodiesel conversion through the non-catalytic trans-esterification reaction is temperature dependent. Non-catalytic biodiesel conversion can be achieved in the presence of a porous material *via* a thermochemical process and a real continuous flow system. In addition, this non-catalytic conversion of biodiesel can be enhanced by the presence of $CO_2$. The transformation of animal fat (beef tallow and lard) into biodiesel was achieved by Eilhann et al. (2012) using charcoal and $CO_2$ under ambient pressure. This methodology for producing biodiesel combines esterification of free fatty acids and transesterification of triglycerides into a single process and leads to a 98.5 conversion efficiency of biodiesel within 1 min at 350–500°C. This new process has very high potential to achieve a breakthrough in minimizing the cost of biodiesel production owing to its simplicity and technical advantages.

(xviii) **Alternative electronic chemicals:** Because of the acute toxicity of arsine (even more toxic than hydrogen cyanide), a search is on for alternatives to arsine in cylinders. One of the possible solutions to the hazard associated with the sudden release of arsine stored in compressed cylinders is containment of the gas at atmospheric pressure. Another solution is the replacement of arsine by completely non toxic reagents. Lethally toxic arsine is made less toxic by alkylation and the fully substituted reagent trimethyl arsine, $As(CH_3)_3$ is practically non toxic according to contemporary chemical standards.

(xix) **Miscellaneous:** Bio-based N-methylpyrrolidone was prepared by the cyclization of γ-aminobutyric acid (obtained from glutamic acid) to 2-pyrrolidone and subsequent catalytic methylation of 2-pyrrolidone with methanol to N-methylpyrrolidone was done in a one-pot procedure (Lammens et al., 2010). Bio-based synthesis of secondary aryl amines from (-)-shikimic acid was reported by Wu et al., (2012). 3-Arylamino-4-hydroxybenzoates and 3,4-dihydroxy-5-alkylaminobenzoates have been synthesized in good yields starting from the biomass-based feedstock (-)-shikimic acid via the tandem cross coupling and aromatization reaction with primary amines.

Starting material should be benign in nature. This is possible by taking biomass and biomass-derived derivatives as the starting materials because these are provided by nature itself. This is not always possible as far as variety of chemical reactions are concerned and therefore, such feedstocks are to be search down, which are either harmless or less toxic as compared to conventional reactants.

## KEYWORDS

- **Biodegradation**
- **Biodiesel**
- **Biomass**
- **Chemical transformation**
- **Sustainable process**

## REFERENCES


1. Akbar, E., Binitha, N., Zahira, Y., Kamarudin, S. K., & Jumat, S. (2009). *Green Chemistry, 11,* 1862–1866.
2. Albini, A., & Fagnoni, M. (2004). *Green Chemistry, 6,* 1–6.
3. Allen, D. J., & Brent, G. F. (2010). *Environment Science Technology,* 44, 12735–2739.
4. Arancon, R. A., Barros, H. R. Jr., Balu, A. M., Vargas, C., & Luque, R. (2011). *Green Chemistry, 13,* 3162–3167.
5. Baghos, V. B., Doss, S. H., & Eskander, E. F. (1993). *Organic Preparations and Procedures International, 25,* 301–307.
6. Balakrishnan, M., Batra, V. S., Hargreaves, J. S. J., & Pulford, I. D. (2011). *Green Chemistry, 13,* 16–24.
7. Balbino, J. M., Menezes, E. W., Benvenutti, E. V., Cataluna, R., Ebeling, G., & Dupont, J. (2011). *Green Chemistry, 13,* 3111–3116.
8. Behr, A., & Henze, G. (2011). *Green Chemistry, 13,* 25–39.
9. Benaglia, M. (2009). *Renewable and recyclable catalysts. Advancing green chemistry series.* West Sussex: Wiley-VCH.

10. Bicker, M., Kaiser, D., Ott, L., & Vogel, H. (2005). *Journal of Supercritical Fluids 36*, 118–126.
11. Binder, J. B., & Raines, R. T. (2009). *Journal of American Chemical Society, 131*, 1979–1985.
12. Bowker, M. (2011). *Green Chemistry, 13*, 2235–2246.
13. Braden, D. J., Henao, C. A., Heltzel, J., Maravelias, C. C., & Dumesic, J. A.(2011). *Green Chemistry, 13*, 1755–1765.
14. Campbell, C. J., & Laherrere, J. H. (1998). *Scientific American*, 60–65.
15. Case, P. A., Heiningen, A. R. P., & Wheeler, M. C. (2012). *Green Chemistry, 14*, 85–89.
16. Chang, R. H., Jang, J., & Wu, K. C. W. (2011). *Green Chemistry, 13*, 2844–2850.
17. Cheng, H., & Gross, R. (Eds). (2011). *Green polymer chemistry: Biocatalysis and biomaterials*. Oxford: Oxford University Press.
18. Chheda, J. N., Leshkov, Y. R., & Dumesic, J. A. (2007). *Green Chemistry, 9*, 342–350.
19. Chia, M., Schwartz, T. J., Shanks, B. H., & Dumesic, J. A. (2012). *Green Chemistry*, 14, 1850–1853.
20. Corma, A. Iborra, S., & Velty, A. (2007). *Chemical Reviews, 107*, 2411–2502.
21. Cukalovic, A., Stevens, C. V. (2010). *Green Chemistry, 12*, 1201–1206.
22. Darensbourg, D. J., & Wilson, S. J. (2012). *Green Chem*istry, *14*, 2665–2671.
23. Demain, A. L. (2000). *Biotechnology Advances, 18*, 499–514.
24. Dewulf, J., & Van Langenhove, H. (Eds.). (2006). *Renewable-based technology: Sustainable assessment*. West Sussex: Wiley-VCH.
25. Eilhann, E. K., Jaegun, S., & Haakrho, Y. (2012). *Green Chemistry*, *14*, 1799–1804.
26. European Parliament and the Council of 23 April 2009, 2009/28/EC, Off. J. Eur. Union, 2009, L140/16.
27. Faugeras, P, A., Elchinger, P, H., Brouillette, F., Montplaisir, D., & Zerrouki, R. (2012). *Green Chemistry*, *14*, 598–600.
28. Fumagalli, C. (1997). In Kroschwitz, J. & Home Grant, M. (Eds.), *Kirk–Othmer encyclopedia of chemical technology* (Vol. 22, p. 1074, 4th ed.). New York: John Wiley and Sons.
29. Galletti, A, M. R., Antonetti, C., Luise, V., & Martinelli, M. (2012). *Green Chemistry, 14*, 688–694.
30. Gandini, A., Belgacem, M, N. (1997). *Progress in Polymer Science, 22*, 1203–1379.
31. Gassama, A., Ernenwein, C., & Norbert, H. (2010). *Green Chem*istry, *12*, 859–865.
32. Gutsche, B., Roβler, H., & Wurkerts, A. (2008). *Heterogeneous Catalysis in Oleochemistry*. West Sussex: Wiley-VCH.
33. Han, J., Sun, H., Ding, Y., Lou, H., & Zheng, X. (2010). *Green Chem*istry, *12*, 463–467.
34. Hengne, A. M., & Rode, C. V. (2012). *Green Chemistry, 14*, 1064–1072.
35. Hester, R. E. (2009). *Carbon capture. Royal Society of Chemistry. Green Chemistry Series*. Cambridge: RSC Publishers.
36. Hick, S. M., Griebel, C., Restrepo, D. T., Truitt, J. H., Buker, E. J., Bylda, C., & Blair, R. G. (2010). *Green Chemistry, 12*, 468–474.
37. Hill, K. (2000). *Pure and Applied Chemistry, 72*, 1255–1264.
38. Holtz, M. H. (2003). Second Annual Conference on Carbon Sequestration: Developing the Technology Base to Reduce Carbon Intensity. Alexandria, Virginia, May 2003.

39. Hu, X. & Li, C. Z. (2011). *Green Chemistry, 13,* 1676–1679.
40. Hwang, D. W., Kashinathan, P., Lee, J. M., Lee, U. H., Hwang, J. S., Hwang, Y. K., & Chang, J. S. (2011). *Green Chemistry, 13*, 1672–1675.
41. International Energy Agency (2012). http://www.iea.org/press/pressdetail.asp Technology Roadmap, Biofuels for Transport, http://www.iea.org/papers/2011/biofuels_roadmap.pdf. Accessed on March 2012.
42. James, O., Maity, S., Adediran, M. M., Ogunniran, K.O., Siyanbola, T.O., Sahu, S., & Chaubey, R. (2011). *Green Chemistry, 13*, 2272–2284.
43. Klass, D. H. (1998). *Biomass for renewable energy, fuels and chemicals. Fossil fuel reserves and depletion* (pp. 10–19). San Diego: Academic Press.
44. Knothe, G. (2011). *Green Chemistry*, *13,* 3048–3065.
45. Kobayashi, H., Ito, Y., Komanoya, T., Hosaka, Y., Dhepe, P. L., Kasai, K., Hara, K., & Fukuoka, A. (2011). *Green Chemistry, 13,* 326-333.
46. Kollonitsch, V. (1970). *Sucrose chemicals: A critical review of a quarter century of research by the Sugar Research Foundation.* Washington, DC: International Sugar Research Foundation.
47. Lammens, T. M., Franssen, M. C. R., Scott, E. L., & Sanders, J. P. M. (2010). *Green Chemistry, 12,* 1430–1436.
48. Leimkuhler, H. J. (Ed). (2010). *Managing $CO_2$ emissions in the chemical industry,* West Sussex: Wiley-VCH.
49. Li, C. Z., Wang, Q., & Zhao, Z. K. (2008). *Green Chemistry, 10,* 177–182.
50. Li, C. Z., & Zhao, Z. K. B. (2007). *Advanced Synthesis and Catalysis, 349,* 1847–1850.
51. Li, N., Tompsett, G. A., Zhang, T., Shi, J., Wyman, C. E., & Huber, G. W. (2011). *Green Chemistry, 13,* 91–101.
52. Lichtenthaler, F. W., Cuny, E., Martin, D., & Ronninger, S. (1991). In Lichtenthaler, F. W., (Ed.) *Carbohydrates as organic raw materials.* New York: VCH.
53. Long, J., Li, X., Guo, B., Wang, F., & Wang, Y. Y. L. (2012). *Green Chemistry, 14,* 1935–1941.
54. Marcotullio, G., & Jong, W. D. (2010). *Green Chemistry, 12,* 1739–1746.
55. Mascal, M., & Dutta, S. (2011). *Green Chem***istry**, *13,* 40–41.
56. Meiyazhagan, A. K., Narayanan, N. T., Reddy, A. L. M., Gupta, B. K., Chandrasekaran, B., Talapatra, S., Ajayan, P. M., & Thanikaivelan, P. **(2012).** *Green Chem*istry, *14,* 1689–1695.
57. Moreau, C., Belgacem, M. N., Gandini, A. (2004). *Topics in Catalysis, 27,* 11–30.
58. Moussa, H. H., Abdel Meguid, S., & Atalla, M. M. (1982). *Pharmazie, 37,* 352–354.
59. Mukund, G. A., Varma, J., & Gokhale, D. V. (2007). *Green Chemistry, 9,* 58–62.
60. Noutsias, D., Alexopoulou, I., Montagnon, T., & Vassilikogiannakis, G. (2012). *Green Chemistry, 14,* 601–604.
61. Okkerse, C., & Bekkum, H. V. (1999). *Green Chemistry*, 107–114.
52. Pentz, W. J. GB. (1984). Patent 2131014,.
63. Purushothaman, R. K., Haveren, J. V., Van, E. D. S., Melián-Cabrera, I., & Heeres H. J. (2012). *Green Chemistry, 14,* 2031–2037.
64. Ragauskas, A. J., Williams, C. K., Davison, B. H., Britovsek, G., Cairney, J., & Eckert, C. A., et al. (2006). *Science,* 311, 484–486.
65. Ran, N., Zhai, L., Chen, Z., Tao, J. (2008). *Green Chemistry, 16*, 361–372.

66. Ravelli, D., Dondi, D., Fagnoni, M., & Albini, A. (2009). *Chemical Society Reviews, 38,* 1999–2011.
67. Ravindranath, N. H., & Hall, D. O. (1995). *Biomass, energy, and environment.* Oxford: Oxford University Press.
68. Relsch, M. S. (1991). *Chemical and Engineering News, 69,* 13–16.
69. Ruhal, R., Aggarwal, S., & Choudhary, B. (2011). *Green Chemistry, 13,* 3492–3498.
70. Serrano-Ruiz, J. C., & Dumesic, J. A. (2009). *Green Chemistry, 11,* 1101–1104.
71. Shapiro, N., Kramer, M., Goldberg, I., & Vigalok, A. **(2010).** *Green Chemistry, 12,* 582–584.
72. Singhvi, M., Joshi, D., Adsul, M., Varma, A., & Gokhale, D. (2010). *Green Chem*istry, *12,* 1106–1109.
73. Soetaert, W., & Vendamme, E. (2009). *Biofuels.* Weinheim: Wiley-VCH
74. Spexripijse, J., Notre, J. L., Haveren, J. V., Scott, E. L. & Sanders, J. P. M. (2012). *Green Chemistry, 14,* 2747–2757.
75. Stein, T. M., Grande, P., Sibilla, F., Commandeur, U., Fischer, R., Leitner, W., & Maria, P. D. (2010). *Green Chemistry, 12,* 1844–1849.
76. Sun, J., & Liu, H. (2011). *Green Chemistry, 13,* 135–142.
77. Tao, J., & Kazlauskas, R. J. (2011). *(Eds). Biocatalysis for green chemistry and chemical process development.* West Sussex: John Wiley and Sons.
78. Tudorache, M., Protesescu, L., Coman, S., & Parvulescu, V. I. (2012). *Green Chemistry, 14,* 478–482.
79. Tukacs, J. M., Kiraly, D., Stradi, A., Novodarszki, G., Eke, Z., Dibo, G., Kegl, T., & Mika, L. T. (2012). *Green Chem*istry, *14,* 2057–2065
80. Umbach, W., in: Eierdanz, H. (Ed.) (1996). Perspektiven Nachwach- sender Rohstoffe in der Chemie (p. XXIX). Weinheim: VCH Publ.
81. Wang, J., Ren, J., Liu, X., Xia, J., Zu, Y., Lu, G., & Wang, Y. (2012a). *Green Chemistry, 14,* 2506–2512.
82. Wang, C. F., Sun, J. Q., Liu, X. D., Sudo, A., & Endo T. (2012b). *Green Chemistry, 14,* 2799–2806.
83. Weissermel, K., & H. J. (2003). *Industrial organic chemistry.* Weinheim: Wiley-VCH.
84. Whittall, J., & Sutton, P. (2009). *Practical methods for biocatalysis and biotransformations.* West Sussex: John Wiley and Sons Inc.
85. Wölfel, R., Nicola, T., Andreas, B., & Peter, W. (2011). *Green Chemistry, 13,* 2759–2763.
86. Wu, W., Zou, Y., Chen, Y., Li, J., Lu, Z., Wei, W., Huang, T., & Liu, X. (2012). *Green Chemistry, 14,* 363–370.
87. Yang, Y., Hu, C., & Abu-Omar, M. M. (2012). *Green Chemistry, 14,* 509–513.
88. Zaccheria, F., Psaro, R., & Ravasio, N. (2009). *Green Chemistry, 11,* 462–465.
89. Zakrzewska, M. E., Bogel-Lukasik, E., & Bogel-Lukasik, R. (2011). *Chemical Reviews, 111,* 397–417.
90. Zhang, Z., & Zhao, Z. K. (2010). *Bioresources Technology, 101,* 1111–1114.
91. Zhang, Z., Wang, Q., Tripathi, P., & Pittman, C. U. Jr. (2011). *Green Chemistry, 13,* 940–949.

**CHAPTER 3**

# ECO-FRIENDLY PRODUCTS

NEELU CHOUHAN, ANIL KUMAR, AJAY SHARMA, and RAMESHWAR AMETA

## CONTENTS

## 3.1 INTRODUCTION

Our atmosphere is in a constant state of chaos or disorderness and is never being static. Internal and external transformations in our planet earth that is either made by Nature or man bring drastic changes in weather pattern, availability of natural resources, and living conditions. Scientific evidences highlighted the role of man in environmental degradation as a result of industrialization exploiting Mother earth insanely. Industrial revolution put momentum to improve the quality of human life at many levels by improving global health, eradicating death rate, scientific gadgets that made our life more comfortable, mode of convenience that save our time by keeping our comfort zones intact, mode of entertainments to relax us, carries our voices over telephone lines through cell phones to keep in touch with dear ones or to expand business activities, quality of potable water, high production of varieties of agriculture products that is, corn, wheat, and rice, paired with fertilizers and pesticides, dramatically increased the food stuff, the lights to enlighten our lives, advancement in satellite research telling us about when and where natural disasters (hurricane, earthquakes, etc) may strike and with how much impact, and many more, all have been brought to us from courtesy of science.

One can say that science and technology becomes an essential and integral part of daily lives of masses, as it serves us 24 x 7 in variety of avatars and eventually or gradually transforming our life style. Hence, the importance of any nation in the 21st century shall be judged not by its economic strength alone, but also by its power to conceptualize inventions and bring their benefits to people taking care of environment. Unfortunately, this development generates a remarkable gap between human race and nature. In past two decades, the size of this gap gaining heights day-by-day. Until recently, almost negligible or no effort has been made to bridge this gap. All the hi-tech progress of human race done during the past two decades has provided us an easy lifestyle that is full of modern facilities on the very high cost of resources consumption and environmental degradation. A great turmoil in the delicate balance of our ecological system, originated the global warming and its associated climate changes, increasing ocean temperature, changes in terrestrial geography, scarcity in the drinking water, rain fall ratio, temperature, and so on. Tremendous critical issues, such as extinction of rare species of flora and fauna from earth, various incurable or semicurable diseases, acid rain, ozone layer depletion, excessive pollution, and photochemical smog, especially in and around the urban areas, arises due to these changes that has been proven dangerous to global life.

Above mentioned concerns, fuels the growing concern of masses and governments about this most eminent need of the time that is, a persistent action for environmental protection. In this reference, great Chinese Philosopher Mencius (20,000 years ago) observed that, "Refraining from over fishing will ensure fishing last forever" and "Cutting wood according to the season will ensure the health of forest" are the means to achieve good harmony between man and nature. In this context, the words of Elsa Reichmanis, the former President of the American Chemical Society, were appropriate, "The days are passed, when we can trade environmental contamination for economical prosperity that is only a temporary bargain and now the cost of pollution on both eco-system and human health is too high"(Ritter, 2003). Even then we still do not wish to quit the comfortable lifestyle, but simultaneously we cannot afford to continue along this path. Therefore, it is the high time to rethink and set our accountability and shake hands with Nature to satisfy our livelihood's demands in an eco-friendly manner. Although, we can't single handedly clean up our environment but we can make choices in our day-to-day life that will benefit the health of the planet and our community. The role of the policy makers also became very crucial in this reference. They have to pay serious attention to the various burning issues that arises due to the modernization and includes developing research agendas, driven by social challenges, engaging citizens through building constituencies, and cultivating scientists with a clear sense of civic responsibility. Therefore, governments, industrialists, and researchers have to put their heads together on this leading issue with their careful concerns about the challenges of modernization and renew the interest in development of eco-friendly alternatives using breakthrough concepts and accelerated application of cutting-edge scientific, engineering, and analytical tools.

Simple changes, such as finding green alternatives for the everyday products we consume, can have a beneficial impact on water and soil quality, reduce energy uses and the amount of pollution or waste by using breakthrough concepts (catalysis by design, biodegradable consumer products) and accelerated application of cutting-edge green technology and products. With the customers getting gradually concerned about the environment friendly products, the world market is now drifting more and more toward the recyclable or decomposable home appliances, hardware equipment and daily life products: (i) Biodegradable bicycle: Marco Facciola has created a bike that is entirely made of glue and wood (ii) Sony presented the world's first eco-friendly camera Odo. This camera is made of biodegradables and do not need any batteries. The power is produced by kinetic means and (iii) Chinese mobile company Je-Hyun Kim has designed an eco-friendly mobile phone that is half recyclable and half biodegradable. The keypad and the screen for the mobile phone are made to be recycled after usage, while the rest of the mobile phone is made of grass like carbonic components, making it fully decomposable.

Efforts of Governments, industrialists, and researchers have been focused primarily to reduce the impact of toxic chemicals and exposure-based initiatives. To meet the goals of pollution prevention, few important legislative decision comes in a form of Water Act-1974, Air Act-1981, Environmental Prevention Act-1986, Hazardous Waste Rules-1989, Pollution Prevention Act-1990, Coastal Regulation Zones-1991, Biomedical Waste Rules-1998, Rules for Recycled Plastics-1999, Fly Ash Notification-2000, Municipal Solid Waste Rules-2000, Battery Managment and Handling Rules-2000, Environment Protection (Amendment) Rules 2012, and so on. One of its important initiatives is "Designing Safer Chemicals". It means designing a chemical that will not affect adversely the normal biochemical and physiological process of any organism. However, considering the complex, diverse and dynamic nature of living organisms, in practice, this becomes a formidable challenge. The design of safer chemicals will require the ready availability of the data and information on the relationship between chemical structure and industrial or commercial function. Chemical toxicity may be reduced by isosteric replacement of a carbon atom, toxic substances (e.g., DDT, etc) can be redesigned such that they will retain their commercial efficacy but will decompose rapidly under physiological conditions to innocuous and readily excretable products.

Chemical substances that are toxic because of their ability to persist in the environment can be redesigned such that they biodegrade readily. Technical, economic, and commercial feasibility of designing safer biocides and paint constituents, the carcinogenic properties of many commercial aromatic amines (e.g., benzidines and anilines) can be greatly reduced by molecular modifications that facilitate excretion in the urine or prevent bioactivation. These initiatives, represents a new approach of designing chemicals that emphasizes safety to human health and the environment as well as efficacy of use. Implementing this concept will require major changes in the current practices of all social, academia, and industry regimes. To accomplish these changes, the concept must be understood, accepted, and practiced by all those associated with the development, manufacture, and use of industrial chemicals.

## 3.2 SOME GREEN PRODUCTS

### *3.2.1 BIOPLASTICS*

Across the nations, there has been a public outcry for action to ban the plastic polybags to get rid of the almost non biodegradable plastic waste. In sunlight, small portion of plastic waste degrades into toxic parts that contaminate soil and water. Accidentally, it can be ingested by animals and thereby enter into the food chain. The death of terrestrial animals such as cow, buffalo, and so on was reported due to consumption of such polythene carry bags. To the innocent marine and terrestrial life, polythene waste is recognized as a major threat, as it could be fatal for fishes, birds, and mammals. As per this report, due to this plastic pollution in the marine environment minimum 267 species are being affected, which includes all mammals, sea turtles (86%), and seabirds (44%) (Derraik, 2002).

Bioplastics are a form of plastics derived from renewable biomass sources, such as vegetable fats and oils, corn starch, cellulose, biopolymers, or microbiota (Hong et al., 1999–2003). They have a high-market potential because of their additional advantage of biodegradability in 10–15 years. Although currently plastics derived from petrochemicals, constitute a sustainable alternative to conventional oil-based plastics, which degrade in 100–150 years. Common plastics rely more on scarce fossil fuels and produce more greenhouse gases. Bioplastics, which are designed to biodegrade, can break down in either anaerobic or aerobic environments, depending on how they were manufactured. There are a variety of bioplastics being made. While aromatic polyesters are almost totally resistant to microbial attack, most aliphatic polyesters are biodegradable due to their potentially hydrolysable ester bonds. Naturally produced bioplastics are Polyhydroxyalkanoates(PHAs), such as poly-3-hydroxybutyrate (PHB), polyhydroxyvalerate (PHV), and polyhydroxyhexanoate (PHH); renewable resource, polylactic acid (PLA), and so on. Some common applications of bioplastics are packaging materials, dining utensils, food packaging, and insulation (Chen and Patel, 2012). Their production is expected to be more than 1.5 million tons per year in 2020. A novel and cost-effective polymerization technology has been developed to produce high-quality bioplastics, with improved thermal stability up to 200°C.

### 3.2.2 GREEN FUEL (HYDROGEN)

Light fuel hydrogen has attracted great attention of environmentalists, scientists, and industrialists as a benign fuel of future because of its capability to produce pollution free energy (no carbon emission and useful byproduct of hydrogen fuel combustion that is only water) with highest energy density, that is, values per mass of 140MJ $kg^{-1}$, are the beauty associated with hydrogen fuel. Hydrogen fuel is quiet contemporary and relevant in present energy scenario, that is, "We are at the peak of the oil age but the beginning of the hydrogen age". Anything else is only an interim solution. The transition will be very messy, and will take many technological paths but the future will be hydrogen-fuel cells". The majority of hydrogen used in industries is derived from fossil fuels or cleavage of water. Currently, majority of industrial hydrogen need is satisfied from conventional sources (coal, oil, and natural gas), which contains about 10% $CO_2$ with hydrogen gas, and only 4% of $H_2$ comes from electrolysis, which is the cheapest method to generate hydrogen ($3.51 $kg^{-1}$). The success of hydrogen technology will depend on the efficient generation of hydrogen from water cleavage powered by renewable sources (such as solar or wind). These are the important steps necessary for overall water cleavage.

(i) Absorption of light near the surface of the semiconductor creates electron-hole pairs.
Photocatalyst + $h\nu \rightarrow$ Photocatalyst ($e^-$) + Photocatalyst ($h^+$)

(ii) Holes (minority carriers) drift to the surface of the semiconductor (the photo-anode), where they react with water to produce oxygen.
$2\,h^+ + H_2O \rightarrow \frac{1}{2}\,O_2\,(g) + 2\,H^+$ (1.23V vs NHE at pH = 0, +0.82V vs NHE pH = 7)

(iii) Electrons (majority carriers) are conducted toward a counter metal electrode (typically Pt), where they combine with $H^+$ ions in the electrolyte solution to generate $H_2$:
$2\,e^- + 2\,H^+ \rightarrow H_2\,(g)$ (0.00V, vs NHE at pH = 0, –0.41V vs NHE pH = 7)

(iv) Transport of $H^+$ from the anode to the cathode through the electrolyte completes the electrochemical circuit. The overall reaction is
$2\,h\nu + H_2O \rightarrow H_2\,(g) + \frac{1}{2}\,O_2\,(g)$ (1.23eV, $\Delta G^0$ = 237kJ/mol)

The use of hydrogen as an energy carrier has the enough potential to reduce energy dependence on gasoline and also reduce pollution and greenhouse gas emissions. Therefore, advancement in hydrogen technologies is under process. Hydrogen economy can provide renewable energy solutions to energy crisis, such as emergency backup power, heating and electricity for commercial and residential purpose, and hybrid electric vehicles (storage of hydrogen still remains a "critical path" barrier, and it is one of the primary focused areas). Market transformation activities aim to promote their adoption in stationary, portable, and specialty vehicle applications, such as forklifts, municipal vehicle, lawn mower (Yvon and Lorenzoni, 2006), and municipal supply of clean fuel hydrogen to run electrical and other utility appliances (heater, air-conditioner, fan, etc). Hydrogen burner is one more interesting example of portable hydrogen appliance that can be used indoors safely without risk and the reaction product is water, which is actually beneficial for the room climate.

A panel of evaluators that confirmed hydrogen roasted meat was undistinguishable in taste from that of propane roasted meat. Likewise eco-friendly portable hydrogen house is another use of PEC hydrogen, where hydrogen is used as an energy source for most of the appliances for example, gas engine driven generator for the seasonal compensation (Wingens et al., 2008). Another most promising application of hydrogen produced by PEC cells is in micro-combined heat and power (CHP) supply of residential houses (Hollmuller et al., 2000) and exploiting both gas-engine generated electric power and waste heat (Matics and Krost, 2007). Hydrogen driven city buses, and so on are some commercial means of transportation, which utilize PEC hydrogen. Highly efficient clean fuel hydrogen is the most famous fuel for space craft. In 1990, the world's first solar-powered hydrogen production plant (a research and testing facility) became operational, at Solar-Wasserstoff-Bayern, in southern Germany.

### 3.2.3 GREEN PESTICIDES

Worldwide one-third harvest losses are from weeds, diseases, and insects, which were gaining heights day-by-day. Every $1 spent by farmers on pesticides saved $3–5 from crop loss, but role of modern pesticides (organochlorines such as DDT, chlorobenzene, etc.) ChE inhibitors, organophosphates, carbamates, phenoxyherbicides, pyrethroids, bromine-based, phenol derivatives, and dipyridyl derivatives are hazardous for our human health and ecosystem. As most of them are non selective, endocrine disrupter, reproductive toxins, neurotoxins (Lindane), CNS toxic, increases hepatocellular tumors, kills more than just the target, are persistent, and move around in the environment. They can cause cancer (lymphoma, leukemia, brain or lung or testicle or breast), sterility, higher rates of miscarriage, greater risk in children with birth defects or stunted limbs, immune system suppression, potential link to Parkinson's, and so on. Green pesticides are derived from organic sources, which are considered environment friendly and causing less harm to human and animal health, and to habitats and the ecosystem (Lai et al., 2006). Pesticides or biocides include germicides, antibiotics, antibacterials, antivirals, antifungals, antiprotozoals, and antiparasites and these are available in the form of sprays and dusts. Biopesticides are generally safer than synthetic pesticides, but they are not always more safe or environment friendly than synthetic pesticides.

Likewise natural pesticides include sulfur, rotenone, mixture of copper, lime and water, nicotine sulfate, strychnine, arsenic containing pesticides, and pyrethrums. These are used as pesticides from ancient time but most of them are banned in organic farming.

Phytoalexin elicitor glucohexatose has been called a green pesticide (Ning et al., 2003) as a new class of insecticides (spinosad), which shows remarkable selectivity in destroying harmful pests and leaving beneficial insects alive. Few of the insecticides are Wormwood extract, Chive extract, Summer tansy dust, Stinging nettle extract, Daffodil extract, Garlic extract, Rhubarb extract, Onion extract, Sambucus extract, Tobacco extract, and Stale beer. Sulfur (organic fungicide, pesticide, and acaricide), Bio-S (sulfur mixture), Pilzvorsoge, Spruzit, Carbolineum, Basalt dusting powder, Gene silencing pesticide, and Steam (Thermal pest control). For the past few decades, there has been a considerable research interest in the area of natural product delivery using particulate system for controlling plant diseases. The secondary metabolites in plants have been used in the formulation of nanoparticles increasing the effectiveness of therapeutic compounds used to reduce the spreading of plant diseases, while minimizing side-effects for being rich source of bioactive chemicals, biodegradable in nature and non polluting (eco-friendly). There are myriad of nanomaterials including polymeric nanoparticles, iron oxide nanoparticles, gold nanoparticles, and silver ion, which can be easily synthesized and exploited as pesticide. Promising results of ZnO nanoparticles antibacterial activity suggest its usage in food systems as preservative agent (Ahmed et al., 2011). Inhibition effects of 100ppm silver nanoparticles against powdery Mildews disease was reported on cucumber and pumpkin (Lamsal et al., 2011).

### 3.2.4 GREEN DRUGS

Pharmaceutical industries became the most dynamic sector of the chemical industries of 21st century as the sales of medicines and other pharmaceutical products have been increased fourfold from 1985. It has been estimated that the number of potential drug targets may be between 5,000 and 10,000 out of the estimated $10^{60}$ possible compounds (Drews and

Ryser, 1997). Their production process generates huge amount of waste, which typically included volatile organic solvents and other hard-to clean-up agents, as the volumes were daunting. According to the systematic data gathered by the Environmental Protection Agency (EPA), only in the United States, around 278 million tonnes of hazardous waste were generated in 1991 at more than 24,000 sites.

More effective, efficient, and elegant solution to this problem in the words of Anastas is "simply better chemistry" or "green chemistry". Hence, pharmaceutical sector has welcomed green chemistry very warm heartedly, perhaps because no company can afford to ignore green chemistry's potential savings in term of money and eco-health. Typical pharmaceutical plants generate 25–100kg of waste per kilogram of product, a ratio known as the environmental factor, or 'E-factor', in comparison to their other industrial counterparts, such as oil refineries (E-factor <0.1), bulk chemicals (E-factor <1–5), and fine chemicals (E-factor ~5–50). Consequently, there is a plenty of room to increase efficiency and cut costs. Re-drug designing in such a way that drugs must have a degradable chemical structure, which results in, acceptable safety profile, stability in synthesis, formulation, storage, and use.

The technical advancement, such as the introduction of apt catalyst (Nagendrappa, 2002), microwave heating (Liu and Zhang, 2011), ultrasonic energy (Mason, 1997; Cintas and Luche, 1999; Cravotto and Cintas, 2006), combinatorial chemistry (Weller et al., 2006), recycling of by products or solvents, use of solvent-free or mild solvents, biocatalyst, and use of ionic liquids in place of conventional solvents (Martínez-Palou, 2007) offers simple, clean, fast, efficient, and economic synthesis of a large number of pharmaceutical products, which provide enough momentum for many chemists to switch from traditional to green chemical pathway. To catalyze the implementation of green chemistry and engineering in the pharmaceutical industry globally, there is the immense need for the modification in current (conventional) manufacturing practices.

Ibuprofen was commercially synthesised by Boot in 1960 as an eminent case study, which belongs to the category of analgesic, non steroidal anti-inflammatory drug with very high sales. The conventional synthesis of ibuprofen was performed in six steps and it has many disadvantages, under the "green" principles, such as production of secondary byproducts and waste, very poor atom economy, and 40% final yield. Afterwards, a new synthetic route with only three steps and increased efficiency was discovered. In both the synthetic routes, the starting chemical is 2-methyl-propylbenzene, a product of petrochemical industry. But in this new innovated synthetic approach, 77% yield was obtained and Raney nickel (Ni or CO or Pt) was used as a catalyst, which can be recycled; thus, decreasing the substantially the steps.

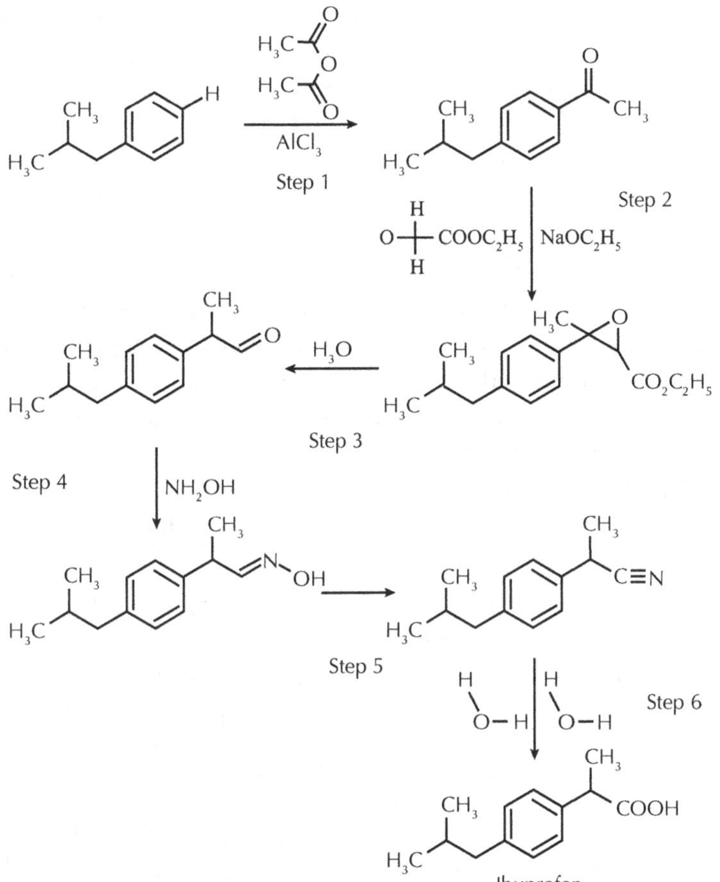

Conventional route

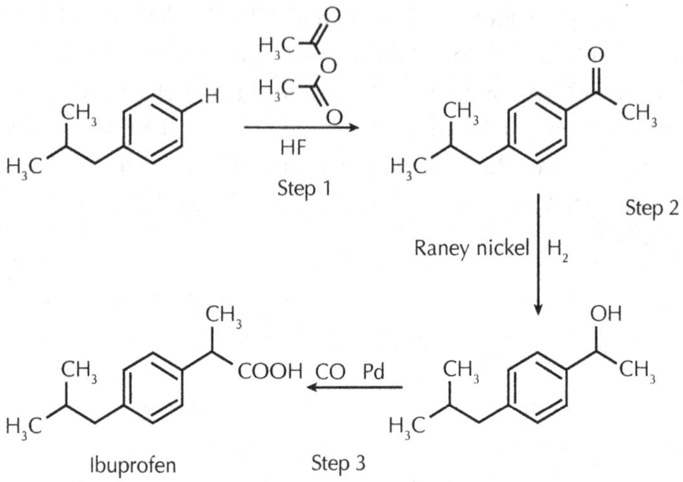

Green chemical route

Similarly, Simvastatin, the second best selling drug for treating high cholesterol, was manufactured from a natural product (lovastatin). The traditional multistep synthesis was wasteful and used large amounts of hazardous reagents (Alberts et al., 1980). Xie and Tang conceived a single step synthesis using an engineered enzyme and a practical low-cost feedstock. Furthermore, Codexis (LovD) optimized both the enzyme and the chemical process. The resulting process greatly reduces E-factor and generates low hazardous waste, cost-effective and meets the needs of customers (2007).

In this context, Pfizer's team rigorously re-examined every step of the synthesis of anti-impotence drug sildenafil citrate (Viagra) and an E-factor of 105 was achieved. All the chlorinated solvents have been replaced with less toxic alternatives, and then measures were introduced to recover and reuse these solvents. The need of using hydrogen peroxide was also eliminated, which can cause burns. They also eliminated any requirement for oxalyl chloride, a reagent that produces carbon monoxide in reactions, a major safety concern. Eventually, later, it cut Viagra's E-factor to 8 (Dunn et al., 2004). After that success, they have also reduced the E-factor of the anticonvulsant pregabalin (Lyrica) from 86 to 9 by biocatalytic mechanism with low (~0.5%) protein loading, conducting all four reactions in water and resolution at first step (wrong enantiomer can be recycled). They have also made similar improvements for the antidepressant sertraline and the non steroidal anti-inflammatory celecoxib. These three drugs altogether eliminated more than half a million metric tons of chemical waste (Sanderson, 2011).

Microwave heating has proven its efficiency in dramatic reduction of reaction time, and high yield product formation almost solventlessly and these are potentially important factors in drug discovery. Various advance techniques introduced to microwave synthesis, such as click chemistry (Simone et al., 2011), green chemistry (Kümmerer and Hempel, 2010), multi-component reactions (Murray et al., 2005), combinatorial synthesis (Lin et al., 2011), parallel synthesis (Zhou et al., 2010), and automated library production (Hsiao et al., 2010), increased the output of pharmaceutically active chemical entities. All these techniques and strategies hold their distinguished advantages as well as shortcomings.

On the other way, microwave heating has proven its efficiency in dramatic reduction of reaction time, which is potentially important in drug discovery. Therefore, one of ideal options to accelerate the synthetic processes is to combine microwave chemistry with these techniques in drug discovery. That is why many pharmaceutical companies are incorporating microwave chemistry into their drug discovery efforts (Kappe and Dallinger, 2006). Tryptamines were synthesized by reduction of glyoxalylamide precursors with lithium aluminium deuteride via microwave enhanced single mode system under elevated pressure and anhydrous tetrahydrofuran as solvent at 150°C for 5min (Brandt et al., 2008). Few microwave assisted green drugs, such as the psychoactive drug N,N-dialkylated tryptamines, dihydropyrimidinones, and dihydropyrimidinethiones as antibacterial, antiviral, antihypertensive, and anticancer agents (used in AIDS therapies) can be produced with improved atom economy, under microwave heating, and fluorous solid-phase extractions (F-SPE) (Piqani and Zhang, 2011).

Microwave-enhanced efficient and convenient combinatorial method was developed for the preparation of 2,4-(1*H*,3*H*)-quinazolinediones and 2-thioxoquinazolinones by substituting methyl anthranilate with various iso(thio)cyanates in DMSO/$H_2O$ without any catalyst or base in 20min (Li et al., 2008).

$+ R_2 - NCX$ — $DMSO/H_2O$, MW, 120°C, 20 min.

X = O or S

$R_1$ = H, 5-F or 4,5-Dimethoxy

$R_2$ = Substituted phenyl or alkyl

Green microwave synthetic approach to diverse fused tricyclic xanthines has been reported as anticonvulsants to treat chemically induced seizures. It has good atom economy and high functional group tolerance (Ye et al., 2009a).

$AuCl(PPh_3)$(10%), $H_2O$, MW, 150°C, 10-60 min.

$AgAsF_6$(10%), $H_2O$, MW, 150°C, 10-60 min.

Biologically interesting compound indole-1-carboxamides were prepared in moderate to high yields in 5min by Au (I) catalyzed 5-endo-dig cyclization in water under microwave irradiation (Ye et al., 2009b). Microwave technology can also address the challenges of the rapid labeling of radiopharmaceuticals (Jones and Lu, 2006). New HIV-1 protease inhibitors with more potency (56 times) and excellent antiviral activities were obtained by introducing microwave irradiation to accelerate Stille or Suzuki cross-couplings at 120°C in 30–50min.

The manufactured drugs were ultimately come to the waterway. Occurrence of a large number of pharmaceuticals and personal care products (PPCP) in the environment is a serious multifaceted issue as many of which are highly bioactive and perpetually present in water locales that badly affects the surface and ground water quality. So far, about 100 different pharmaceuticals have been detected in the aquatic resources, usually in STW effluents (Vanderford and Snyder, 2006; Batt et al., 2008; Kasprzyk-Hordern, et al., 2008). Those pharmaceuticals cover many different therapeutic classes, including analgesics, beta-blockers, selective serotonin reuptake inhibitors (SSRIs), fibrates, anti-epileptics, and steroids. Indirect continuous exposure of multiple PPCP to human and environment (even at low dose), needs wide range of proactive action, which could be implemented in near and long term to minimize the introduction of these PPCP into our eco-system.

If we still do not realize the risk paradigm that arises from the magnitude of this issue, and the size of the problem, then the future generations will not going to forgive us; although, there are very few evidences of these pharmaceuticals in the environment, which are visualized to result in acute effects as organ damage. There are two cases, where drugs have had drastic effects (i) where the anti-inflammatory drug diclofenac has virtually wiped out the 97% of vulture population of Asia in 2004 (Rattner et al., 2008), and (ii) steroid estrogens, both natural (e. g., estradiol) and man-made (ethinyl estradiol; EE2), which has caused feminization of male fish worldwide in surface water receiving domestic wastewater treatment plant effluent (Sumpter and Johnson, 2008). Similarly, the presence of the most popular analgesic drug ibuprofen show antibacterial activity in water; furazolidone-medicated fish feed causes acute toxicity on crustaceans or copepods; antibiotic drugs streptomycin and chlortetracycline or oxytetracycline prevent growth of blue-green algae and pinto beans, respectively.

Tinidazole prescribed for protozoa infections results bacterial mutagens in the urine of patients in therapeutic treatment. The volume of pharmaceuticals that are used is expected to continue increasing worldwide as population density increases, per capita incomes rise, and new disease target groups as well as more potent compounds are identified (Jjemba, 2008). Therefore, the unintended consequences of pharmaceuticals in the environment cannot continue to be overlooked, and minimizing their potential of impact is an enormous task that is not going to be accomplished easily. It could also improve medical healthcare outcomes for customers and reduce healthcare cost.

Sincere research on this major thrust area is needed to improve the risk assessment from pharmaceuticals at different trophic levels. Drinking water is a cocktail of chemical compounds (more than 95 organic wastewater contaminants, including pharmaceuticals, personal care products, and other extensively used chemicals, such as detergent metabolites and insecticides). However, studies have been only focused on the effects of individual substance. Therefore, the mixtures of substances in water could be studied with respect to their chronic exposure to low doses. Nowadays, there is no regulation for pharmaceuticals. Thus, there is an urgent need to set a national or area wise primary drinking water regulation or policies for pharmaceuticals because it is highly important to make health care providers and patients made aware of the medical and environmental consequences of our medication practices, including overprescribing. It also aims at minimizing pharmaceutical use by creating awareness about the linkages between human health and ecological health.

### *3.2.5 GREEN DETERGENTS*

Synthetic detergents are surfactant or a mixture of surfactants, which lowered surface tension in water and broke down fatty materials. In other words, they decreased the fabric's hold on the dirt and they also dissolved dirt particles. Detergents with cleaning properties in dilute solutions were manufactured in the United States from the early 1930s. The core component of the detergents was usually alkylbenzenesulfonates, sodium tripolyphosphate, and so on (Hampel and Blasco et al., 2002; Leon et al., 2006; Diederik et al., 2007) in particular, obtained from petroleum. A benzene ring can be added to the tetramer of propylene dodec-1-ene (petroleum product), via Fridal-Crafts reaction in presence of hydrogen chloride or aluminium chloride catalyst. This dodecylbenzene is sulphonated by refluxing it with concentrated sulphuric acid and gives 4-dodecylbenzene sulphonic acid, which generates detergent after neutralization with sodium hydroxide.

$$CH_3(CH_2)_9CH{=}CH_2 + C_6H_6 \xrightarrow[HCl]{AlCl_3} CH_3(CH_2)_{11}{-}C_6H_5$$

1-Dodecene Benzene Dodecylbenzene

$$\downarrow H_2SO_4$$

$$CH_3(CH_2)_{11}{-}C_6H_4{-}SO_3H + H_2O$$

4-Dodecylbenzene sulphonic acid

$$\downarrow NaOH$$

$$CH_3(CH_2)_{11}{-}C_6H_4{-}SO_3Na + H_2O$$

4-Dodecylbenzene sulphonic (Detergent)

Water hardness is a significant factor of modern detergents. The harder is the water, more detergent is required. The ions in hard water, particularly of calcium and magnesium, bind to the surfactant components, so manufacturers added builders, which bind to and essentially remove these ions and soften the water. Phosphates have been most commonly used as the builder in detergents that creates freely available phosphates. However, excess phosphates can cause problems in our waterways, such as excessive toxic algal growth (blue-green algae) or decomposer organisms that require oxygen may increase, which can deplete the amount of oxygen dissolved in the water. Therefore, some detergent manufacturers have developed phosphate-free detergents. As an alternative to phosphates, manufacturers can use a builder, or combination of builders, including zeolites (aluminosilicates), sodium citrate and nitrilotriacetate (NTA). Detergent wastewaters containing alternative builders also have environmental impacts and must be treated by sewage treatment works. Some of them (alkyl phenols) are oestrogen mimics that can have serious detrimental effects on populations of aquatic animals, such as decreasing their ability to reproduce. Even after treatment, the environmental impacts of some alternative builders remain.

Most of the detergents, no matter, if they are with phosphate builder or not, found in the market are chemical-based and they are resistant to the action of the biological agents (bacteria, and the decomposers). It is very difficult to eliminate them from municipal waste waters. They can pose a serious problem to aquatic life. These problems are overcome by biodegradable detergents (Jan, 2007). The Australian standard for biodegradability of surfactants (AS1792) requires 80% of the mixture to be degraded within 21 days if the product is to carry the label 'biodegradable'. There is a growing market for green products now and these detergents are preferred by several environmental lovers. As they are effective, low cost, can remove all types of stains without damaging paint fastness, do not harm hands and fabric, non toxic, and safe for environment. Earth friendly detergents are plant-based and use absolutely zero petrochemicals, ammonia bleach, animal ingredients (or testing), paraben, phosphates, dyes, artificial fragrances, and so on. Instead, most of the ingredients would include things such as water, natural salts, essential oils, and plant-based cleaners.

Vegetable oil derived detergents contain straight chain linear alkyl sulfonate (LAS), making them suitable for detergent production. Therefore, because of availability and the environmental friendliness of these oils, detergents produced from neem seed oil (Alewo et al., 2010) (low surface tension as 0.00523N/m, Foamability 3cm/10min.), and castor oil (Isah, 2006) are preferred. Recently, it has been reported the possible anaerobic degradation of LAS under methanogenic conditions (Mogensen et al., 2003; Angelidaki et al., 2004; Lobner et al., 2005). The experiments were performed mainly with UASB reactors operating either in mesophilic (37°C) or in thermophilic (55°C) conditions. Removal of 40–80% of the initial LAS concentration was reported. Although, doubts still remains about the actual abatement of LAS through biological reactions under anaerobic conditions, because of the LAS-induced inhibition of autotrophic nitrification in soil and water, which can be monitored by the account of ammonia-oxidizing bacteria (Nielsen et al., 2004). Moreover, synthetic surfactants or oil in marine aerosol magnify the toxic effects of pollutants (ozone, polyaromatic hydrocarbons, phthalates, n-alkanes, etc.) on plants, especially in winter season, when degradation of LAS in seawater is considerably inhibited at lower temperatures (Leon et al., 2004). Some recent data indicates that an improvement of the coastal vegetation (in Spain) is observed in areas where wastewater is adequately treated by waste water treatment plants (WWTP).

### 3.2.6 GREEN DYES

Until the discovery of first synthetic dye mauveine, the plant and animal kingdom provided all the materials for dyeing textiles, paints and cosmetics, and colored food stuffs for making food more attractive, visually. Use of synthetic dyes involves use and release of large amount of hazardous chemicals in the environment during their production and subsequent use and creating worker safety concerns. Petroleum is the starting material for all synthetic dyes; thus, the price of dyes is sensitive to the petroleum price. Synthetic dyes are clearly superior with better fastness properties, good fixation and shade reproducibility. Preparation of synthetic dyes has substantial drawbacks, such as relatively strenuous reaction conditions namely and refluxing reactants for several hours in organic solvents. A lot of organic solvents are not friendly to environment and complexity of isolation of the products and their non-biodegradability is also a problem. Recently, the awareness of environment as well as increasing disputes about the risks of synthetic dyes (more than 4400 organic dyes are available) resulted in growing interest in natural coloring agents for textiles, leather, plastic newspaper, magazines, decorative items, films, every day utility products and to satisfy other human need.

Presently, the dye industry has start paying great attention to newer products, which has nice blend of fashion trends as well as environmental specifications. Recent ban on the use of azo dyes, due to their carcinogenic properties, has led all major dye stuff manufacturers to search for benign alternatives of toxic dyes. The safest way is the use of eco-friendly natural dyes. From the sustainable point of view, environmentally benign or natural dyeing stuff is becoming a top priority in recent years because of the medium in which, these plant cells or fungi or bacteria grow contain no expensive or toxic chemicals, low processing temperature (around 30°C), neutral pH of synthetic process, high yield, high purity and their biodegradablity. Along this line, dyeing by extract of green tea is particularly of interest due to its antibacterial and UV protection nature. Black tea (*Ca-*

*milla sinensis* variety *assamica*) is used as source of colorant to dye cotton and nylon fabrics.

Chitosan and commercially available cationic fixing agent to control the fastness and dye uptaking capacity were used as the potential substitute of the heavy metal salts, which are not eco-friendly. Chitosan and the cationic fixing agents are utilized as the linkers between tea catechin and cellulose of cotton and polyamide of nylon via the formation of hydrogen bonding and ionic bonding (Nuramdhani et al., 2011). Tamarind seed coat tannin extracted and employed as a natural mordant with 0.5–1% metal mordant namely copper sulphate for cotton, wool and silk fabrics dyeing using natural dyes namely turmeric and pomegranate rind. This resulted in good antibacterial activity up to 20 washes (Prabhu et al., 2011). The cranberry fruit contains anthocyanin and flavonol pigments (Boulanger and Singh, 1998), which play an important protective function against damaging effects of UV radiation on widely used polyamide fabrics. Nylon-66 and wool (Takahashi et al., 1991), Canadian golden rod plant (Bechtold et al., 2007a), barberry, madder, hollyhock, privet, walnut, sticky alder tree (Bechtold et al., 2003), ash tree(Bechtold et al., 2007b), *Hibiscus mutabilis* (Shanker and Vankar, 2007), *Terminalia arjuna*, *Punica granatum*, *Rheum emodi* (Vankar et al., 2007), *Coffea arabica* L. (Lee, 2007), *Garciniamangostana* L. (Chairat et al., 2007), *Rhizoma coptidis* (Ke et al., 2006), curcumin (Erdawati, 2013), and *Rubia cordifolia* (Vankar et al., 2008), plants were used as natural dyes for silk, wool and cotton fabrics with different mordant. Natural dyes are extracted from some abundantly occurring plant materials of forest origin. These dyes may be used for imparting different shades on silk, wool, and cotton using common mordant's like alum, salts of iron, tin, and chrome. Environmental-friendly syntheses such as microwave irradiation, sonochemical or ultrasonic techniques are the powerful tools toward organic synthesis of dyes. Solvent free microwave irradiation is well known as environmentally benign method, which offers several advantages including shorter reaction times, cleaner reaction profiles, and simple experimental or product isolation procedures (Loupy, 2007). For example, styryl dyes are widely used as sensitizers, additives in the photographic industry, a biologically active compound in pharmaceutical industry, and novel successful fluorescent probes in bioanalytical methods.

However, conventional synthesis of such dyes is often carried out by the reaction of 2- or 4-methyl quaternary salts and aromatic aldehydes at high temperature for a relatively longer reaction time 8hr, which usually leads to the formation of undesired side products, low yields, and considerable power consumption. Therefore, environmentally benign procedure under solvent free conditions and microwave irradiation in the presence of different basic or acidic reagents was adopted to synthesize the series of styryl dyes (Vasilev et al., 2008).

R, CHO + $H_3C$, $N^+$, A, $CH_3$ — Solvent / Base reflux → R, $N^+$, A, $CH_3$

A = $I^-$, $ClO_4^-$

R = Alkyl or Substituted alkyl

$CH_3$, $N^+$, $CH_3$ + O, H, R, N, R — Base (or p-TsOH) / 1. 3 drops $H_2O$ / 2. rt or MW / 3. $KI/H_2O$ or $NaClO_4/H_2O$ → $N^+$, $CH_3$, R, N, R

9 Examples

R = Alkyl, Y = $I^-$; $ClO_4$

In this revolutionary way, environmentally benign yellow rare earth inorganic pigments that is, $NaCe_{0.5}(MoO_4)$ (Sreeram et al., 2007), $Ce(MoO_4)_2$, and $Ce_{1-(x+y)}Zr_xTa_yO_2^{+\delta}$ ($x$ ranges from 0.15 to 0.2 and $y$ ranges from 0 to 0.05) (Vishnu et al., 2009) displaying colors ranging from white to yellow have been synthesized by design yellow pigments consisting of non toxic elements as alternatives to lead, cadmium and chromium pigments.

Acute toxicity of dyes and the numerous efforts undertaken to avoid or reduce risks, dye manufacturers, and tanneries are confronted with effluents, wastes and contaminated containers or packaging material that require carefully thought out disposal. The contamination of soil and water by dye containing effluents is a serious environmental concern. Due to the increasing awareness and concern of the global community over the discharge of synthetic dyes into the environment and their persistence there, much attention has been focused on the remediation of these pollutants. Among the current pollution control technologies, biodegradation of synthetic dyes by different microbes is emerging as an effective and promising approach. The biodegradation of synthetic dyes is an economic, effective, biofriendly, and environmentally benign process (Ali, 2010). Synthetic dyes Bisphenol A, Bromophenol blue, Remazol brilliant blue R, Methyl orange, Relative black 5, Congo red, and Acridine orange were decolorized by effective enzyme *Trametes laccase* obtained from *Trametes polyzon*a and the percentage of decolorization increased, when 2mM HBT was added in the reaction mixture (Chairin et al., 2013).

Although government, industry, and advocacy groups have taken significant actions to solve related problems, including restricting the use of certain substances, but the response remains inadequate. The EEC has promulgated at "EC Control of Substance Hazardous to Health Act, 1989" and published a red list enumerating a number of chemicals, the presence of which in any kind of fabric has been banned. An ordinance in Germany stipulates that no garment or any other article that comes into contact with the skin shall contain any of the twenty aromatic amines named there. Similar restrictions have been imposed in many other countries. In India, the ban on the use of azo dyes has been imposed by the Union Ministry of Environment and Forest under section 6(2) (d) of the Environment (Protection) Act, 1986 read with the Rule 13 of the Environment (Protection) Rules, 1986.

After textile dyeing, hair coloring is another popular application of colors. Some experts estimate that 65–75% of women dye their hair, and it is believed that the practice goes all the way back to ancient Egyptians. Kingsley says that vegetable dyes do not adhere to the hair and cause more damage because they must be used more frequently, while more effective formula simply uses less irritating and odorless substitute involving different chemicals. Synthetic hair dyes contain ammonia, which binds the color (blue/green); p-phenylenediamine (PPD), which increase the risk of cancer and cell mutation that is linked to problems with the immune and nervous systems with increased risk of diseases such as non-Hodgkin's lymphoma parabens; hydrogen peroxide, which lightens existing color, plastics, sulphites; sodium lauryl suflates, and other baddies, as common ingredients that causes allergic reactions, scalp sensitivity and inhaled fumes can lead to respiratory issues. Healthier hair color is certainly on the minds of industry professionals, as even the best organic hair colors are going to have some chemicals that we may not like.

After all this notorious salons outcomes, a promising news is coming from research laboratories that a new hair bleach may be on the way, and it will be gentler for both people and Earth. Japanese scientists are on their way to creating an eco-friendly hair bleach whose, roots are in fungus. An enzyme from a strain of forest fungus naturally breaks down melanin, the substance that gives hair its color. Compared bleaching with hydrogen peroxide, this enzyme would probably be easier on hair and skin. But researchers are still pretty far from turning theories into a product. More studies are required to make this dream into a reality.

### *3.2.7 ECO-WAXES*

Waxes are commercially produced at large scale for use in cosmetics, polishes, surface coatings, and many other applications. Waxes are mainly derived from three major plant sources. These are Jojoba, carnauba, and beeswax. The composition of the waxes varies greatly according to the plant species and the site of wax deposition (leaf, flower, fruit, etc.). Plant waxes usually provide a hydrophobic coating that reduces water loss and protects the surface. A greener approach to obtain waxes is to utilize the cheaper and abundant byproducts, such as cereal straw, carnauba, and sheep wool as a raw material and to use benign extraction techniques such as supercritical $CO_2$ for their selective removal from the plant matrix with lipids and pigments. Selecting the temperature and pressure of the supercritical $CO_2$ used in the extraction can vary their physical properties. Importantly the waxes from straws have been shown to have a microcrystalline structure and this property is important for many cosmetic uses. These waxes can be used as replacements for a wide range of existing products.

### 3.2.8 GREEN BUILDING CONSTRUCTION MATERIALS

Green cements or concrete are resource saving structures with reduced environmental impact in term of energy saving, $CO_2$ emissions and waste water. Otherwise, the entire cement making is environmentally destructive process as it includes extraction and mining of limestone, transportation of materials and energy intensive. It was reported in Chronicle that each ton of cement emits about 1,763 pounds of $CO_2$ during manufacture. Combustion in kilns sources air pollution (worse when burning tires, 37% in U.S. or hazardous waste), and resulting toxic ash (cement kiln dust i.e., 9 tons per 100 tons of clinker). Cement industries pollution generate carbon dioxide (global warming gas), acid gases ($H_2SO_4$, HF, HCl), nitrogen oxides, sulfur dioxide, particulate matter (dioxins and furans), 19 heavy metals including lead, mercury, cadmium, and chromium, as the products of incomplete combustion (PICs), including dioxins, furans, and polycyclic aromatic hydrocarbons (PAHs). Over 7% of all green house gas emissions worldwide are caused from the manufacturing of Portland cement.

To develop a new green cements and binding materials, the use of alternative raw materials and energy saving strategies are in practice (Phair, 2006) that includes fly ash (Yazici and Hasan, 2012), waste glass (Nassar and Soroushian, 2011), waste fibreglass (Wang et al., 2000), blast furnace slag (Videla and Gaedicke, 2004), volcano ash, metakaolin or calcined clay, calcined shale, rice hull ash (Paya et al., 2000), calcined shale, municipal solid waste incinerated product (Horiguchi and Saeki, 2004), and alternative fuels (sewage sludge, etc) (Zabaniotou and Theofilou, 2008), to develop or improve cement with low energy consumption. Residual products from the concrete industry, that is, stone dust (from crushing of aggregate) and concrete slurry (from washing of mixers and other equipment) and ceramic wastes are used as green aggregates. These new types of cement with reduced environmental impact are more cost effective, and it provides tremendous environmental benefits. Few more ecological benign cementitious materials are

(i) **Geopolymer Concrete**

Geopolymers are amorphous alumino-silicate binding materials, unlike natural crystalline zeolitic materials and these can be synthesized by polycondensation reaction of geopolymeric precursor, and alkali polysilicates. Basic ingredients of geopolymers include, fly ash C (FA), sand aggregates (SA), alkaline liquid (sodium silicate and sodium hydroxide solution (AL), water, and super plasticizer (SP). They contain the molar ratio of Si-to-Al about 1 to 3. Since no limestone is used in geopolymer cement, they possess excellent properties to survive within the tough environments (both acid and salt). Compared with Portland cement, the geopolymers has a relative higher strength, excellent volume stability with better durability and low cost. Geopolymer concrete based on pozzolana is a new material that does not need the presence of Portland cement as a binder. The polymerization process involves a substantially fast chemical reaction under alkaline condition on Si-Al minerals that results in a three dimensional polymeric chain and ring structure consisting of Si-O-Al-O bonds. A geopolymer can take one of these three basic forms (Figure 3.1).

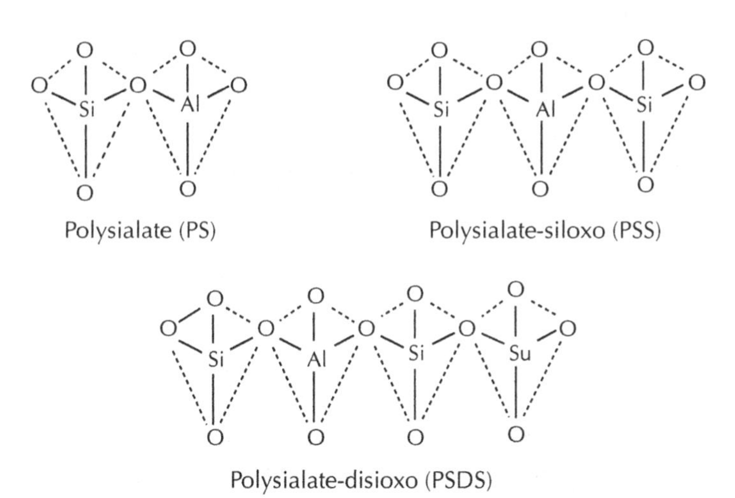


**FIGURE 3.1** Three basic forms of geopolymer (a) Si/Al = 1, PS: Polysialate, (b) Si/Al = 2, PSS: Polysialate-sioxo, and (c) Si/Al = 3, PSDS: Polysialate-disioxo.

This technology makes concrete more corrosion resistant and durable than Portland-based concrete. Sea water can be used for the blending of the geopolymer cement, which was impossible with Portland cement.

(ii) **Novacem's Cement**

Novacem is carbon negative cement. It is a composite of hydrated MgO and silicate. Its strength developed through the formation of magnesium silicate hydrates (M-S-H) rather than carbonation with atmospheric $CO_2$. Novacem technology is strongly protected by more than four patents. Non carbonate feedstock (uses magnesium silicates) is used in Novacem, so that no $CO_2$ is emitted from the raw material. No absorption of $CO_2$ in cement production and lower process temperature (700°C) can make us able to utilize biomass as fuel. 0–150kg $CO_2$ is created per ton of cement, depending on fuel mix used, whereas, conventional cement produces almost 800kg $CO_2$ per ton of cement. Cement composition includes a carbonate created during production process by absorbing $CO_2$.

(iii) **Mineral Admixtures or Supplementary Cementitious Materials (SCMs)**

SCMs are alumino-siliceous materials that possess pozzolanic (materials containing reactive silica and/or alumina, which on their own have little or no binding property) reactivity and/or latent hydraulic reactivity. CaO, silicates, and $H_2O$ are the main constitutes of SCMs.

$$2\,C_3S + 6\,H \rightarrow C_3S_2H_3 + 3\,CH$$

$$2\,C_2S + 4\,H \rightarrow C_3S_2H_3 + CH$$

where, C = CaO; S = $SiO_2$; H = $H_2O$ and C-S-H; molar ratios can vary with strength giving phase. CH has no cementitious properties (does not contribute to strength), easily leached and prone to chemical attack. SCMs can improve concrete properties by many of the beneficial effects. SCM observed the pore structure effect such as micro-filler effect that increased packing of cementitious particles, as the porous CH replaced with C-S-H, wall effect that densify the ITZ (interfacial transition zone) at the cement aggregate interface, causes pore blocking, which occurs because of a combination of these factors. These effects refine the pore structure and reduce the permeability of concrete; thereby making it more resistant to the penetration of deleterious agents. SCMs are of two types (i) Natural (ASTM C 618 Class N); it is produced from natural mineral deposits (e. g., volcanic ash or pumicite, diatomaceous earth, opaline cherts, and shales). It also requires heat treatment (e. g., metakaolin or calcined clay, calcined shale, rice hull ash, and calcined shale) and (ii) Processed / Manufactured, which contains silica fume (ASTM C 1240), fly ash (ASTM C 215), slag (ASTM C 989), and so on. Sustainable construction practices also includes blended cements, a combination of Portland cement and SCMs, which has improved strength and workability, durability. It is less expensive than pure cement. These are currently more common in Europe and South America than in the US. Partial replacement for cement reduces energy consumption and $CO_2$ emissions when used as productive use of industrial waste, which may be land filled. Contractors can derive some financial benefits to green construction.

### *3.2.9 BIO-BASED MATERIAL STARBONS*

Starbons are a novel family of mesoporous materials derived from polysaccharides, which retained their organized structure during pyrolysis, with tunable surface properties.

$$\text{Starch} \xrightarrow{\text{Expansion}} \text{Porous gel block} \xrightarrow{\text{Drying}} \text{Mesoporous starch} \xrightarrow{\text{Pyrolysis}} \text{Starbon}$$

Starbon products deliver a step change in performance over existing systems for example, acid resins in catalysis, porous graphitic carbons in separation, and activated carbon in adsorption. They are more selective, efficient, and effective, because of their unmatched high mesoporosity (0.4–0.7cm$^3$/g), high surface areas (150–500m$^2$/g), readily functionalisable (range of heteroatoms, acid/base, functionality, metal complexation), adjustable surface properties (different energies, hydrophilic to hydrophobic), excellent solvent stability, good chemical and heat resistance, controllable electrical conductivity, availability of starbons in different forms, from homogeneous micronized powders to beads to monoliths, and so on.

Starbons are used for various applications such as catalysis of bio-refinery downstream processes including esterification reactions in aqueous systems.

(a) Starbon > DArco > $SO_3H$ > β-25 > $ZrO_2$ > KSF

(b) Starbon > Darco > Norit > Blank

HO, O, O, OH ⇌ ETOH (aq) ⇌ HO, O, O, $OC_2H_5$ ⇌ ETOH (aq) ⇌ $H_5C_2O$, O, O, $OC_2H_5$

Biomass fermentation produces a wide range of organic acids, which can be utilized as platform molecules in various applications such as polymers and higher value intermediates. Esterification is one of the key upgrading steps for these acids. The fermentation process is carried out in aqueous media. Therefore, it requires intensive separation steps before the acids can be upgraded. Here, Starbon supported sulphonic acid catalysts developed by the Green chemistry centre overcome this problem and are able to perform esterification reactions with very high conversion yields and rates.

Other reactions where these catalysts have excelled in, includes aromatic amidiations and acylations. Starbon or nanometals perform well in other aqueous phase reactions including reductions with $H_2$ and oxidations with $H_2O_2$.

- Starbon C Series: Catalysis/catalyst support (solid acid/base, immoblised metal or active in aqueous media).
- Starbon P Series: Gas trapping and water purification, in particular, removal of harmful organics and heavy metals to purify water and clean up waste streams.
- Starbon S Series: Separation of complex mixtures for production and analysis with Starbon as the stationary phase in chromatographic systems.
- Starbon R Series: Recovery of precious metals through reductive adsorption.

These have also been used as biomedical devices, separation media, absorbency, remediation, effluent treatment, fuel cells, and so on (Clark et al., 2012).

### 3.2.10 BIODESIEL

Green chemistry tries to utilize benign and renewable feedstocks as raw materials, whenever it is possible. Therefore, combustion of fuels obtained from renewable feedstocks would be more preferable than combustion of the fossil fuels from depleting finite sources. Worldwide, there are many vehicles fuelled with diesel oil, and the production of biodiesel oil is a promising green option. Fat embedded plant's oil can be converted into the biodiesel via a transesterification reaction by using methanol and caustic or acid catalysts. During these reactions, the triglycerides are converted into the methyl ester and glycerol, which is a valuable raw material for soap production. Same can be achieved by utilizing supercritical methanol without a catalyst (Bunyakiat et al., 2006). This has the advantage of allowing a greater range of feedstocks (in particular, used cooking oil), the product does not need to be washed to remove catalyst, and is it easier to design as a continuous process.

$$\begin{array}{l} CH_3(CH_2)_{16}\overset{O}{\overset{\|}{C}}OCH_2 \\ CH_3(CH_2)_{16}\overset{O}{\overset{\|}{C}}OCH \\ CH_3(CH_2)_{16}\overset{O}{\overset{\|}{C}}OCH_2 \end{array} + \underset{\text{Methanol}}{3\ CH_3OH} \xrightarrow{KOH} \underset{\text{Biodiesel}}{3\ CH_3(CH_2)_{16}\overset{O}{\overset{\|}{C}}OCH_2} + \underset{\text{Glycerine}}{HOCH_2\overset{O}{\overset{\|}{C}}OCH_2OH}$$

### 3.2.11 ENVIRONMENTAL BENIGN SUPERCRITICAL FLUIDS

The use of supercritical fluids (SCFs) in chemical processes became more and more rampant (Jessop and Leitner, 1999). Planets of our solar system are good examples of systems that possess SCFs. Out of these Sun is the best, which is mainly composed of the hydrogen and helium at temperatures well above their critical points. The gaseous outer atmospheres of

Jupiter and Saturn transit smoothly into the interior SCFs. The term supercritical fluids comprises of the liquids and the gases at temperatures and pressures higher than their critical temperatures and pressures. It makes easy to adjust density and solution ability by a small changes in temperature or pressure. Due to this, the supercritical fluids are able to dissolve many compounds with different polarity and molecular mass. Therefore, SCFs are suitable as a substitute for organic solvents in a wide range of industrial and laboratory processes. Among them, green reagents carbon dioxide (sc$CO_2$) and water (sc$H_2O$) are quite popular. Easily available (from natural sources) and low cost, supercritical carbon dioxide is a new surfactant with high surface activity, which has opened a way to new processes in textile and metal industries and for dry cleaning of clothes (Wardencki et al., 2005). Supercritical carbon dioxide can be used instead of PERC (perchloroethylene) or other undesirable solvents for dry cleaning. Moreover, $CO_2$ at high pressures has antimicrobial properties (Fraser, 1951).

Room-temperature ionic liquids are considered to be environmentally benign reaction media because they are low viscosity liquids with no measurable vapor pressure. However, the lack of sustainable techniques for the removal of products from the room temperature ionic liquids has limited their application. Fortunately, sc$CO_2$ dissolves in the ionic liquid at appreciable extent to facilitate their extractions. Moreover, no measurable cross-contamination of the sc$CO_2$ by the ionic liquids is there, so that the product can be recovered in pure form. Hence, it can be used on a large scale for the decaffeination of green coffee beans, the extraction of hops for beer production, the production of essential oils, and pharmaceutical products from plants.

Supercritical water can be used to decompose biomass via supercritical water gasification of biomass. This type of biomass gasification can be used to produce hydrocarbon and hydrogen fuels. Supercritical fluid chromatography (SFC) can be used on an analytical scale, where in a few cases such as chiral separations and analysis of high molecular weight hydrocarbons. Impregnation is, in essence, the converse of extraction. A substance is dissolved in the supercritical fluid; the solution flowed past a solid substrate, and is deposited on or dissolves in the substrate. Dyeing of polyester fiber can also be performed using SCFs. Formation of the small particles (nano systems, quantum dots, etc.) of a substance is an important process in the pharmaceutical and many other industries. Supercritical fluids provide a number of ways for their production by rapidly exceeding saturation point of a solute by dilution, depressurization, or a combination of these that promotes nucleation or spinodal decomposition over crystal growth and harvests very small and regular sized particles.

Recent supercritical fluids have shown the capability to reduce particles up to a range of 5–2,000nm (Yeob and Kirana, 2005). Furthermore, supercritical fluids can be used to deposit functional nanostructured films and nanometer size particles of metals onto surfaces (Bart, 2005). Supercritical drying of the archeological and biological samples is a method of removing solvent from main products without surface tension effects. As a liquid dries, the surface tension drags on small structures within a solid, causing distortion and shrinkage. Under supercritical conditions, there is no surface tension, and the supercritical fluid can be removed without distortion.

Supercritical water oxidation uses supercritical water along with molecular oxygen as oxidizing agent that gives up oxygen atoms, which oxidizes the hazardous waste and eliminate the production of the toxic combustion products that the burning can produce (Oakes et al., 1999). The efficiency of a heat engine (via raise in operating temperature of the power station) for subcritical operation is raised from 39 to 45%, when $scH_2O$ is used as the working fluid. The efficiency of the power stations will be improved by raising the operating temperature. Supercritical water or carbon dioxide reactors (SCW/CRs) are the promising advanced nuclear systems that offer the thermal efficiency gains. Supercritical water or $scCO_2$ reactors are promising advanced nuclear systems that offer thermal efficiency gains to power stations (Dostal et al., 2004). Supercritical carbon dioxide is being used in domestic heat pumps as an emerging new refrigerant material for low-carbon solutions by small inputs of electric power by moving heat into the system from their surroundings.

### *3.2.12 NANOPARTICLES*

Routine techniques for nanoparticle production such as photochemical reduction (Eustis et al., 2005), laser ablation (Mafune et al., 2002), electrochemistry (Rodríguez-Sánchez et al., 2000), lithography, or high energy irradiation (Zhang and Wang, 2008; Treguer et al., 1998), either remain expensive or employ hazardous substances, such as organic solvents, and toxic reducing agents like sodium borohydride and N,N-dimethylformamide The biosynthetic procedures involve living organisms, such as bacteria (Joerger et al., 2000), fungi (Bhainsa and D'Souza, 2006), plants (Gardea-Torresday et al., 2002), plants extracts (Vilchis-Nestor et al., 2008), biocatalyst (enzymes or engineered enzyme), and a practical low cost natural feedstock are used to perform many chemical reactions of great importance. Resulting process greatly reduces hazard and waste, cost effective and meets the needs of customers. Biological synthetic processes have emerged as a simple and viable alternative to more complex physicochemical approaches to obtain nanomaterials with adequate control of size and shape (Shankar et al., 2004).

### *3.2.13 ANTIFOULANTS*

Rohm and Haas Company received a Presidential Green Chemistry Challenge Award for designing the environmentally safe marine antifoulant called Sea-Nine™. The unwanted growth of plants and animals on a ship's surface causes fouling and it costs the shipping industry approximately $3 billion a year. The main compounds used worldwide to control this fouling are organotin antifoulants, such as tributyltin oxide (TBTO). This agent has widespread environmental problems due to its persistence in the environment and the side effects. They cause acute toxicity, bioaccumulation, decreased reproductive viability, and increased shell thickness in shellfish.

This company selected 4,5-dichloro-2-*n*-octyl-4- isothiazolin-3-one as an alternative. TBTO bioaccumulates as much as 10,000 times, while Sea-Nine™ bioaccumulation is essentially zero. Both TBTO and Sea-Nine™ were acutely toxic to marine organisms, but TBTO had widespread chronic toxicity, but Sea-Nine™ antifoulant showed no chronic toxicity.

### 3.2.14 OTHER GREEN CHEMICALS

Term green chemistry was first coined to design chemicals and chemical processes that will be less harmful to human health and environment by implementing sustainable development in chemistry and chemical technologies. But the irony is that the analytical methods used to assess the state of environmental pollution may in fact be the great source of pollutants emission, which influences the environment, adversely. Therefore, the basic principles of green chemistry were proposed to protect the environment from pollution. These principles helps to increase the efficiency of synthetic methods, use of less toxic and renewable solvents or starting materials, reduce the stages of the synthetic routes, lower the energy used, minimizing waste as far as practically possible, and more biodegradable byproducts (Tundo and Anastas, 2000; Sheldon and Arends, 2006; Poliakoff and Licence, 2007). In this way, organic synthesis will be part of our effort for sustainable development. Green chemistry along with the green engineering provides us the potential tools, alternative materials, processes and systems. Figure 3.2 shows not only the sustainability of the chemicals/materials production, but also their environmental credentials by reducing toxicity and increasing recyclability.

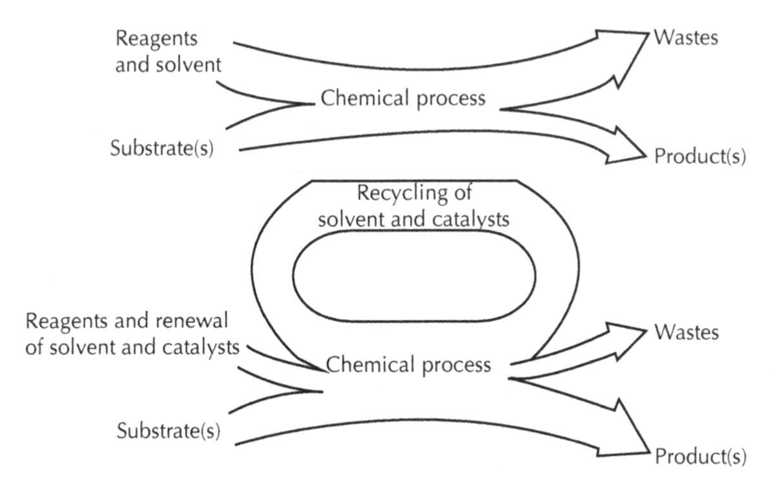


**FIGURE 3.2** Chemical process (i) Conventional way and (ii) Green chemical method and (iii) green process of material synthesis.

One interesting example is of adipic acid formation. Adipic acid is a very popular starting material for nylon-6,6 and catechol synthesis (widely used in the pharmaceutical and pesticide industries). Traditionally, the adipic acid was produced by the oxidation of cyclohexanone or cyclohexanol with nitric acid in solvent benzene and catalyst copper or vanadium (Cu 0.1–05% and V 0.02–0.1%).

$$C_6H_{12}O + 2\ HNO_3 + H_2O \rightarrow C_6H_{10}O_4 + NO_x\ (NO, NO_2, N_2O, N_2)$$

Benzene is mainly obtained from petrochemical industry and is also known for its carcinogenic properties. Oxidant nitric acid, also generate toxic fumes of nitric oxides ($NO_x$), which is one of the contributors to the greenhouse effect and the destruction of the ozone layer in the stratosphere. The yield of this reaction and its reaction mass efficiency (RME) is found 93% and 55.7%, respectively. Finally, a greener chemical route with green oxidizing agent $H_2O_2$, without solvent, using a new generation of catalysts, phase transfer catalyst (PTC), was presented (Anastas et al., 2001). The starting chemical is cyclohexene and its oxidation was performed by 30% hydrogen peroxide ($H_2O_2$). The tungsten salts ($Na_2WO_4$/$KHSO_4$/Aliquat 336 (Stark's catalyst) was dissolved in a special organic solvent (Aliquat 336). With cyclohexene or 1,2-cyclohexanediol, the yield is 45-86% (Sato et al., 1998; Usui and Sato, 2003).

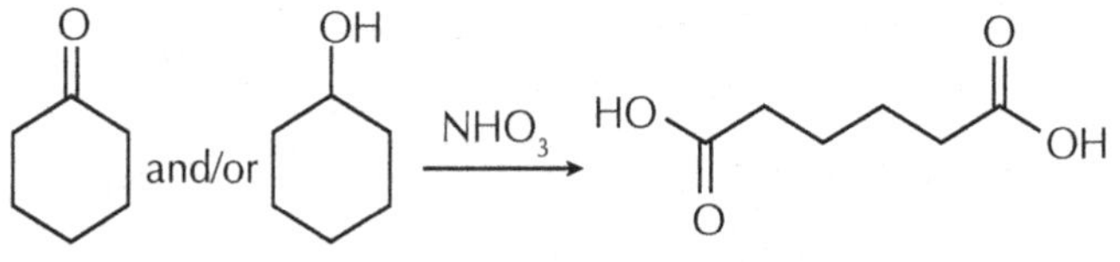

O OH
and/or
$NHO_3$
HO O OH O

Conventional route Adipic acid

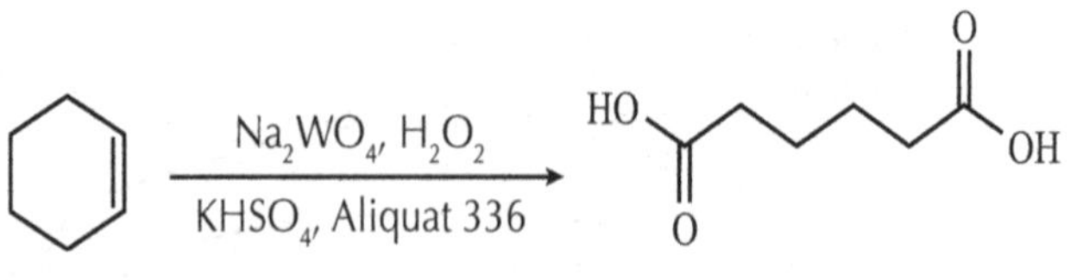

$Na_2WO_4$, $H_2O_2$
$KHSO_4$, Aliquat 336
HO O OH O

Green chemical route Adipic acid

Solvent Aliquat 336 is a mixture of octyl C8 and decyl C10 chains (with C8 predominating). It is a quaternary ammonium salt used as PTC and metal extraction reagent able to dissolve metal complexes. Here, one can also use W as oxpoeroxo tungsten complexes with molybdenium.

$$C_6H_{10} + 4\ H_2O_2 \rightarrow \text{Catalysts} \rightarrow C_6H_{10}O_4 + 4\ H_2OC$$

This eco-friendly method did not produce toxic waste. Moreover, its yield and reaction mass efficiency is 90 and 67%, respectively. RME of this, green chemical method is 11% higher than the traditional one. Recently, biocatalytic method for the synthesis of adipic acid from D-glucose has also been promoted, where genetically transgenic bacteria *Klebsiella Pneumoniae*, a non-toxic strain of *Escherichia coli* and *Enterobactiriaceae*, used as biocatalyst. (Draths and Frost, 1994).

Maleic anhydride (MA or cis-butenediol acid) synthesis is used as a starting material for the production of polyimides, polyester resins, surface coatings, lubricant additives, phthalic-type alkyd, plasticizers, and copolymers. It is also used as an important intermediate in the synthesis of 1,4-butanediol (in the industry of polyurethane and butyrolactone). Traditionally, maleic anhydride was produced using benzene, butene or butane as a starting material, and air as an oxidizing gas, in presence of the catalyst, which was composed of oxides of vanadium and molybdenum, $V_2O_5$, and $MoO_3$ (fixed bed reactor) under 3–5 bar pressure and 350–450°C temperature.

$$2\ C_6H_6 + 9\ O_2\ (\text{air}) \xrightarrow{\text{Catalysts}} 2\ C_4H_2O_3 + H_2O + 4\ CO_2$$

Reaction has a yield of 95% and RME equals to 44.4%. In the 1990s, two very big industrial enterprises UCB chemicals (Belgium) and BASF (Germany) started producing this as a byproduct of the oxidation of naphthalene into phthalic acid and phthalic anhydride. Another method was proposed with no use of solvent and starting material n-butane and catalyst $(VO)_2\ P_2O_5$ or new catalysts (special complexes of vanadium-phosphorous; fixed bed reactor, which is now converted into circulating fluidized bed reactor) in air, at temperatures 0–200°C (RME = 57.6% and yield = 60%) (Contractor et al., 1994). Therefore, these methods are much greener as compared with the original which uses benzene and the atom economy of the reaction was better without much waste. The new greener method did not produce toxic waste and its yield is 90% with RME 67%.

$$C_4H_{10} + 3.5\ O_2\ (\text{Air}) \xrightarrow{\text{Catalyst}} C_4H_2O + 4\ H_2O$$

There are few more examples of eco-benign chemical reactions, such as, mild N-formylation in the presence of indium metal as a catalyst under solvent free conditions. It is a chemoselective reaction of amines and $\alpha$-amino acid esters without epimerization (Kima and Jang, 2010).

$$RR'NH + 3\ \text{eq.}\ HCO_2H \xrightarrow[\text{Neat, } 100°C,\ 1.5 - 24\ h]{0.1\ \text{eq. In}} RR'N\text{-}CHO$$

R = Ar, Alkyl
R′ = H, Alkyl

Similarly, the use of phosgene and methylene chloride in the synthesis of polycarbonates has been replaced by diphenylcarbonate. Borono–Mannich reactions can be performed in solvent free conditions under microwave irradiation with short reaction time. Full conversion of the starting materials toward the expected product was achieved, starting from stoichiometric quantities of reactants, avoiding column chromatography. No purification step other than an aqueous washing was required (Nun et al., 2010).

$$Ar\text{-}B(OH)_2 + \text{(2-hydroxybenzaldehyde: CHO, OH)} + HNR_2 \xrightarrow[\text{MW, } 120°C,\ 2\ h]{\text{Neat}} \text{(2-(Ar)(NH}_2\text{)CH-phenol: Ar, NH}_2\text{, OH)}$$

R = Alkyl, Bn

Intermolecular addition of perfluoroalkyl radicals on electron rich alkenes and electron deficient alkenes with in water, mediated by silyl radicals gives perfluoroalkyl substituted compounds in good yields. The radical triggering events employed consists of thermal decomposition of 1,1′-azobis(cyclohexanecarbonitrile) (ACCN) or dioxygen initiation (Barata-Vallejo and Postigo, 2010).

Oxidation is the most polluting reaction in industries. Implementation of green chemistry has led to alternative less polluting reagents viz., molecular $O_2$ as the primary oxidant with extremely high oxidation state transition metal complexes. A convenient green synthesis of acetaldehyde is by Wacker oxidation of ethylene with $O_2$ in presence of a catalyst, instead of its synthesis by oxidation of ethanol or hydration of acetylene with $H_2SO_4$. Conventional methylation reactions employ toxic alkyl halides or methylsulfate that leads to environmental hazards. These are replaced by dimethylcarbonate with no deposit of inorganic salts. Microwave-assisted eco-friendly syntheses have also become a novel approach in green synthesis of chemicals. Organic synthesis under the microwave irradiation has many advantages as compared with the conventional reactions, which need very high temperatures and hazardous solvents. Microwave assisted reactions are "cleaner"; last only for very few minutes, no solvent is used, have high yield and produce minimum waste (Lidstrom et al., 2001).

Ultrasound-assisted organic syntheses were also found as other green synthetic routes with great advantages for high efficiency, low waste, and low energy requirements. Ultrasonic region of 20kHz–1MHz has many applications due to its high energy and the ability to disperse reagent in small particles and accelerate reactions. Acoustic cavitations occasionally occurs (growth, and implosive collapse of bubbles). These cavitations can create extreme physical and chemical conditions (bubbles have temperatures around 5,000K and/or pressures of roughly 1000atm) in otherwise cold liquids (Mason, 1997; Cintas and Luche, 1999; Cravotto and Cintas, 2006). Sonochemical engineering is a newly emerging field involving the application of sonic and ultrasonic waves to chemical processing. Sonochemistry enhances or promotes chemical reactions and mass transfer. It offers the potential for shorter reaction cycles, cheaper reagents, and less extreme physical conditions. The traditional multistep synthesis was wasteful and used large amounts of hazardous reagents.

Chemistry has provided us comfort from all angles but it has also added to environmental pollution. One cannot change his life style without using these chemicals in some or the other way. However, these products can be replaced by some other chemicals or products, which are eco-friendly.

In this context, biodegradable materials are welcome. Nature has provided numerous examples, such as corn husk wrapper, beautiful network of pipelines for water supply in leaves and so on which are biodegradable. It is presumed that most of the chemical products will be replaced by some or the other biodegradable or less toxic materials in years to come.

## KEYWORDS


- **Bioplastics**
- **Eco-waxes**
- **Geopolymer concrete**
- **Green chemicals**
- **Nanoparticles**


## REFERENCES


1. Ahmed, A. T., Wael, F. T, Shaaban, M. A., & Mohammed, F. S. (2011). *Journal of Food Safety, 31,* 211–218.
2. Alberts, A. W., Chen, J., Kuron, G., Hunt, V., Huff, J., & Hoffman, C. et al. (1980). *Proceedings of National Academy of Sciences. U S A*, *77,* 3957–3961.
3. Alewo, O. A., Muhammed, T. I., & Ebenezer, K. U. (2010). *Leonardo Electronic Journal of Practices and Technologies*, *16,* 69–74.
4. Ali, H. (2010). *Water, Air, & Soil Pollution, 213,* 251–273.
5. Anastas, P. T., Kirchhoff, M. M., & Williamson, T. C. (2001). *Applied Catalysis A: General., 221,* 3–13.
6. Angelidaki, I., Torang, L., Waul, C. M., & Schmidt, J. E. (2004). *Water Science Technology*, *49,* 115–22.
7. Barata-Vallejo, S., & Postigo, A. (2010). *Journal of Organic Chemistry, 75,* 6141–6148.
8. Bart, C. J. (Ed.). (2005). *Additives in polymers: Industrial analysis and applications.* Chichester: John Wiley & Sons.
9. Batt, A. L., Kostich, M. S., & Lazorchak, J. M. (2008). *Analytical Chemistry, 80,* 5021–5030.
10. Bechtold, T., Mahmud-Ali, A., & Mussak, R. (2007a). *Dyes Pigments, 75,* 287–293.
11. Bechtold, T., Mahmud-Ali, A., & Mussak, R. A. M. (2007b). *Coloration Technology, 123,* 271–279.

12. Bechtold, T., Turcanu, A., Ganglberger, E., & Geissler, S. (2003). *Journal Cleaner Production, 11,* 499–509.
13. Bhainsa, K. C., & D'Souza, S. F. (2006). *Colloids and Surfaces B: Biointerfaces, 47,* 160–164.
14. Boulanger, R. R. Jr., & Singh, B. R. (1998). Light regulation of anthocyanin and flavonol biosyntheis in cranberry plants. *The Nucleus (Northeastern Section American Chemical Society), 76,* 14–18.
15. Brandt, S. D., Tirunarayanapuram, S. S., Freeman, S., Dempster, N., Barker, S. A., & Daley, P. F. et al. (2008). *Journal of Labelled Compounds and Radiopharmaceuticals, 51,* 423–429.
16. Bunyakiat, K., Makmee, S., Sawangkeaw, R., & Ngamprasertsith, S. (2006). *Energy and Fuels, 20,* 812–817.
17. Chairat, M., Bremner, J. B., & Chantrapromma, K. (2007). *Polymers, 8,* 613–619.
18. Chairin, T., Nitheranont, T., Watanabe, A., Asada, Y., Khanongnuch, C., & Lumyong, S. (2013). *Applied Biochemistry and Biotechnology, 169,* 539–545.
19. Chen, G., & Patel, M. (2012). *Chemical Reviews, 112,* 2082–2099.
20. Cintas, P., & Luche, J-L. (1999). *Green Chemistry, 1,* 115–125.
21. Clark, J. H., Budarin, V. L., Macquarrie, D. J., & Breeden, S. W. (2012). *Nanotechnology Conference and Trade Show* (pp. 493–494). *Santa Clara, USA.*
22. Contractor, R. M., Gemett, D. I., Horowitz, H. S., Bergna, H. E., Patience, G. S., Schwartz, J. T., & Sisler, G. M. (1994). *Studies in Surface Science and Catalysis, 82,* 233–242.
23. Cravotto, G., & Cintas, P. (2006). *Chemical Society Reviews, 35,* 180–196.
24. Derraik, J. G. B. (2002). *Marine Pollution Bulletin, 44,* 842–852.
25. Diederik, S., Helen, D., Rosa, F., Jeremy, H., Holger, K., & Paul, H. K., et al. (2007). *Regulatory Toxicology and Pharmacology, 49,* 245–259.
26. Dostal, M., Driscoll, J., & Hejzlar, P. (2004). *A supercritical carbon dioxide cycle for next generation nuclear reactors.* MIT-ANP-TR-100, MIT-ANP-Series.
27. Draths, C. M., & Frost, J. W. (1994). *Journal of. American Chemical Society, 116,* 399–400.
28. Drews, J., & Ryser, St. (Eds.). (1997). *Human Disease from genetic causes to biochemical effects* (pp. 5–9). Berlin: Blackwell.
29. Dunn, P. J., Galvin, S., & Hettenbach, K. (2004). *Green Chemistry, 6,* 43–48.
30. Erdawati, N. H. F. (2013). *Journal of Basicand Applied Science Research, 3,* 5–14.
31. Eustis, S., Hsu, H. Y., & El-Sayed, M. A. (2005). *Journal of Physical Chemistry B., 109,* 4811–4815.
32. Fraser, D. (1951). *Nature, 167,* 33–34.
33. Gardea-Torresday, J. L., Parsons, J. G., Gomez, E.; Peralta-Videa, J., Troiani, H. E., Santiago, P., & Yacaman, J. (2002). *Nano Letters, 2,* 397–401.
34. Hampel, M., & Blasco, J. (2002). *Ecotoxicoogy and Environmental Safety, 51,* 53–59.
35. Hollmuller, P., Joubert, J.-M., Lachal, B., & Yvon, K. (2000). *International Journal of Hydrogen Energy, 25,* 97–109.
36. Hong, C., Yu, P. H. F., & Chee, K. M. (1999–2003). *Applied Biochemistry and Biotechnology, 78,* 389–399.

37. Horiguchi, I., & Saeki, N. (2004). Compressive strength and leachate characteristics of new green CLSM with eco-cement and melted slag from municipal solid waste. *Special Publication, 221,* 529–558.
38. Hsiao, Y., Yellol, G. S., Chen, L., & Sun, C. (2010). *Journal of Combinatorial Chemistry, 12,* 723–732.
39. Isah, A.G. (2006). *Leonardo Electronic Journal of Practices and Technologies, 9,* 153–160.
40. Jan, W.G. (2007). *Encylopedic dictionary of polymers* (pp. 108–109). New York: Springer Science.
41.`Jessop, P. G., & Leitner, W. (Eds). (1999). *Chemical synthesis using supercritical fluids.* Weinheim: Wiley-VCH.
42 Jjemba, P. (2008). *Pharma-Ecology: The occurrence and fate of pharmaceuticals and personal care products in the environment.* New Jersey: Wiley.
43. Joerger, R., Klaus, T., & Granqvist, C. G. (2000). *Advanced Materials, 12,* 407–409.
44. Jones, J. R., & Lu, S. (Eds.). (2006). *Microwaves in organic synthesis* (2nd edn., pp. 820–859). Weinheim: Wiley-VCH.
45. Kappe, C. O., & Dallinger, D. (2006). *Nature Reviews Drug Discovery, 5,* 51–63.
46. Kasprzyk-Hordern, B., Dinsdale, R. M., & Guwy, A. J. (2008). *Water Research, 42,* 3498–3518.
47. Ke, G., Yu, W., & Xu, W. (2006). *Journal of Applied Polymer Science, 101,* 3376–3380.
48. Kima, J.-G., & Jang, D. O. (2010). *Synlett,* 1231–1234.
49. Kümmerer, K., & Hempel, M. (Eds.). (2010). *Green and sustainable pharmacy.* Berlin: Springer-Verlag.
50. Lai, F., Wissing, S. A., & Müller, R. H. (2006). *AAPS PharmSciTech, 7,* 10–18.
51. Lamsal, K., Kim, S.-W., Jung, J. H., Kim, Y. S., Kim, K. S., & Lee, Y. S. (2011). *Microbiology, 39,* 26–32.
52. Lee, Y. H. (2007). *Journal Applied Polymer Sciences, 103,* 251–257.
53. Leon, V. M., Gomez-Parra, A., & Gonzalez-Mazo, E. (2004). *Environmental Science and Technology, 38,* 2359–2367.
54. Leon, V. M., Lopez, C., Lara, M., Prats, D., Varo, P., & Gonzalez-Mazo, P. (2006). *Chemosphere, 64,* 1157–1166.
55. Li, Z., Huang, H., Sun, H., Jiang, H., & Liu, H. (2008). *Journal of Combinatorial Chemistry, 10,* 484–486.
56. Lidstrom, P., Tierney, J., & Wathey, B. (2001). *Tetrahedron,* 57, 9225–9283.
57. Lin, P., Salunke, D. B., Chen L., & Sun, C. (2011). *Organic & Biomolecular Chemistry, 9,* 2925–2937.
58. Liu, H., & Zhang, L. (Eds.). (2011). *Microwave heating.* New York: InTech.
59. Lobner, T., Torang, L., Batstone, D. J., Schmidt, J. E., & Angelidaki, I. (2005). *Biotechnology Bioengineering, 89,* 759–65.
60. Loupy, A. (2007). *Bulletin of Korean Chemical Society, 28,* 83–88.
51. Mafune, F., Kohno, J., Takeda, Y., & Kondow, T. (2002). *Journal of Physical Chemistry B., 106,* 7575–7577.
62. Martínez-Palou, R. (2007). *Journal of the Mexican Chemical Society, 51,* 252–264.
63. Mason, T. J. (1997). *Chemical Society Reviews, 26,* 443–451.

64. Matics, J., & Krost, G. (2007). *Prospective and adaptive management of small combined heat and power systems in buildings*, Proceedings of 9th REHVA World Congress CLIMA, Helsinki Finland, 2007.
65. Draths, C. M., & Frost, J. W. (Eds.). (1994). *Microbial catalysis: Synthesis of adipic acid from D-glucose,* ACS Symposium Series, 577. Washington DC: American Chemical Society.
66. Mogensen, A. S., Haagensen, F., & Ahring, B. K. (2003). *Environmental Toxicology and Chemistry, 22,* 706–711.
67. Murray, J. K., Farooqi, B., Sadowsky, J. D., Scalf, M.,Freund, W. A., Smith, L. M., Chen, J., & Gellman, S. H. (2005). *Journal of American Chemical Society, 127,* 13271–13280.
68. Nagendrappa, G. (2002). *Resonance, 7,* 64–77.
69. Nassar, R.-ud-D., & Soroushian, P. (2011). *Journal of Solid Waste Technology and Management, 37,* 307–319.
70. Nielsen, K. B., Brandt, K. K., Jacobsen, A. M., Mortensen, G. K., & Sorensen, J. (2004). *Environmental Toxicology and Chemistry, 23,* 363–370.
71. Ning, J., Kong, F., Lin, B., & Lei, H. (2003). *Journal of Agricultural and Food Chemistry, 51,* 987–91.
72. Nun, P., Martinez, J., & Lamaty, F. (2010). Microwave-assisted neat procedure for the petasis reaction. *Synthesis,* 2063–2068.
73. Nuramdhani, I., Widodo, M., Harnirat, H., & Juhana, J. (2011). *The application of chitosan and cationic fixing agent for environmentally benign process of black tea dyeing cf cotton and polyamide fabrics* (pp. 347–352). Proceedings of the 2nd International Seminar on chemistry, Jatinangor.
74. Oakes, R. S., Clifford, A. A., Bartle, K. D., Pett, M. T., & Rayner, C. M. (1999). *Chemical Communication, 3,* 247–248.
75. Paya, J., Monzo, J., Borrachero, M. V., Peris-Mora, E., & Ordonez, L. M. (2000). *Waste Management, 1,* 493–503.
76. Phair, J. W. (2006). *Green Chemistry, 8,* 763–780.
77. Piqani, B., & Zhang, W. (2011). *Beilstein Journal of Organic Chemistry, 7,* 1294–1298.
78. Poliakoff, M., & Licence, P. (2007). *Nature, 450,* 810–812.
79. Prabhu, K. H., Teli, M. D., & Waghmare, N. (2011). *Fibre Polymers, 12,* 753–759.
80. Rattner, B. A., Whitehead, M. A., Gasper, G., Meteyer, C. U., Link, W. A., & Taggart, M. A. et al. (2008). *Environmental Toxicology and Chemistry, 27,* 2341–2345.
81. Ritter, S. K. (2003). *Chemical & Engineering News, 81,* 31–33.
82. Rodríguez-Sánchez, L., Blanco, M. C., & López-Quintela, M. A. (2000). *Journal of Physical Chemistry B, 104,* 9683–9688.
83. Sanderson, K. (2011). *Nature, 469,* 18–20.
84. Sato, K., Aoki, M., & Noyori, R. A. (1998) *Science, 281,* 1646–1647.
85. Shankar, S. S., Rai, A., Ankamwar, B., Singh, A., Ahmad, A., & Sastry, M. (2004). *Nature Materials, 3,* 482–488.

86. Shanker, R., & Vankar, P. S. (2007). *Dyes Pigments, 74,* 464–469.
87. Sheldon, R. A., & Arends, I. (2006). *Green chemistry and catalysis.* Indianapolis: Wiley–VCH.
88. Simone, R. D., Chini, M. G., Bruno, I., Riccio, R., Mueller, D., Werz, O., & Bifulco, G. (2011). *Journal of Medical Chemistry, 54,* 1565–1575.
89. Sreeram, K. J., Srinivasan, R., Devi, J. M., Nair, B. U., & Ramasami, T. (2007). *Dyes Pigments, 75,* 687–692.
90. Sumpter, J. P., & Johnson, A. C. (2008). *Journal of Environmental Monitoring, 10,* 1476–1485.
91. Takahashi, A., Takeda, K., & Ohnishi, T. (1991). *Plant Cell Physiology, 32,* 541–548.
92. Treguer Cointet, C., Remita, H., Khatouri, J., Mostafavi, M., Amblard, J., & Belloni, J. (1998). *Journal of Physical and Chemistry B, 102,* 4310–4321.
93. Tundo, P., & Anastas, P. T. (Eds). (2000). *Green chemistry: Challenging perspectives.* Oxford: Oxford University Press.
94. Usui, Y., & Sato, K. A. (2003). *Green Chemistry, 5,* 373–375.
95. Vanderford, B. J., & Snyder, S. A. (2006). *Environmental Science and Technology, 40,* 7312–7320.
96. Vankar, P. S., Shanker, R., Mahanta, D., & Tiwari, S. C. (2008). *Dyes Pigments, 76,* 207–212.
97. Vankar, P. S., Shanker, R., & Verma, A. (2007). *Journal of Cleaner Production, 15,* 1441–1450.
98. Vasilev, A., Deligeorgiev, T., Gadjev, N., Kaloyanova, S., Vaquero, J. J., Alvarez-Builla, J., & Baeza, A. G. (2008). *Dyes Pigments, 72,* 550–551.
99. Videla, C., & Gaedicke, C. (2004). *ACI Material Journal, 101,* 365–375.
100. Vilchis-Nestor, A. R., Sánchez-Mendieta, V., Camacho-López, M. A., Camacho-Ló pez, M. A., & Arenas-Alatorre, J. A. (2008). *Material Letters, 62,* 3103–3105.
101. Vishnu, V. S., George, G., Divya, V., Reddy, M. L. P. (2009). *Dyes Pigments, 82,* 53–57.
102. Wang, Y., Wu, H. C., & Li, V.C. (2000). *Journal of Materials in Civil Engineering, 12,* 314–319.
103. Wardencki, W., Cury, J., & Namieoenik, J. P. (2005). *Journal of Environmental Studies, 14,* 389–395.
104. Weller, H. N., Nirschl, D. S., Petrillo, E. W., Poss, M. A., Andres, C. J., & Cavallaro, C. L. et al. (2006). *Journal of Combinatorial Chemistry, 8,* 664–669.
105. Wingens, J., Krost, G., Ostermann, D., Damm, U., & Hess, J. (2008). *Hydrogen production for autonomous solar based electricity supply.* DRPT2008, Nanjing, China.
106. Xie, X., & Tang, Y. (2007). *Applied Environmental Microbiology, 73,* 2054–2060.
107. Yazici, S., & Hasan¸ S. A. (2012). *Sadhana, 37,* 389–403.
108. Ye, D., Zhang, X., Zhou, Y., Zhang, D., Zhang, L., & Wang, H. et al. (2009a). *Advanced Synthesis, 351,* 2770–2778.
109. Ye, D., Wang, J., Zhang, X., Zhou, Y., Ding, X., & Feng, E. et al. (2009b). *Green Chemistry, 11,* 1201–1208.

111. Yeob, S. D., & Kirana, E. (2005). *Journal of Supercritical Fluids, 34,* 287–308.
112. Yvon, K., & Lorenzoni, J. L. (2006). *International Journal of Hydrogen Energy, 31,* 1763–1767.
113. Zabaniotou, A., & Theofilou, C. (2008). *Renew Sustainable Energy Reviews, 12,* 531–541.
114. Zhang, G., & Wang, D.Y. (2008). *Journal of American Chemical Society, 130,* 5616–3617.
115. Zhou, Y., Zhai, Y., Li, J., Ye, D., Jiang, H., & Liu, H. (2010). *Green Chemistry, 12,* 1397–1404.

## CHAPTER 4

# GREEN CATALYSTS

SHIKHA PANCHAL, YUVRAJ JHALA, ANURADHA SONI, and SURESH C. AMETA

## CONTENTS

## 4.1 INTRODUCTION

Many organic reactions of synthetic importance are very slow and it is quite important to enhance their reaction rate. The rate of the reactions can be enhanced by using a catalyst. This catalyst may be toxic in nature and it is important to find out some alternate catalyst, which is harmless or less toxic. Just to avoid environmental pollution, such a job can also be done by any enzyme. These enzymes are called biocatalyst or in general, green catalyst. Enzymes are used in chemical industries, when extremely specific catalysts are required. However, use of enzymes is limited because of their stability in organic solvents and higher temperatures. Searching or creating new enzymes with novel properties, either through rational designing or *in vitro* evolution (Hult and Berglund, 2003; Renugopalakrishnan et al., 2005) is a challenging task for chemists. A few enzymes have now been designed from scratch to catalyzed reactions that do not occur in nature (Jiang et al., 2008).

## 4.2 TYPES OF BIOCATALYTS

### *4.2.1 BIOCATALYSTS*

Lipase catalyzed double enantioselective transesterification of racemic carboxylic esters and cyclic *meso*-diols to give the hydroxy esters have been investigated by Theil et al. (1994). The penicillin acylase catalyzed synthesis of ampicillin, *via* acylation of 6-aminopenicillanic acid with D-phenylglycine amide, is accompanied by the formation of the hydrolysis product D-phenylglycine (Langen et al., 2001). The immobilized resting-cell of *Geotrichum candidum* was used as a catalyst for the reduction of a ketone in a semi-continuous flow process using supercritical carbon dioxide (Matsuda et al., 2003) while Shao et al. (2002) carried out biocatalytic synthesis of uridine 5′-diphosphate *N*-acetylglucosamine by multiple enzymes co-immobilized on agarose beads.

Using Baker's yeast as a biocatalyst, the chemoselective reduction of aromatic nitro compounds bearing electron-withdrawing groups gave the corresponding hydroxylamines with good to excellent conversion under mild conditions (Li et al., 2004). Bohn et al. (2007) investigated that caffeine affects the stereoselectivity of microbial high cell density reductions with commercial grade *Saccharomyces cerevisiae* (Baker's yeast) while a novel Baker's yeast (*Saccharomyces cerevisiae*) catalyzed protocol for Knoevenagel condensation of aldehydes and active methylene compounds including 2,4-thiazolidinedione in an organic solvent at ambient temperature has been developed (Pratap et al., 2011).

New bacterial alcohol dehydrogenases with high and complementary enantioselectivity for the reduction of ethyl 3-keto-4,4,4-trifluorobutyrate and methyl 3-keto-3-(3′-pyridyl)-propionate have been investigated by Zhang et al. (2004). Stereoselective reductase catalyzed asymmetric deoxygenation of racemic alkylaryl, dialkyl, and phenolic sulfoxide was observed by Boyd et al. (2004). Cytochrome $P450_{BM3}$, from *Bacillus megaterium*, catalyzes the epoxidation of linolenic acid yielding 15,16-epoxyoctadeca-9,12-dienoic acid with complete regio- and moderate enantio-selectivity (60% ee) (Çelik et al., 2005). Wu et al. (2005) have reported a new enzymatic process, where penicillin G acylase from *Escherichia coli* displays a promiscuous activity in catalyzing the Markovnikov addition of allopurinol to vinyl ester.

A mediatorless microbial fuel cell based on the direct biocatalysis of *Escherichia coli* shows significantly enhanced performance by using bacteria electrochemically evolved in fuel cell environments through a natural selection process and a carbon/PTFE composite anode with an optimized PTFE content (Zhang et al., 2006). The enantioselective synthesis of (2*S*)-2-phenylpropanol and (2*S*)-2-(4-*iso*-butylphenyl)propanol ((*S*)-Ibuprofenol) has been achieved by horse liver alcohol dehydrogenase through dynamic kinetic resolution (Giacomini et al., 2007) Ríos et al. (2007) investigated green method for Baeyer–Villiger oxidation of substituted cyclohexanones *via* lipase-mediated perhydrolysis utilizing urea–hydrogen peroxide in ethyl acetate.

DNAzyme cascades activated by $Pb^{2+}$- or L-histidine-dependent DNAzymes yield the horseradish peroxidase-mimicking catalytic nucleic acids that enable the colorimetric or chemiluminescence detection of $Pb^{2+}$ or

L-histidine (Elbaz et al., 2008). A green procedure for the kinetic resolution of chiral amines *via* enzymatic acylation and deacylation has been demonstrated by Ismail et al. (2008). Asymmetric dihydroxylation of aryl olefins has been carried out by sequential enantioselective epoxidation and regioselective hydrolysis with tandem biocatalysts (Xu et al., 2009). Dupont et al. (2009) discussed multiphase conditions using classical acid or base catalysts as well as biocatalysts with some recent catalytic approaches and achievements, such as alcoholysis of triglycerides. The flavoprotein catalyzed reduction of aliphatic nitro-compounds represents a biocatalytic equivalent to the Nef-reaction (Durchschein et al., 2010).

$R^1R^2CH{-}NO_2$ —Reductase [H]→ —Tautomerization (Spont.)→ $R^1R^2C{=}N{-}OH$ —Reductase [H]→ —Hydrolysis $H_2O$ (Spont.)→ $R^1R^2C{=}N + NH_3$

Three enzymes, *Rhizopus oryzae* lipase (ROL), lysozyme, and phytase are reported to catalyze the condensation of the model compound, trimethylsilanol formed *in situ* from trimethylethoxysilane and produced hexamethyldisiloxane in aqueous media at 25°C and pH 7 (Abbate et al., 2010). A straightforward, high-yielding, chemoenzymatic total synthesis of enantiopure (*S*)-rivastigmine was developed by Fuchs et al. (2010) using various ω-transaminases for the asymmetric amination of appropriate acetophenone precursors. The sponge-restricted enzyme silicatein-α catalyzes *in vivo* silica formation from monomeric silicon compounds from sea water (*i.e.* silicic acid) and plays the pivotal role during synthesis of the siliceous sponge spicules (Wolf et al., 2010). Znabet et al. (2010) synthesized a very important drug candidate telaprevir, featuring a biocatalytic desymmetrization and two multicomponent reactions as the key steps.

Mitochondria, considered as the "powerhouse" of the living cell, can do bioelectrocatalysis of pyruvate, fatty acids, and amino acids at electrode surfaces for biofuel cell applications (Bhatnagar et al., 2011). Li et al. (2011) discovered the unnatural ability of nuclease p1 from *Penicillium citrinum* to catalyze asymmetric aldol reactions between aromatic aldehydes and cyclic ketones under solvent free conditions.

Nuclease pl
15°C

up to 99% ee
up to $>$ 99 : 1(Anti/Syn)

An enzymatic oxidation of met hanol combined with a lyase for the hydroxymethylation of aldehydes, coupled with a further enzymatic reduction affords enantiopure diols in one pot (Shanmuganathan et al., 2012).

Enzymatic orchestra

+ $CH_3OH$ → Oxidation + Ligation + Reduction → or

75 – 99%yield; 99% ee

Stereoselective benzylic hydroxylation of alkylbenzenes and epoxidation of styrene derivatives catalyzed by the peroxygenase of *Agrocybe aegerita* has been reported (Kluge et al., 2012). Balke et al. (2012) reported that Baeyer–Villiger monooxygenases (BVMOs) are useful enzymes for organic synthesis as they are enable in the direct and highly regio- and stereoselective oxidation of ketones to esters or lactones simply with molecular oxygen. A hybrid biocatalyst, where a synthetic rhodium complex is covalently inserted into a $\beta$-barrel protein scaffold, undergoes a stereoregular polymerization of poly(phenylacetylene) containing *trans* stereostructure (Onoda et al., 2012). Yara-Varón et al. (2012) synthesized poly(ethyl acrylate-*co*-allyl acrylates) from acrylate mixtures prepared by a continuous solvent-free enzymatic process.

Enzymatic synthesis of amoxicillin by penicillin G acylase in the presence of ionic liquids has been reported (Pereira et al., 2012). Ni et al. (2012) reported biohydrogenation (under mild reaction conditions) of carboxylic acids to the corresponding alcohols or aldehydes using *Pyrococcus furiosus*. Lozano et al. (2012) have developed a clean biocatalytic approach for producing flavor esters in switchable ionic liquid or solid phases by using an iterative cooling or centrifugation protocol. Highly selective semisynthetic lipases have been prepared by site-specific incorporation of tailor-made peptides on the lipase-lid site (Romero et al., 2012). Strategies have been reported for reducing the side-reactions of chemoenzymatic DKR tuning the organometallic catalyst and entrapping the biocatalyst (Pollock et al., 2012). Dioxygenase catalyzed stereoselective dihydroxylation of benzo[*b*] thiophenes and benzo[*b*]furans yielded *cis* and *trans* diols having synthetic potential (Boyd et al., 2012).

## 4.2.2 ORGANOCATALYSTS

Ranu et al. (1999) developed environment-friendly procedure for acylation of ferrocene with direct use of carboxylic acid in the presence of trifluoroacetic anhydride on the solid phase of alumina. Tanaka et al. (2000) reported that solvent-free condensation of cyclohexanone and diethyl succinate in the presence of t-BuOK at room temperature gives cyclohexylidenesuccinic acid, while heating mixture and t-BuOK at 80°C gives only cyclohexenylsuccinic acid. Jiang et al. (2000) carried out Wacker reaction in supercritical carbon dioxide or ROH/supercritical carbon dioxide and they observed that both $scCO_2$ and co-solvent can remarkably affect the selectivity toward methyl ketone and the presence of ROH accelerates the reaction.

Tetramethyladipic acid (TMAA), a starting monomer for several technically important polymers (polyester resins and polyamide fibers) was synthesized by direct carbon–carbon bond formation between the saturated primary carbon atoms of pivalic acid using a sonoelectrochemical Fenton process (Bremner et al., 2001). Solvent-free intermolecular and intramolecular Thorpe reactions proceeded efficiently to give acyclic and cyclic enamines (Yoshizawa et al., 2002). A simple and convenient system for benzylic bromination of toluenes has been developed by Mestres and Palenzuela using a two-phase mixture (sodium bromide, aqueous hydrogen peroxide/carbon tetrachloride, or chloroform) under visible light. Substitution of the chlorinated solvents by other more environmentally benign organic solvents has been attempted and good results were obtained for methyl pivalate (2002).

Perchlorinated aryl compounds were efficiently dechlorinated and dearomatised in hydrogen during 0.5–2 hr at 50–90 °C (Yuan et al., 2003).

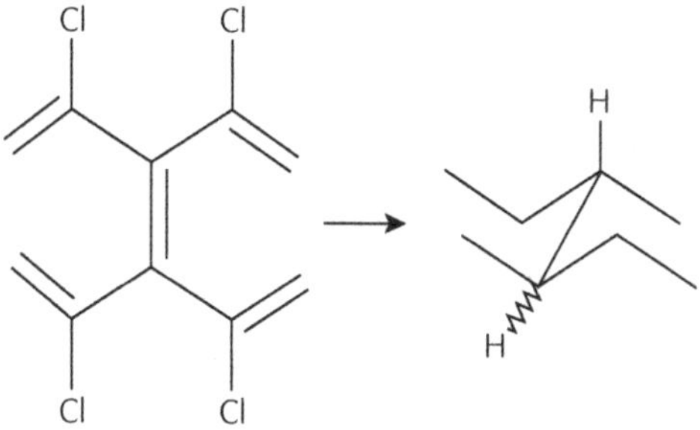

Cl Cl

H

H

Cl Cl

Zhu et al. (2003) reported that esterification of carboxylic acids occurs with alcohols in the ionic liquid $[Hmim]^{+}BF_4^{-}$ without any organic solvents and the ionic liquid could be reused over eight times.

$$\mathrm{RCOOH} + \mathrm{R'OH} \xrightarrow[110°\mathrm{C}]{[\mathrm{Hmim}]^{+}\,\mathrm{BF_4^{-}}} \mathrm{RCOOR'}$$

Laitinen et al. (2004) carried out Ene reaction of allylbenzene and N-methylmaleimide in subcritical water and ethanol. A novel emulsifier-free copolymerization can be achieved by the ultrasonic radiation (Yan et al., 2004). Comisar and Savage (2005) carried out base-catalyzed benzil rearrangement, at conventional conditions, which proceeds in high temperature water without added base. Acid catalysis also occurred under these conditions.

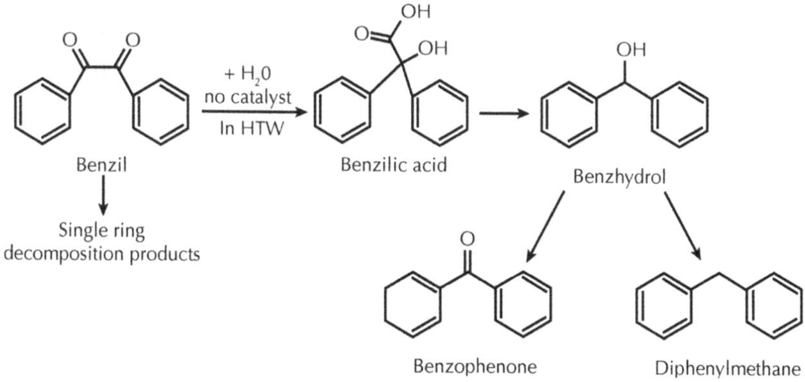

O O

+ $H_2O$
no catalyst
In HTW

OH
O
OH

OH

Benzil

Benzilic acid

Benzhydrol

Single ring
decomposition products

O

Benzophenone

Diphenylmethane

Zhang et al. (2005) reported that biotin methyl ester can be synthesized using catalytic carbonylation to generate urea, avoiding the traditional phosgene and phosgene derivative methodology.

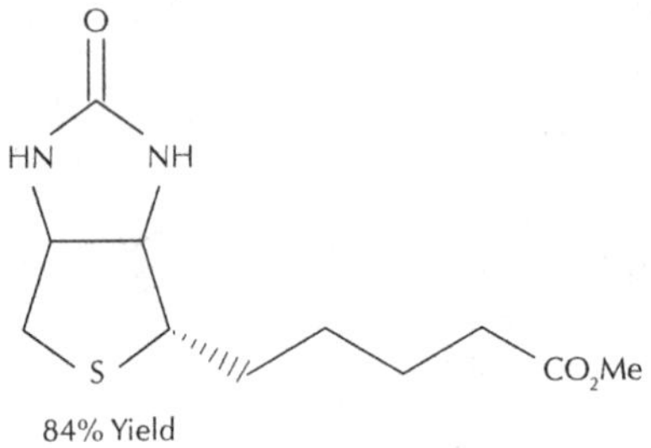

O
HN NH
S
$CO_2Me$
84% Yield

Bonnet et al. (2006) investigated a single step air oxidation of cyclohexane based on a new lipophilic catalytic system, which leads to the production of adipic acid with excellent results.

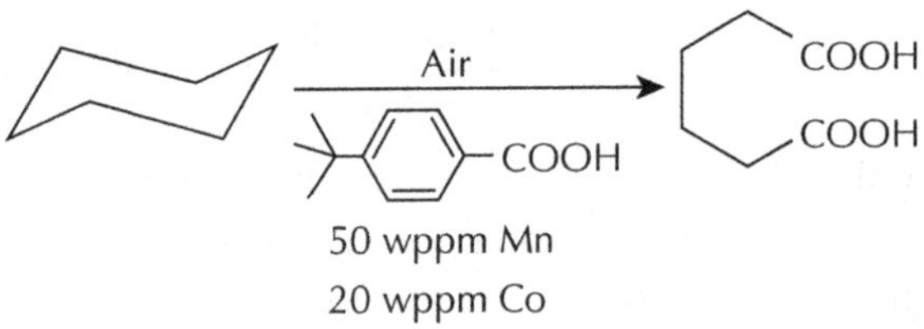

Air
COOH
50 wppm Mn
20 wppm Co
COOH
COOH

Guo et al. (2006) reported Beckmann rearrangement of cyclohexanone oxime to afford caprolactam in a novel caprolactam-based Brønsted acidic ionic liquid as catalyst and this reaction proceeded with high conversion and selectivity at 100°C. Baylis–Hillman products were produced in 98% yield in as little as 30min. by solvent-free mechanochemistry. This represents one of the fastest methods of Baylis–Hillman reactions under neat conditions in presence of 20% DABCO (Mack and Shumba, 2007).

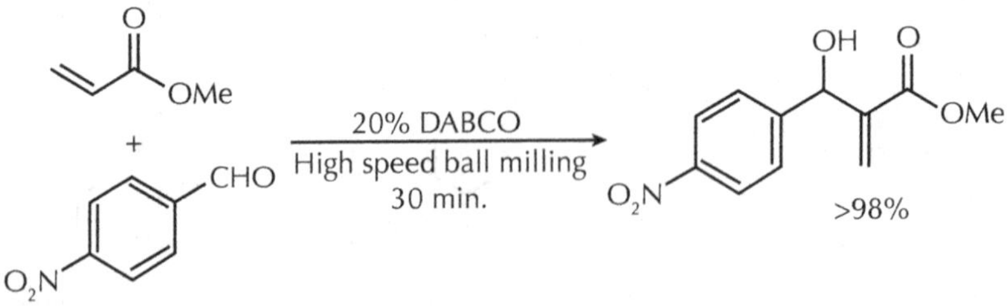

O
OMe
+
CHO
$O_2N$
20% DABCO
High speed ball milling
30 min.
OH O
OMe
$O_2N$
>98%

Chahbane et al. (2007) described orange II oxidation by peroxides catalyzed by $Fe^{III}$–TAMLs at pH 9–11 leads to $CO_2$, CO, phthalic, and smaller aliphatic acids as non toxic major mineralization products while Jin et al. (2008) reported hydrothermal conversion of carbohydrate biomass into formic acid in an excellent yield at mild temperatures.

$$\text{Glu cos e} \xrightarrow[250^{\circ},60\text{s}]{O_2+OH-} \underset{75\%}{\text{HCOOH}}$$

KI and $\beta$-cyclodextrin ($\beta$-CD) show excellent synergetic effect in promoting cycloaddition of $CO_2$ with epoxides to produce cyclic carbonates (Song et al., 2008).

$$\text{R-epoxide} + CO_2 \xrightarrow{\text{KI, }\beta\text{–CD}} \text{R-cyclic carbonate}$$

Tee et al. (2008) investigated activity of effects of ten ionic liquids on cytochrome P450 BM-3 by evaluating the influence of hydrophobicity and ion pairs on P450 BM-3. Stevens et al. (2009) have explored the aldol reactions of propionaldehyde and butyraldehyde in supercritical carbon dioxide over a variety of heterogeneous acidic and basic catalysts.

Kong et al. (2009) developed a green way to synthesize allyl phenols. Quantitative yield of 2-allyl-4-methoxyphenol was obtained via a fast Claisen rearrangement in a microreactor system without solvent and work-up. Polshettiwar and Varma (2009) reported an economical and sustainable transfer hydrogenation for aldehydes and ketones. The general protocol is mild, chemoselective, and important. It uses neither precious nor non precious metals or even ligands.

$$R^1C(O)R^2 \xrightarrow[\text{KOH, iPrOH, 85°C}]{\text{No metal catalyst}} R^1CH(OH)R^2$$

$R^1$ – Alkyl, Aryl, Heteroaryl
$R^2$ = H, $R^1$

Baj et al. (2009) reported the Baeyer–Villiger oxidation of ketones with bis(trimethylsilyl) peroxide in the presence of ionic liquids as the solvent and catalyst.

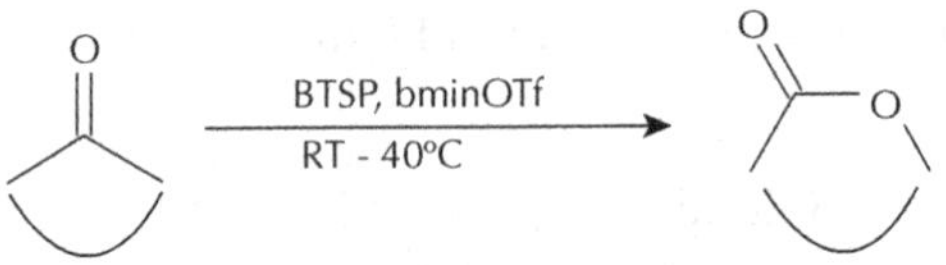

BTSP, bminOTf
RT - 40°C

Huertas et al. (2009) used dicyclopentadiene as a source of *in situ* generated cyclopentadiene for Diels–Alder reactions under solvent-free conditions.

Reflux
15-22 min.
37% 35%

Phadtare and Shankarling (2010) have described that biodegradable solvents provide an effective green method for the bromination of 1-aminoanthra-9,10-quinone under mild conditions and these are recyclable.

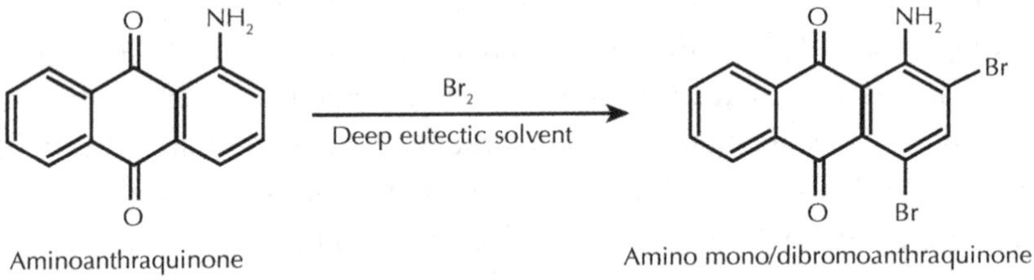

$Br_2$
Deep eutectic solvent

Aminoanthraquinone Amino mono/dibromoanthraquinone

The aerobic oxidation of alcohols was performed by Yang et al. (2010), immobilizing a mixture of palladium-guanidine complex and guanidine in the nanocages of SBA-16, an efficient solid catalyst for Suzuki coupling. Highly efficient, one-pot and three component reactions of amines and carbon disulfide with alkyl vinyl ethers via Markovnikov addition reaction were carried out by Halimehjani et al. (2010) in water under a mild and green procedure with excellent yields and complete regiospecificity. Duval and Lener (2010) reported that mildly acidic aqueous conditions are suitable for Fischer syntheses of naltriben, naltrindole, and naltrindole analogs. Methyltetrahydrofuran is a useful bio-based (co)solvent for benzaldehyde-catalyzed reactions, affording a straightforward work-up with high yields and enantioselectivities (Shanmuganathan et al., 2010).

ThDP-Lyase
Buffer/(CO) solvent
up to 99% ee
up to 60 g/L
Renewable
Resources
Abiotically degraded

A rational design of phosphonium ionic liquid for ionic liquid coated-lipase (IL1-PS)-catalyzed reaction has been investigated by Abe et al. (2010). A very rapid transesterification of secondary alcohols was accomplished, when IL1-PS was used as a catalyst in 2-methoxyethoxymethyl(tri-n-butyl)phosphonium bis(trifluoromethanesulfonyl)amide ($[P_{444MEM}][NTf_2]$) as solvent while perfect enantioselectivity was maintained. Free hemoglobin (Hb) in water at pH 5 was able to oxidize 11 polycyclic aromatic hydrocarbons (PAH) (300 nM each) in the presence of $H_2O_2$ amounting to 75% PAH removal. PAH are carcinogenic, mutagenic and xenobiotic pollutants found in wastewaters of oil refineries (Laveille et al., 2010). Balbino et al. (2011) reported that dicyclohexylguanidine group covalently attached on silica gel is an efficient basic heterogeneous catalyst for the production of biodiesel in a continuous flow reactor. Malonic acid half esters were used as the equivalent of ester carbanions for the practical one-step synthesis of $\beta$-hydroxy and $\beta$-amino esters from aryl aldehydes and arylimines via decarboxylative aldol and Mannich type reactions. Two mechanisms were unveiled depending on the substitution of the malonyl substrate (Baudoux et al., 2010).

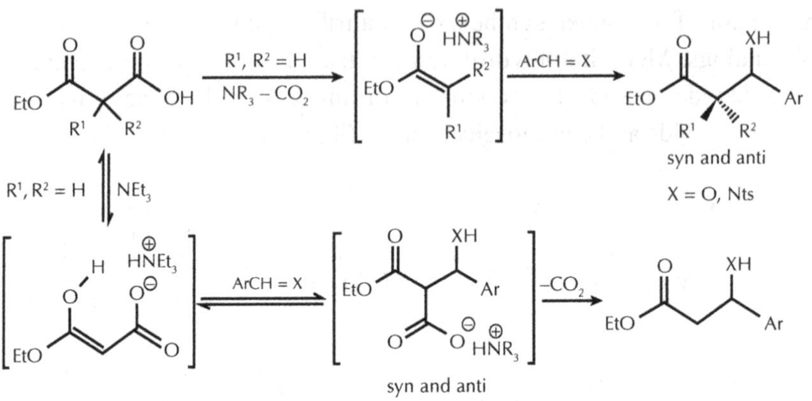


Bromination of industrially-important aromatics using an aqueous $CaBr_2$–$Br_2$ system as an instant and renewable brominating agent was carried out by Kumar et al. (2011).

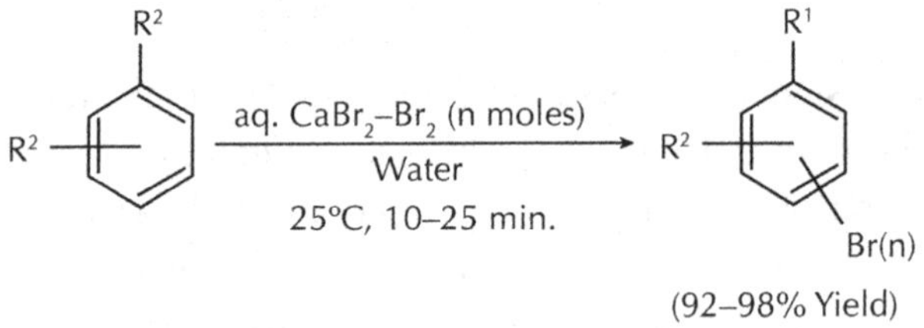


$R^1$ = OH, $NH_2$, NHCOMe, NHCOPh, CHO, COOH

$R^2$ = H, OH, $NO_2$, $SO_2NH_2$

Kotlewska et al. (2011) used hydrogen peroxide and a lipase dissolved in ionic liquids for epoxidation and Baeyer–Villiger oxidation.

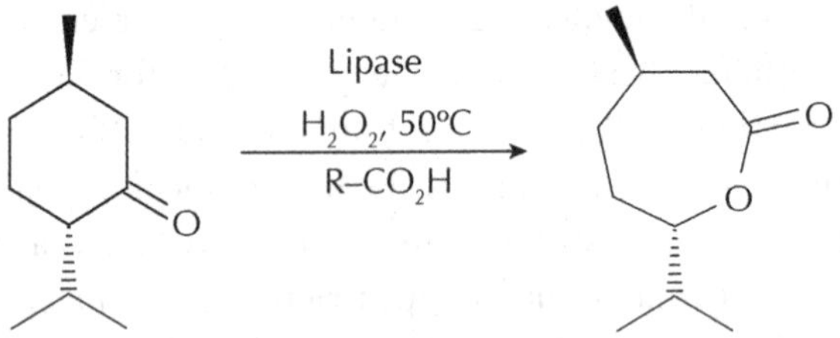


Kamimura et al. (2011) investigated treatment of waste nylon-6 with supercritical MeOH resulted in smooth depolymerization giving methyl 6-capronate and methyl 5-hexenoate in good yields.

Nylon-6 —Supercritical MeOH, 330°C→ HO~~~~~$CO_2Me$ + ~~~~$CO_2Me$

A highly efficient Knoevenagel condensation was catalyzed by a tertiary amine functionalized polyacrylonitrile fiber with excellent recyclability and reusability (Li et al., 2011).

ArCHO + $H_2C(R)CN$ —Ethanol→ ArCH=C(R)CN

R = –COOEt, –CN

A simple and scalable organocatalytic aldol reaction of acetol and aromatic aldehydes has been developed by Czarnecki et al. (2011). Ando and Yamada investigated solvent free Horner–Wadsworth–Emmons reaction catalyzed by DBU in the presence of $K_2CO_3$ or $Cs_2CO_3$, which gave olefins with high *E*-selectivity (2011).

$(EtO)_2P(O)CH_2CO_2Et$ —RCHO, $K_2CO_3$–DBU→ RCH=CH$CO_2Et$ Yield 89–99% E/Z = 98:2 – 99:1

$(EtO)_2P(O)CH_2CO_2Et$ —Me(Ar)C=O, $Cs_2CO_3$–DBU, No solvent→ R(Me)C=CH$CO_2Et$ Yield 50–81% E/Z = 95:5 – 99:1

Hajimohammadi et al. (2011) used energy of sunlight and air in the presence of porphyrins for a highly selective, green, and economical conversion of alcohols into aldehydes and ketones. Radical benzyl bromination in diethyl carbonate under microwave assisted reaction condition was performed by Pingali et al. (2011). Both the solvent and the reagent (NBS) are recyclable.

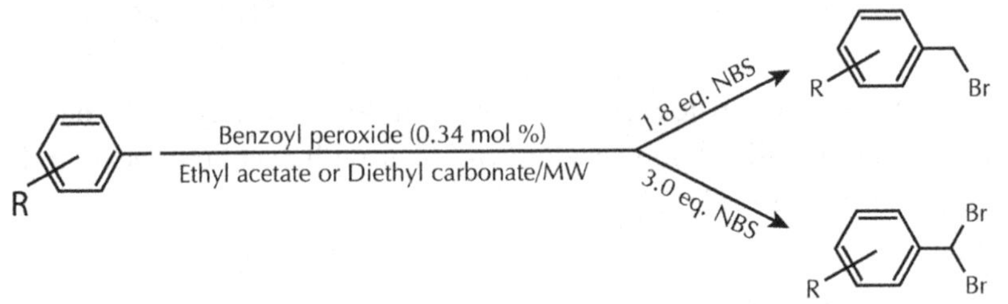

2,4-Disubstituted-1,2-dihydroquinazolines and quinazolines can be readily obtained from 2-aminobenzophenone, aldehyde, and urea under microwave irradiation in absence of any solvent or catalyst. The reaction is simple, clean and excellent yields are obtained within minutes (Sarma and Prajapati, 2011).

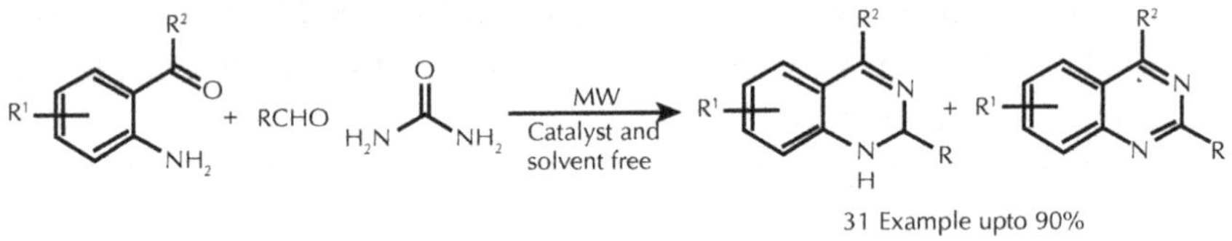

Ntainjua et al. (2012) observed that Au–Pd-ion exchanged heteropolyacid catalysts are considerably more effective in achieving high $H_2O_2$ yields in the absence of promoters than previously reported catalysts. According to Bellomo et al. (2012) organocatalyzed direct aldol reactions were efficiently performed in aqueous solutions of facial amphiphilic carbohydrates with high diastereoselectivity and yields. Mitsudome et al. (2012) reported titanium-exchanged montmorillonite ($Ti^{4+}$-mont) as an efficient heterogeneous catalyst for the etherification of various alcohols under mild reaction conditions. Solvent free brominations of 1,3-dicarbonyl compounds, phenols, and alkenes were achieved by employing sodium bromide and oxone under mechanical milling conditions (Wang and Geo, 2012).

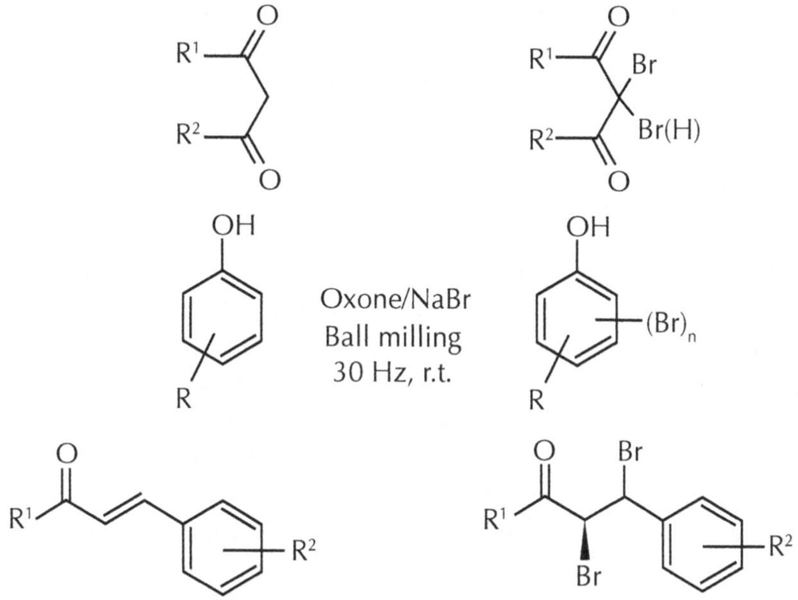

Oxone/NaBr
Ball milling
30 Hz, r.t.

Wang et al. (2012) reported water as a green additive, which enhances the ring opening and contraction reactions of aromatics. A key pharmaceutical intermediate for production of edivoxetine·HCl was prepared in >99% ee via a continuous Barbier reaction, which improves process greenness relative to a traditional Grignard batch process (Kopach et al., 2012). Shen et al. (2012) used ionic liquid supported imidazolidinone catalyst (an efficient and recyclable organocatalyst) for mediating highly enantioselective Diels–Alder reactions involving *α,β*-unsaturated aldehydes and cyclopentadiene.

Efficient, continuous flow methylation was effected with dimethylcarbonate using a basic ionic liquid as catalyst (Glasnov et al., 2012). Synthesis of *α*-acyloxy amides has been developed by Cui et al. (2012) by ultrasound-promoted sterically congested Passerini reactions under solvent free conditions.

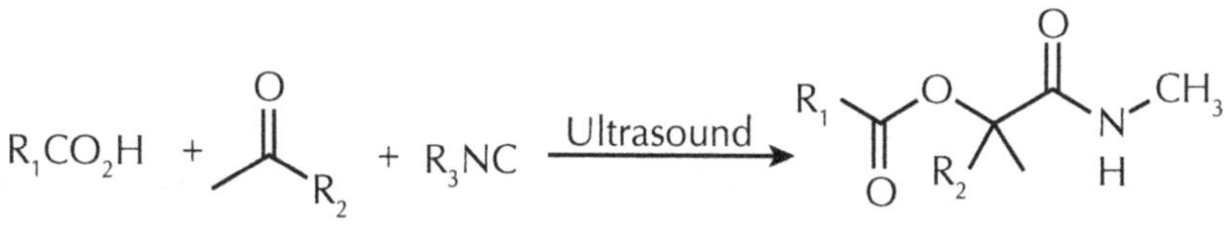

$R_1CO_2H$ + $R_2$-ketone + $R_3NC$ —Ultrasound→ product

A novel magnetic nanoparticle supported acidic catalyst was prepared and used as a highly efficient and magnetically recoverable catalyst for the one-pot synthesis of benzoxanthenes (Zhang et al., 2012).

### 4.2.3 METALLOCATALYSTS

Palladium(II)-catalyzed Heck arylation of both electron-poor and electron-rich olefins with arylboronic acids as arylpalladium precursors were conducted under air (Enquist et al., 2006). A novel Pd/Ph–Al-MCM-41 catalyst was designed by Li et al., (2007), which exhibited excellent activity, selectivity, and hydrothermal stability in an aqueous-medium Ullmann reaction. This may be due to the promoting effects of the ordered mesoporous structure, and the Ph- and Al-modifications. The Pd catalyst supported on 1,1,3,3-tetramethylguanidinium (TMG)-modified molecular sieve SBA-15 is a very active and stable catalyst for the Heck coupling reaction in solvent-free conditions (Ma et al., 2008).

R–X + $R_1$, $R_2$ — N(Et)3, SBA-TMG-Pd / 140°C, without solvent → R, $R_1$, $R_2$

Wang et al. (2009) used ethylene carbonate as a unique solvent for the Wacker oxidation of higher alkenes and aryl alkenes. They have successfully used molecular oxygen as the sole oxidant, when colloidal Pd nanoparticles stabilized in ethylene carbonate facilitate its reoxidation under cocatalyst-free conditions.

R (Higher alkene, aryl alkene) — $PdCl_2$/Ethylene carbonate, $O_2$, NaOAc / Cocatalyst and ligand free → R, O (51-92% Yield, 88-99% Selectivity)

Waddell and Mack (2009) reported Tishchenko reaction using high speed ball milling and a sodium hydride catalyst for aryl aldehydes in high yields in 0.5hr.

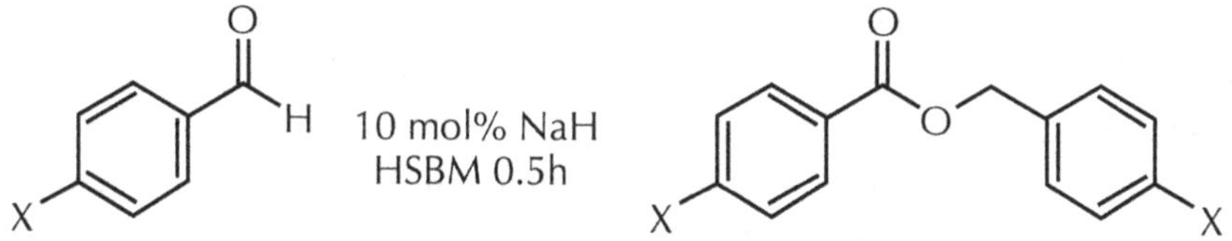

10 mol% NaH
HSBM 0.5h

Cu–Mn spinel oxide catalyst has been used by Yousuf et al. (2010) for the ligand-free Huisgen [3+2] cycloaddition. Zhang et al. (2011) described the reductive cleavage of the C–O bond of aromatic and aliphatic acetals to ethers, catalyzed by $Cu(OTf)_2$ or $Bi(OTf)_3$ in excellent yields and selectivity.

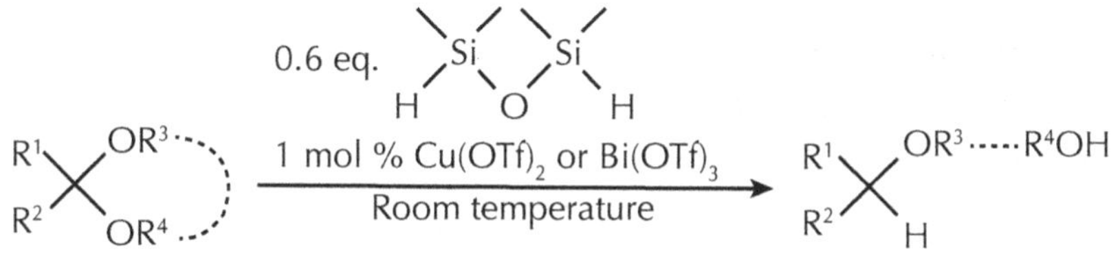

0.6 eq.
1 mol % $Cu(OTf)_2$ or $Bi(OTf)_3$
Room temperature

A green process using nickel-loaded $La_xNa_{1-x}TaO_3$ prepared by hydrogen peroxide-water based solvent for hydrogen production has been reported by Husin et al. (2011). Direct transformation of cellulose and a lignocellulosic biomass raw material into 5-hydroxymethyl-2-furfural (HMF) using a specific combination of Cr(II) and Ru(III) metals in [emim] Cl (Kim et al., 2011). The Ru/ZnO–$ZrO_x(OH)_y$ catalyst is very efficient for the selective hydrogenation of benzene to cyclohexene, and the yield of cyclohexene can reach 56% without using any additive (Liu et al., 2011).

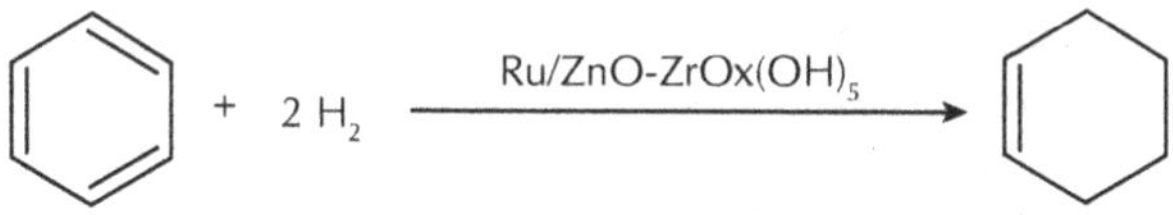

+ $2 H_2$ Ru/ZnO-ZrOx$(OH)_5$

Füldner et al. (2011) reported that colored metal oxides ($PbBiO_2X$) are efficient visible light photocatalysts for nitrobenzene reduction. A fluorous oxime-based palladacycle, which promotes carbon-carbon bond formation reactions (Suzuki–Miyaura, Sonogashira and Stille) in aqueous media under microwave irradiation was developed by Susanto et al. (2012).

Most of the chemical reactions require some or the other catalyst to enhance this rate, but some of these are toxic in nature as these contain some metal, metal derivatives or harmful compounds; of course, leaving aside biochemical reaction involving enzymes as catalyst. Although various green catalysts have been used by some workers, but search is still on for environmentally friendly catalyst, which can serve the purpose of a catalyst taking care of environment.

## KEYWORDS

- **Biocatalysts**
- **Metallocatalysts**
- **Organocatalysts**
- **Solvent free conditions**
- **Wacker reaction**

## REFERENCES

1. Abbate, V., Bassindale, A. R., Brandstadt, K. F., Lawson, R., & Taylor, P. G. (2010). *Dalton Transactions, 39,* 9361–9368.
2. Abe, Y., Yoshiyama, K., Yagi, Y., Hayase, S., Kawatsura M., & Itoh, T. (2010). *Green Chemistry, 12,* 1976–1980.
3. Ando, K. & Yamada, K. (2011). *Green Chemistry, 13,* 1143–1146.
4. Baj, S., Chrobok, A., & Słupska, R. (2009). *Green Chemistry, 11,* 279–282.
5. Balbino, J. M., Menezes, E. W. D., Benvenutti, E. V., Cataluña, R., Ebeling, G., & Dupont, J. (2011). *Green Chemistry, 13,* 3111–3116.
6. Balke, K., Kadow, M., Mallin, H., Saß, S., & Bornscheuer, U. T. (2012). *Organic and Biomolecular Chemistry, 10,* 6249–6265.
7. Baudoux, J., Lefebvre, P., Legay, R., Lasne, M., & Rouden, J. (2010). *Green Chemistry, 12,* 252–259.
8. Bellomo, A., Daniellou, R., & Plusquellec, D. (2012). *Green Chemistry, 14,* 281–284.
9. Bhatnagar, D., Xu, S., Fischer, C., Arechederra, R. L., & Minteer, S. D. (2011). *Physical Chemistry, 13,* 86–92.
10. Bohn, M., Leppchen, K., Katzberg, M., Lang, A., Steingroewer, J., Weber, J., Bley, T., & Bertau, M. (2007). *Organic and Biomolecular Chemistry, 5,* 3456–3463.
11. Bonnet, D., Ireland, T., Fache, E., & Simonato, J. (2006). *Green Chemistry, 8,* 556–559.

12. Boyd, D. R., Sharma, N. D., Brannigan, I. N., Evans, T. A., Haughey, S. A., McMurray, B. T., Malone, J. F., McIntyre, P. B. A., Stevenson, P. J., & Allen, C. C. R. (2012). *Organic and Biomolecular Chemistry, 10,* 7292–7304.
13. Boyd, D. R., Sharma, N. D., King, A. W. T., Shepherd, S. D., Allen, C. C. R., Holt, R. A., Luckarift, H. R., & Dalton, H. (2004). *Organic and Biomolecular Chemistry, 2,* 554–561.
14. Bremner, D. H., Burgess, A. E., & Li, F. (2001). *Green Chemistry, 3,* 126–130.
15. Çelik, A., Sperandio, D., Speight, R. E., & Turner, N. J. (2005). *Organic and Biomolecular Chemistry, 3,* 2688–2690.
16. Chahbane, N., Popescu, D., Mitchell, D. A., Chanda, A., Lenoir, D., Ryabov, A. D., Schramm, K., & Collins, T. J. (2007). *Green Chemistry, 9,* 49–57.
17. Comisar, C. M. & Savage, P. E. (2005). *Green Chemistry, 7,* 800–806.
18. Cui, C., Zhu, C., Du, X., Wang, Z., Li, Z., & Zhao, W. (2012). *Green Chemistry, 14,* 3157–3163.
19. Czarnecki, P., Plutecka, A., Gawroński, J., & Kacprzak, K. (2011). *Green Chemistry, 13,* 1280–1287.
20. Dupont, J., Suarez, P. A. Z., Meneghetti, M. R., & Meneghetti, S. M. P. (2009). *Energy and Environmental Science, 2,* 1258–1265.
21. Durchschein, K., Silva, B. F., Wallner, S., Macheroux, P., Kroutil, W., Glueck, S. M., & Faber, K. (2010). *Green Chemistry, 12,* 616–619.
22. Duval, R. A., & Lever, J. R. (2010). *Green Chemistry, 12,* 304–309.
23. Elbaz, J., Shlyahovsky, B., & Willner, I. (2008). *Chemical Communications,* 1569–1571.
24. Enquist, P., Lindh, J., Nilsson, P., & Larhed, M. (2006) *Green Chemistry, 8,* 338–343.
25. Fuchs, M., Koszelewski, D., Tauber, K., Kroutil, W., & Faber, K. (2010). *Chemical Communications, 46,* 5500–5502.
26. Füldner, S., Pohla, P., Bartling, H., Dankesreiter, S., Stadler, R., Gruber, M., Pfitzner, A., & König, B. (2011). *Green Chemistry, 13,* 640–643.
27. Giacomini, D., Galletti, P., Quintavalla, A., Gucciardo, G., & Paradisi, F. (2007). *Chemical Communications, 13,* 4038–4040.
28. Glasnov, T. N., Holbrey, J. D., Kappe, C. O., Seddon, K. R., & Yan, T. (2012). *Green Chemistry, 14,* 3071–3076.
29. Guo, S., Du, Z., Zhang, S., Li, D., Li, Z., & Deng, Y. (2006). *Green Chemistry, 8,* 296–300.
30. Hajimohammadi, M., Safari, N., Mofakham, H., & Deyhimi, F. (2011). *Green Chemistry, 13,* 991–997.
31. Halimehjani, Z. A., Marjani, K., & Ashouri, A. (2010). *Green Chemistry, 12,* 1306–1310.
32. Huertas, D., Florscher, M., & Dragojlovic, V. (2009) *Green Chemistry, 11,* 91–95.
33. Hult, K., & Berglund, P. (2003). *Current opinion in biotechnology, 14,* 395–400.
34. Husin, H., Su, W., Chen, H., Pan, C., Chang, S., Rick, J., Chuang, W., Sheu, H., & Hwang, B. (2011). *Green Chemistry, 13,* 1745–1754.
35. Ismail, H., Lau, R. M., Langen, L. M. V., Rantwijk, F. V., Švedas, V. K., & Sheldon, R. A. (2008). *Green Chemistry, 10,* 415–418.
36. Jiang H., Jia, L., & Li, J. (2000). *Green Chemistry, 2,* 161–164.

37. Jiang, L., Althoff, E. A., & Clemente, F. R. (2008). *Science, 319,* 1387–91.
38. Jin, F., Yun, J., Li, G., Kishita, A., Tohji, K., & Enomoto, H. (2008). *Green Chemistry, 10,* 612–615.
39. Kamimura, A., Kaiso, K., Suzuki, S., Oishi, Y., Ohara, Y., Sugimoto, T., Kashiwagi, K., & Yoshimoto, M. (2011). *Green Chemistry, 13,* 2055–2061.
40. Kim, B., Jeong, J., Lee, D., Kim, S., Yoon, H., Lee, Y., & Cho, J. K. (2011). *Green Chemistry, 13,* 1503–1506.
41. Kluge, M., Ullrich, R., Scheibner, K., & Hofrichter, M. (2012). *Green Chemistry, 14,* 440–446.
42. Kong, L., Lin, Q., Lv, X., Yang, Y., Jia, Y., & Zhou, Y. (2009). *Green Chemistry, 11,* 1108–1111.
43. Kopach, M. E., Roberts, D. J., Johnson, M. D., Groh, J. M., Adler, J. J., Schafer, J. P., Kobierski, M. E., & Trankle, W. G. (2012). *Green Chemistry, 14,* 1524–1536.
44. Kotlewska, A. J., Rantwijk, F. V., Sheldon, R. A., & Arends, I. W. C. E. (2011). *Green Chemistry, 13,* 2154–2160.
45. Kumar, L., Mahajan, T., & Agarwal, D. D. (2011). *Green Chemistry, 13,* 2187–2196.
46. Laitinen, A., Takebayashi, Y., Kylänlahti, I., Yli-Kauhaluoma, J., Sugeta, T., & Otake, K. (2004). *Green Chemistry, 6,* 49–52.
47. Langen, L. M. V., Vroom, E. D., Rantwijk, F. V., & Sheldon, R. A. (2001). *Green Chemistry, 3,* 316–319.
48. Laveille, P., Falcimaigne, A., Chamouleau, F., Renard, G., Drone, J., Fajula, F., Pulvin, S., Thomas, D., Bailly, C., & Galarneau, A. (2010). *New Journal of Chemistry, 34,* 2153–2165.
49. Li, F., Cui, J., Qian, X., & Zhang, R. (2004). *Chemical Communications, 20,* 2338–2339.
50. Li, G., Xiao, J., & Zhang, W. (2011). *Green Chemistry, 13,* 1828–1836.
51. Li, H., Chen, J., Wan, Y., Chai,W., Zhang, F., & Lu, Y. (2007). *Green Chemistry, 9,* 273–280.
52. Li, H., He, Y., Yuan, Y., & Guan, Z. (2011). *Green Chemistry, 13,* 185–189.
53. Liu, H., Jiang, T., Han, B., Liang, S., Wang, W., Wu, T., & Yang, G. (2011). *Green Chemistry, 13,* 1106–1109.
54. Lozano, P., Bernal, J. M., & Navarro, A. (2012) *Green Chemistry, 14,* 3026–3033.
55. Ma, X., Zhou, Y., Zhang, J., Zhu, A., Jiang, T., & Han, B. (2008). *Green Chemistry, 10,* 59–66.
56. Mack, J. & Shumba, M., (2007). *Green Chemistry, 9,* 328–330.
57. Matsuda, T., Watanabe, K., Kamitanaka, T., Harada, T., & Nakamura, K. (2003). *Chemical Communications, 14,* 1198–1199.
58. Mestres, R. & Palenzuela, J. (2002). *Green Chemistry, 4,* 314–316.
59. Mitsudome, T., Matsuno, T., Sueoka, S., Mizugaki, T., Jitsukawa, K., & Kaneda, K. (2012). *Green Chem*istry, *14,* 610–613.
60. Ni, Y., Hagedoorn, P., Xu, J., Arends, I. W. C. E., & Hollmann, F. (2012). Chemical Communications, *48,* 12056–12058.
61. Ntainjua, E. N., Piccinini, M., Freakley, S. J., Pritchard, J. C., Edwards, J. K., Carley, A. F., & Hutchings, G. J. (2012). *Green Chemistry, 14,* 170–181.
62. Onoda, A., Fukumoto, K., Arlt, M., Bocola, M., Schwaneberg, U., & Hayashi, T. (2012). *Chemical Communications, 48, 9756–9758.*

63. Pereira, S. C., Bussamara, R., Marin, G., Giordano, R. L. C., Dupont, J., & Giordano, R. D. C. (2012). *Green Chemistry, 14,* 3146–3156.
64. Phadtare, S. B. & Shankarling, G. S. (2010). *Green Chemistry, 12,* 458–462.
65. Pingali, S. R. K., Upadhyay, S. K., & Jursic, B. S. (2011). *Green Chemistry, 13,* 928–933.
66. Pollock, C. L., Fox, K. J., Lacroix, S. D., McDonagh, J., Marr, P. C., Nethercott, A. M., Pennycook, A., Qian, S., Robinson, L., Saunders, G. C., & Marr, A. C. (2012). *Dalton Transactions, 41,* 13423–13428.
67. Polshettiwar, V. & Varma, R. S. (2009). *Green Chemistry, 11,* 1313–1316.
68. Pratap, U. R., Jawale, D. V., Waghmare, R. A., Lingampalle, D. L., & Mane, R. A. (2011). *New Journal of Chemistry, 35,* 49–51.
69. Ranu, B. C., Jana, U., & Majee, A. (1999). *Green Chemistry, 1,* 33–34.
70. Renugopalakrishnan, V., Garduno-Juarez, R., Narasimhan, G., Verma, C. S., Wei X., & Li P. (2005). *Journal of Nanoscience and Nanotechnology, 5,* 1759–1767.
72. Ríos, M. Y., Salazar, E., & Olivo, H. F. (2007). *Green Chemistry, 9,* 459–462.
73. Romero, O., Filice, M., Rivas, B. D. L., Carrasco-Lopez, C., Klett, J., Morreale, A., Hermoso, J. A., Guisan, J. M., Abian, O., & Palomo, J. M. (2012). *Chemical Communications, 48,* 9053–9055.
74. Sarma, R. & Prajapati, D. (2011). *Green Chemistry, 13,* 718–722.
75. Shanmuganathan, S., Natalia, D., Greiner, L., & María, P. D. D. (2012). *Green Chemistry, 14,* 94–97.
76. Shanmuganathan, S., Natalia, D., Wittenboer, A. V. D., Kohlmann, C., Greiner, L., & María, P. D. D. (2010). *Green Chemistry, 12,* 2240–2245.
77. Shao, J., Zhang, J., Nahálka, J., & Wang, P. G. (2002). *Chemical Communications, 21,* 2586–2587.
78. Shen, Z., Cheong, H., Lai, Y., Loo, W., & Loh, T. (2012). *Green Chemistry, 14,* 2626–2630.
79. Song, J., Zhang, Z., Han, B., Hu, S., Li, W., & Xie, Y. (2008). *Green Chemistry, 10,* 1337–1341.
80. Stevens, J. G., Bourne, R. A., & Poliakoff, M. (2009). *Green Chemistry, 11,* 409–416.
81. Susanto,W., Chu, C., Ang, W. J., Chou, T., Lo, L., & Lam, Y. (2012). *Green Chemistry, 14,* 77–80.
82. Tanaka, K., Sugino, T., & Toda, F. (2000). *Green Chemistry, 2,* 303–304.
83. Tee, K. L., Roccatano, D., Stolte, S., Arning, J., Jastorff, B., & Schwaneberg, U. (2008). *Green Chemistry, 10,* 117–123.
84. Theil, F., Kunath, A., Ramm, M., Reiher, T., & Schick, H. (1994). *Journal of the Chemical Society, Perkin Transactions, 1,* 1509–1516.
85. Waddell, D. C. & Mack, J. (2009). *Green Chemistry, 11,* 79–82.
86. Wang, G. & Gao, J. (2012). *Green Chemistry, 14,* 1125–1131.
87. Wang, J., He, L., Miao, C., & Li, Y. (2009). *Green Chemistry, 11,* 1317–1320.
88. Wang, Q., Fan, H., Wu, S., Zhang, Z., Zhang, P., & Han, B. (2012). *Green Chemistry, 14,* 1152–1158.
89. Wolf, S. E., Schlossmacher, U., Pietuch, A., Mathiasch, B., Schröder, H, Müller, W. E. G., & Tremel, W. (2010). *Dalton Transactions, 39,* 9245–9249.

90. Wu, W., Wang, N., Xu, J., Wu, Q., & Lin, X. (2005). *Chemical Communications,* 2348–2350.
91. Xu, Y., Jia, X., Panke, S., & Li, Z. (2009). *Chemical Communications,* 1481–1483.
92. Yan, L., Wu, H., & Zhu, Q. (2004). *Green Chemistry, 6,* 99–103.
93. Yang, H., Han, X., Ma, Z., Wang, R., Liu, J., & Ji, X. (2010). *Green Chemistry, 12,* 441–451.
94. Yara-Varón, E., Joli, J. E., Balcells, M., Torres, M., & Canela-Garayoa, R. (2012). *RSC Advances, 2,* 9230–9236.
95. Yoshizawa, K., Toyota, S., & Toda, F. (2002). *Green Chemistry, 4,* 68–70.
96. Yousuf, S. K., Mukherjee, D., Singh, B., Maity, S., & Taneja, S. C. (2010). *Green Chemistry, 12,* 1568–1572.
97. Yuan, T., Majid, A., & Marshall, W. D. (2003). *Green Chemistry, 5,* 25–29.
98. Zhang, J., Duetz, W. A., Witholt, B., & Li, Z. (2004). *Chemical Communications,* 2120–2121.
99. Zhang, Q., Su, H., Luo, J., & Wei, Y. (2012). *Green Chemistry, 14,* 201–208.
100. Zhang, T., Cui, C., Chen, S., Ai, X., Yang, H., Shen, P., & Peng, Z. (2006). *Chemical Communications,* 2257–2259.
101. Zhang, Y., Forinash, K., Phillips, C. R., & McElwee-White, L. (2005). *Green Chemistry, 7,* 451–455.
102. Zhang, Y., Dayoub, W., Chen, G., & Lemaire, M. (2011). *Green Chemistry, 13,* 2737–2742.
103. Zhu, H., Yang, F., Tang, J., & He, M. (2003). *Green Chemistry, 5,* 38–39.
104. Znabet, A., Polak, M. M., Janssen, E., Kanter, F. J. J., Turner, N. J., Orru, R. V. A., & Ruijter, E. (2010). *Chemical Communications, 46,* 7918–7920.

## CHAPTER 5

# IONIC LIQUIDS: PROMISING SOLVENTS

ARPIT PATHAK, NIRMALA JANGID, RAKSHIT AMETA, and P. B. PUNJABI

## CONTENTS

## 5.1 INTRODUCTION

The first ionic liquid was reported by Gabriel and Weiner (1888), but it was debated for a long time. It was ethanol ammonium nitrate (m.p. = 52–55°C). Room temperature ionic liquid ethyl ammonium nitrate ($C_2H_5NH_3^+NO_3^-$) was synthesized by Walden (1914), which melts at 12°C. In the 1970s and 1980s, some more alkyl substituted imidazolium and pyridinium cations with halide or tetrahaloaluminate anions were developed and used in batteries as electrolytes (Chum et al., 1975; Wilkes et al. 1982).

Solvents are required for the synthesis of organic compounds, but its vapor creates air pollution. Therefore, efforts are being made to use solvents with high boiling points or to avoid solvent (solvent free reaction). A new class of solvents has emerged, which are fluid at room temperature. The thermodynamics and kinetics of reaction was also carried out, which is however different from the traditional molecular solvents. As these solvents have high boiling point, it means lower vapor pressure of that solvent and hence, no volatile organic compounds (VOCs) are escaped from these liquids at lower temperatures.

It makes these solvents interesting to chemists. An important quality of these green solvents is their physical property like melting point, viscosity, low combustibility, good solvating property, and so on, which could be adjusted by varying the cation and anion. As these solvents can be designed to accommodate majority of conditions and therefore, these are named as "Designer Solvents" by Freemantle (1998). But there remain two major disadvantages and these are, sensitivity toward moisture and acidity or basicity. In early 1990s, Wilkes and Zavarotko prepared anionic liquid, which can be used in a variety of applications (1992). They used hexafluorophosphate ($PF_6^-$) and tetrafloroborate ($BF_4^-$) as anions. Less toxic cations can be used for better ionic liquids. Search is still on for various cations and anions, which can improve the desired properties of ionic liquids. The field of ionic liquids has been excellently reviewed by several authors time to time (Earle and Seddon, 2000; Holbrey and Seddon, 1999; Welton, 1999).

These ionic liquids are synonymously known as liquid salts, liquid electrolytes, ionic fluids, ionic melts, fuse salts, or ionic glass. Since 1942, the term ionic liquid is most commonly used. Thus, ionic liquids contain large ions (cation and/or anions) and the cations with a low degree of symmetry. As a result, the lattice energy of the crystalline form of ionic salts is decreased resulting into low melting points. Most of the ionic liquids are important because of the following reasons:

(i) They have negligible vapor pressure than conventional solvents.
(ii) They have higher thermal stability than conventional organic molecular solvents.
(iii) They have wide range of solubilities and miscibilities. Some ionic liquids are hydrophobic while others are hydrophilic in nature.
(iv) They have quite wider liquid range than molecular solvents.
(v) They can be easily recycled.
(vi) They are non flammable as compared with other organic solvents.
(vii) They are useful in reaction media and catalyst for various types of chemical reactions.
(viii) They have wide electrochemical window.
(ix) They are also used for the separation and extraction of a chemical from aqueous solutions as well as organic solvents.
(x) They also contribute toward development of green chemistry technology because it is replacing flammable volatile and toxic conventional solvent and also reduce chemical wastage.

Ionic liquids are excellent solvent to volatile and hazardous organic compounds because these are having low vapor pressure, thermal and chemical stability, non corrosive, act as catalyst and are non flammable. The ionic liquids are ionic in nature that is, they are salt like materials. There is one organic cation and one anion. These exist in liquid form below 100°C (Deetlefs and Seddon, 2003). The discovery of binary ionic liquids was made from mixture of aluminum (III) chloride and N-alkyl pyridinium (Pegot and Loupy, 2004). Ionic liquids are of two types. First type is simple salts, which are made by combination of simple cations and anions and second type is binary ionic liquid salts, where an equilibrium is involved for example, [$EtNH_3$] [$NO_3$] is simple ionic liquid whereas 1,3-dialkyl imidazolium chloride is a binary ionic liquid.

## 5.2 SYNTHESIS AND APPLICATIONS

### 5.2.1 *IONIC LIQUID AS GREEN SOLVENT*

In this sense, in present time, many studies describe the use of ionic liquid as a green solvent to develop more eco-friendly and efficient chemical synthesis in both; academic and industrial area. Ionic liquids are excellent alternative solvents to volatile organic solvents.

#### *DESIGNER SOLVENTS FOR A CLEANER WORLD*

Ionic liquids have been described as designer solvents (Freemantle, 1998), as these are made-up of two components (i) cations and (ii) anions, which vary with different types of groups. Property of ionic liquids depends on this group. Hence, the term "designer solvents" has been justified for ionic liquids. The nature of the cations and anions has large influence on the properties of these ionic liquids. The most employed ionic liquid anions are polyatomic inorganic species, halogens, organic and perfluronited anions, such as $[BF_4]$, $[PF_6]$, $[SbF_6]$, $[NO_3]^-$, $[AcO]^-$, $Cl^-$, $Br^-$.

#### *NOTATIONS*

Name of these ionic liquids includes various cations and anions, but particularly containing alkyl chains attached to nitrogen, or phosphorous and therefore, in scientific literature various short notations are used for such cations, 1-ethyl-3-methyl imidazolium cation in ionic liquids is represented as $[emim]^+$, $[EtMeim]^+$ or $[C_2C_1im]^+$ with or without bracket. Ethyl pyridinium cation is represented as $[C_2Py]^+$ or $[EtPy]^+$ cation. Anions are normally denoted in square bracket, if these are polyatomic like $[PF_6]^-$, $[BF_4]^-$ or $[NO_3]^-$ while monoatomic anions are denoted in simple manner, some of the examples are

*ANIONS*

- Halides: bromide $Br^-$, chloride $Cl^-$
- Nitrate $NO_3^-$
- Hexafluoro phosphate $[PF_6]^-$
- Tetrafluoroborate $[BF_4]^-$
- Chloroaluminate $[AlCl_4]^-$
- Hydrogen sulphate $[HSO_4]^-$
- Alkyl sulphate $[RSO_4]^-$
- Acetate $[CH_3COO]^-$
- Tosylate $[OTs]^-$
- Tetrafluoroantimonate $[SbF_6]^-$

*CATIONS*

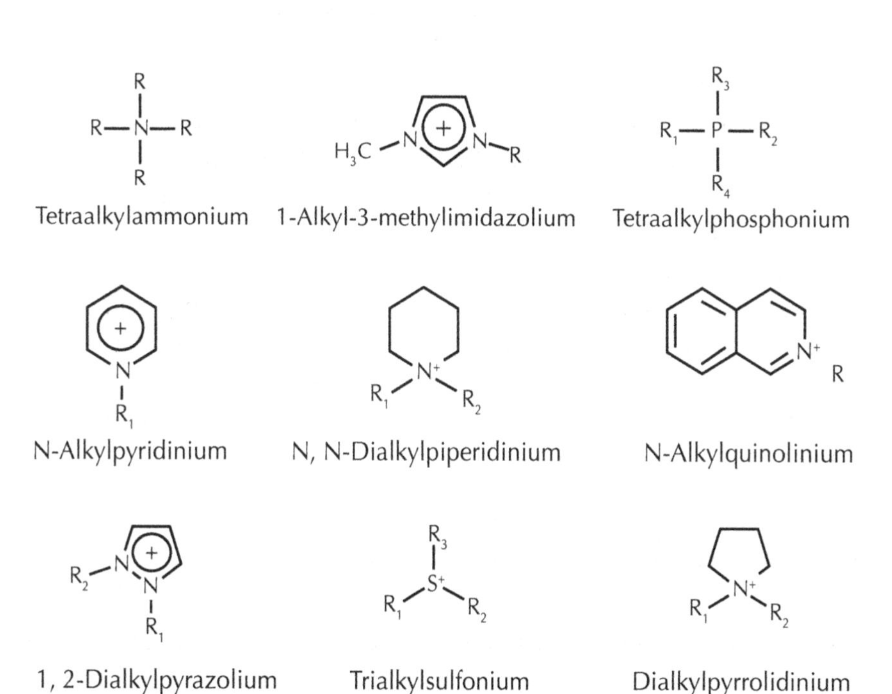


Due to favorably solvating properties by the selection of a proper cation and an anion, it can dissolve a wide range of polar and non polar compounds and due to higher viscosity the ionic liquids are not emitted as vapor like other volatile organic solvents. Due to these different properties, ionic liquids are attracting the attention of synthetic organic chemists as well as electrochemists, chemical engineers, and so on. These ionic liquids find use in biotechnology (Wasserschied and Keim, 2000), chemical engineering (Chum et al., 1975), fuel cells (Wilkes et al., 1982), and so on.

### *IONIC LIQUID GENERATION*

Ionic liquids are classified into three generations. First one is haloaluminate room temperature ionic liquids. Haloaluminate ionic liquids have been studied as solvent and catalyst in Friedel–Craft and other organic reactions.

Second generation ionic liquids are non haloaluminate ionic liquids, which are used as the best solvent in organic synthesis. First example of second generation was described by Wilkes and Zaworotko (1992). Other examples of this generation include dialkylimidazolium cation coordinated with hexafluorophosphate $[PF_6]^-$ and tetrafluoroborate $[BF_4]^-$.

The third generation of ionic liquids is known as Task Specific Ionic Liquid [TSIL] and chiral ionic liquid. It is also known as functionalized ionic liquid. These ionic liquids are designed with functionalized cation or anion with special property (Davis, 2004). Examples of this type of ionic liquid are imidazolium cation functionalized with sulphonic acid group ($-SO_3H$). Many types of ionic liquids are synthesized with chiral cation or chiral anions (Baudequin et al., 2005). Chiral ionic liquids are most useful solvent or catalyst in the asymmetric synthesis. Other use of this ionic liquid is in resolution of racemate.

## 5.2.2 SYNTHESIS OF IONIC LIQUID

Synthesis of ionic liquid consists of two steps (i) synthesis of the desired cation by the simple alkylation or quarterization of nitrogen or phosphorus containing compounds and (ii) anion resulting from the alkylation can be exchanged by metathesis reaction or by direct combination with Lewis acid or ion exchange resin or acid–base procedure. A general preparation of imidazolium ionic liquid is

Me–N(imidazole)N $\xrightarrow{R\text{-}X}$ [Me–N(imidazolium)$N^+$–R] $X^-$

R = Alkyl

X = $Cl^-$, $Br^-$

+ MY  Metathesis

– MX  M = Metal as Ag

In the first step, preparation of ionic liquid has two disadvantages (i) it is time consuming and (ii) an excess of haloalkane is required to gain high yield. As high boiling alkyl halides are difficult to remove and the quarterization reaction is dirty, especially when higher chain derivatives are prepared. So to reduce this problem, ionic liquids are prepared by using microwave radiation.

The second step in synthesis of ionic liquid is the preparation by metathesis process or acid–base reaction. In the metathesis reaction, metal salt is used such as silver nitrate, silver thiocyanate or acid–base neutralization reaction is carried out producing byproduct MX.

Now many alkyl ammonium halides are commercially available or they are made by reaction of alkyl halide and amine. The use of salt with longer chain substituents such as [bmim]Cl has become popular, which can be prepared by conventional method under reflux (Dyson et al. 1997).

## *USE OF MICROWAVE IRRADIATION*

The use of microwave oven as a tool in synthetic chemistry is one of the fastest growing areas of research (Blasius et al., 1985; Gloe et al., 1982). Microwave irradiation for the synthesis of ionic liquids offers some advantages as compared with conventional heating method. In conventional method, synthesis of ionic liquid requires several hours of reaction (10–80h) at high temperature and large amounts of reactants are required while use of microwaves in the synthesis of ionic liquids reduced dramatically the time required to obtain these compounds.

The first synthesis of ionic liquids under microwave was reported by Varma and Namboodiri (2005) for the synthesis of 1-alkyl-3-methylimidazolium halide ($Cl^-$ and $Br^-$). An ionic liquid was obtained in 2min with yields higher than 70%. Because of an increasing interest in ionic liquid (Deetlefs and Seddon, 2003; Fu and Liu, 2006), scientists have started focusing on chiral ionic liquids to use their potential in chiral synthesis (Pegot and Loupy, 2004; Thang, 2004).

Many amino acids based ionic liquids were synthesized by using microwaves in recent year (Ohno and Fukumoto, 2007). Some new members of amino acid based ionic liquids were recently synthesized under microwave irradiation by neutralization of L-glutamic acid with nitric sulphuric and hydrochloric acid. Recently, 1, 3-dialkylimidazolium tetrachloroindates were prepared by Kim and Varma (2005a, 2005b), and these ionic liquids were used in the protection of alcohol, diols, and synthesis of cyclic carbonates.

Ionic liquids have abilities to dissolve organic as well as inorganic substances. Due to these properties, ionic liquids offer numerous opportunities for the modification of existing and also for the development of new extraction processes.

### *5.2.3 USE OF IONIC LIQUIDS IN ORGANIC REACTIONS*

Ionic liquids are attracting a great deal of attention as an alternate solvent replacing conventional molecular solvents for many organic reactions.

#### *(I) OXIDATION*

Although ionic liquids are highly stable and have been evaluated as media for oxidation reactions (Bonhote et al., 1996), surprisingly little attention has been focused on carrying out catalytic oxidations in ionic liquids. Asymmetric epoxidations were performed with NaOCl in [bmim][$PF_6$] catalyzed by a chiral Mn complex (Jacobsen's catalyst) (Song and Roh, 2000). An improvement in the catalytic activity was observed by adding the ionic liquid to the dichloromethane solvent. The ionic liquid containing the catalyst was reused in four consecutive runs without significant loss in yield; however, after the 5th run, the conversion dropped from 83 to 53%. This drop in conversion is believed to be due to degradation of the [Mn (III) (salen)] complex.

A more exciting study by Gaillon and Bedioui (2001) was carried out utilizing a chiral Mn(salen) complex in [bmim][$PF_6$] for the electroassisted biomimetic activation of molecular oxygen. It was observed that a highly reactive oxomanganese (V) intermediate could transfer its oxygen to an olefin, which hints at a promising future for clean oxidations with molecular oxygen in ionic liquid media.

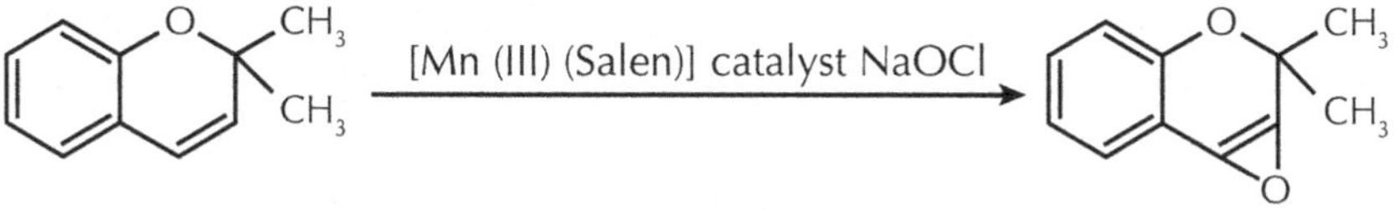

$CH_3$ $CH_3$ [Mn (III) (Salen)] catalyst NaOCl $CH_3$ $CH_3$ O

An efficient method for lactone synthesis was reported by Baj et al. (2009) that utilizes bis(trimethylsilyl) peroxide as an oxidant and ionic liquid 1-butyl-3-methylimidazolium trifluoromethanesulfonate as both; the solvent and catalyst.

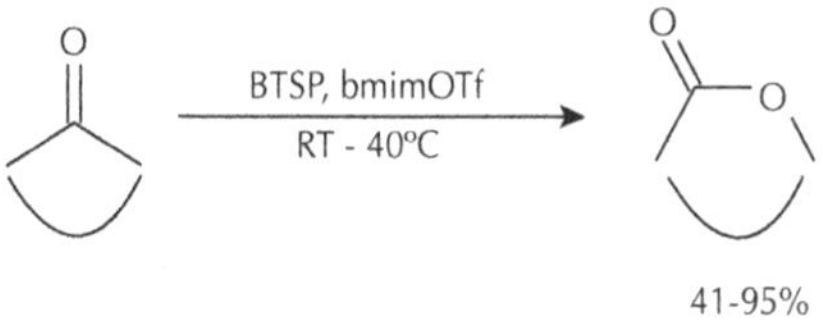

O BTSP, bmimOTf RT - 40°C O O

41-95%

A spectroscopic investigation of the complexes involved in the cobalt catalyzed oxidation of lignin model compounds in ionic liquids was conducted using *in situ* ATR-IR, Raman, and UV-Vis spectroscopy (Zakzeski et al., 2011). Hydrophilic N,N-dimethylpyrrolidinium and N,N-dimethylpiperidinium based ionic liquids provide a highly effective and selective environment for styrene oxidation (Chiappe et al., 2011).

### *(II) OXIDATIVE CARBONYLATION OF ANILINE*

Isocyanates and 4,4′-diphenylmethyldiisocyanate, are currently manufactured by phosphogenation of the corresponding amines with toxic phosgene, which may cause serious environmental pollution and also equipment corrosion. Therefore, these are produced either by oxidative carbonylation of amines or by reductive carbonylation of nitro compounds in the presence of an alcohol for the synthesis of isocyanates with non phosgene routes and have been extensively studied for the last two decades involving carbamates as intermediates. Pd, Ru, Rh, Au, and other transitional metal complexes were employed as the catalysts, but the corresponding catalytic turnover frequency is still not high enough for industrial applications. Palladium complexes coordinated with N-containing compounds is normally insoluble in most of the conventional organic solvents, which played an important role in the homogeneous catalysis. Moreover, organic solvents, which are available to establish a suitable homogenous catalyst system for the carbonylations of N-containing compounds, are very limited. The diversity of ionic liquids may form an optimal homogeneous catalyst system with a specific organometallic complex towards a specific reaction, for example, carbonylation of amine (Feng et al., 2004).

$$C_6H_5{-}NH_2 + MeOH + CO \xrightarrow[\text{Complex}]{\text{Palladium}} C_6H_5{-}NH{-}COOCH_3$$

### (III) REDUCTION

Howarth et al. (2001) reported the use of ionic liquid in the reduction of aldehyde and ketone. They used $NaBH_4$ as reducing agent in [bmim][$PF_6$]. This ionic liquid can be recycled.

$$R_1{-}C(=O){-}R_2 \xrightarrow[\text{[Bmim][PF}_6\text{]}]{NaBH_4} R_1{-}CH(OH){-}R_2$$

$R_1$ = Ph
$R_2$ = H, Ph, PhCH(OH)

Kabalka and Malladi (2000) also reported the reduction of carbonyl group into alcohol by using trialkylborane catalyst in the presence of [bmim][$BF_4$], [emim][$BF_4$] and [emim][$PF_6$] ionic liquids as solvents.

CHO — Tributylborane / [emim][$PF_6$] → OH

Recently, synthesis of (S)-naproxen has been reported by Moneiro et al. (1997) in presence of [bmim][$BF_4$] ionic liquid involving asymmetric hydrogenation. Trihexyl(tetradecyl) phosphonium ionic liquids were found to support the formation of Pd (0) nanoparticles without the addition of reducing agents such as $NaBH_4$. The resulting particles are highly crystalline and the particle shape is highly dependent on the anion of the ionic liquid (Kalviri et al., 2011).

## *(IV) BECKMANN REARRANGEMENT*

The rearrangement of a ketoxime to the corresponding amide is known as Beckmann rearrangement. It is utilized in manufacture of caprolactam in chemical industries in presence of excess or stoichiometric amounts of concentrated sulfuric acid or hydrogen chloride in a mixture of acetic acid and acetic anhydride. The application of inorganic acids usually causes large amount of byproducts and serious corrosion resulting in environment problems. Development of clean and highly efficient catalytic process for Beckmann rearrangement is still necessary. In a recent work, it was reported that ionic liquid could be a catalytic reaction media for Beckmann rearrangement with high conversion and selectivity.

N–OH — Phosphorated compound / Ionic liquids/ 343-363 K → O, NH

A novel task specific ionic liquids use in the Beckmann rearrangement was reported by Gui et al. (2004).

### *(V) DIELS–ALDER REACTION*

The Diels–Alder reaction of cyclopentadiene and methyl acrylate ester has been reported. Diels–Alder reaction is one of the most important tools for carbon–carbon bond formation. In this reaction, 1-ethyl-3-methyl-imidazolium or chloroaluminate was used ionic liquid and the ratio of exo or endo products depends on the ratio of emimCl/$(AlCl_3)_x$ (Lee, 1999). Diels–Alder reactions in neutral ionic liquids (such as 1-butyl-3-methy-limidazolium trifluoromethanesulfonate, 1-butyl-3-methylimidazolium hexafluorophosphate, 1-butyl-3-methylimidazolium tetrafluoroborate, and 1-butyl-3-methyl-imidazolium lactate) were reported by Earle et al. (1999).

The Diels–Alder reaction in the presence of pyridinium-based ionic liquid has also been reported (Xiao and Malhotra, 2004). Two ionic liquids for example, 1-ethyl-pyridinium tetrafluoroborate ([EtPy][$BF_4$]) and 1-ethyl-pyridinium trifluoroacetate ([EtPy][$CF_3COO$]) were used in the Diels–Alder reaction between isoprene and acrylonitrile, acrylic acid, and methacrylic acid. Ionic liquid supported imidazolidinone catalyst was found to be an efficient and recyclable organocatalyst for mediating highly enantioselective Diels–Alder reactions involving *α,β*-unsaturated aldehydes and cyclopentadiene (Shen et al., 2012).

R
O
+
Catalyst 1 (10 mol %)
$CF_3COOH$ (10 mol %)
$CH_3CH$ or $CH_3NO_2$, $H_2O$
r.t. 21 h
R
CHO
+
CHO
R
45-99% Yield
71-90% ee (exo)
73-94% ee (endo)

## *(VI) MANNICH REACTION*

Mannich reaction is one of the most important C–C bond forming reactions in the organic synthesis. Mannich bases are versatile synthetic intermediates, which are traditionally synthesized via acid (base) catalyzed reaction. Bronsted acidic ionic liquid can be used as the catalyst and solvent for this type of reaction. In this case, high yield is achieved and the ionic liquid can be easily reused (Zhao et al., 2004).

RCHO + $R_1NH_2$ + (O, $R_3$, $R_2$) → 25°C, Acidic ionic liquid → ($R_1$, NH, O, R, $R_3$, $R_2$)

The Mannich reaction of the aldehyde, ketone, and amine in the presence of Bronsted acidic ionic liquid has also been reported by Sahoo et al. (2006).

O + $NH_2$ + CHO → Ionic liquid, r.t → O, NH

## *(VII) HECK REACTION*

The first use of ionic liquids as reaction media for the palladium catalyzed Heck coupling was reported by Kaufmann et al. (1996). In the Heck reaction, palladium catalyst, polar solvent, and an aryl iodide were used. 1-Butyl-3-methylimidazolium bromide ionic liquid [bmim]Br was used as solvent, when aryl bromide reacts with styrene to produce stilbenes in high yield without adding phosphine ligand (Xu et al., 2000).

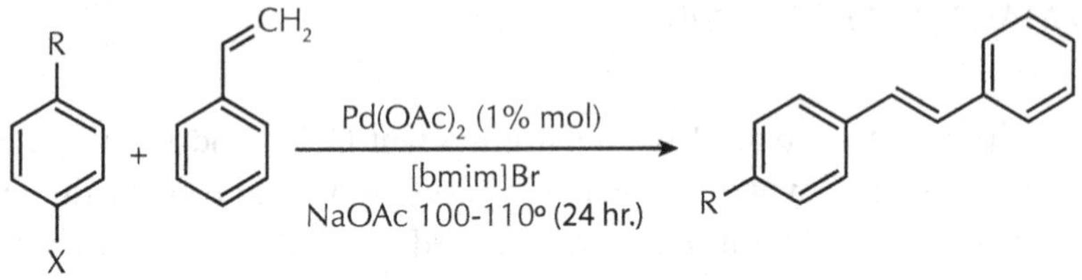

## *(VIII) FRIEDEL–CRAFTS REACTION*

The Friedel–Crafts acylation of benzene has been conducted in acidic chloroaluminate (III) ionic liquid (Boon et al., 1986). Other example is the benzoylation of anisole in the presence of copper triflate in $bmimBF_4$, where methoxybenzophenone was obtained (Ross and Xiao, 2002).

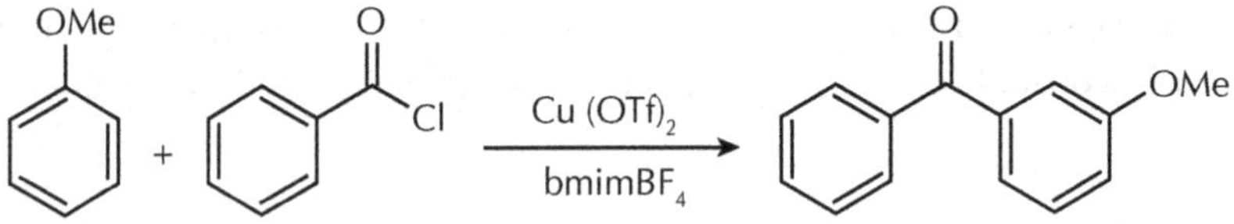

Naphthalene has been acetylated in ionic liquid with highest known selectivity for the 1-position (89%) as compared with 2-position (2%) (Adams et al., 1998). Another interesting development is the use of [bmim] [chloroaluminate] as Lewis acid catalyst for the Friedel–Crafts sulfonylation of benzene and substituted benzenes with TsCl (Nara et al., 2001).

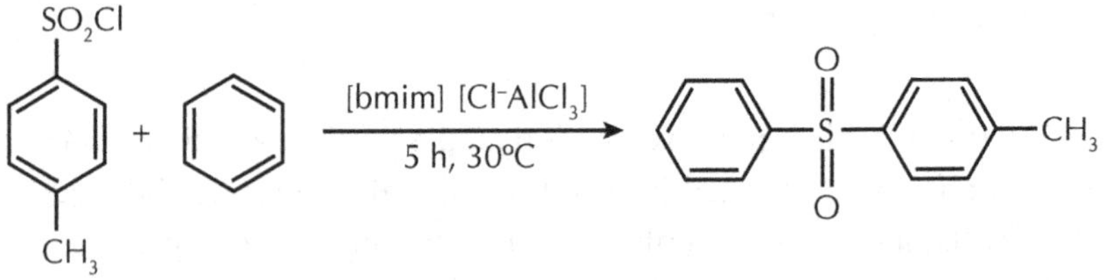

Gordon and Ritchie investigated the application of ionic liquids as solvents for indium and tin mediated allylation of carbonyl compounds (2002).

## (IX) WITTIG REACTION

Wittig reaction is a useful reaction for C=C formation. In the Wittig reaction, separation of product from the triphenylphosphine oxide is problematic. But when the reaction is carried out in the ionic liquid as solvent, the product is easily separated from the triphenylphosphine oxide by ether extraction (Boulaire and Gree, 2000).

$PPh_3$, $H_3C$, O + CHO → ($bmimBF_4$, 60°C, 2.5 h) Ph, O, $CH_3$

## (X) MICHAEL ADDITION REACTION

Zare et al. (2007) reported the Michael addition reaction of sulphonamide to $\alpha$, $\beta$-unsaturated esters using ZnO in [bmim] Br as a recyclable solvent.

O, Ph—S—$NH_2$, O + $R_1$, $R_2$, $COOR^3$ → ([bmim]Br, MW) $R_1$, $R_2$, HN, $COOR_3$, O=S=O, Ph

## (XI) FISCHER INDOLE SYNTHESIS

Fischer indole synthesis in ionic liquid was reported using chloroaluminate ionic liquid as a catalyst and solvent (Rebeiro et al., 2001).

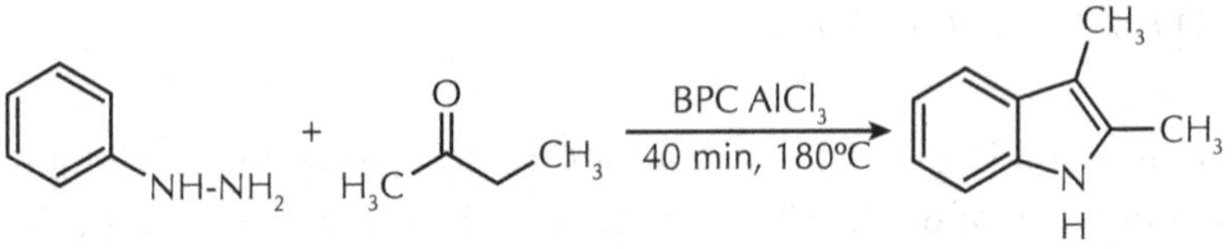

n-BPC = 1-(N-Butyl)pyridinium chloride

*(XII) CONDENSATION REACTION*

Benzoin condensation is one of the oldest C–C bond forming reaction in organic chemistry and in this reaction, cyanide ions are used as a catalyst. Benzoin condensation in imidazolium based room temperature ionic liquids has been reported by Fang et al. (2005).

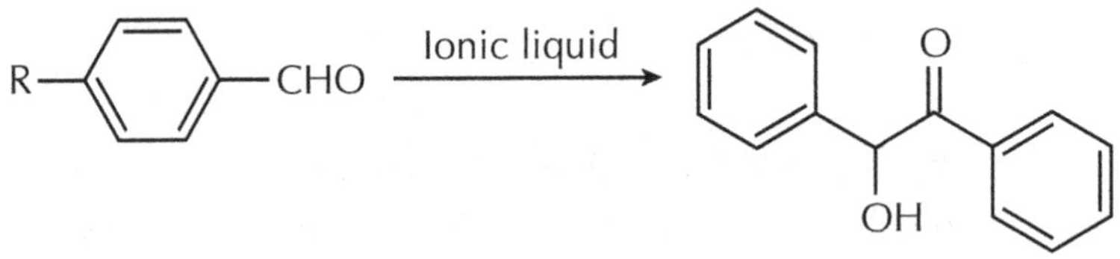

Xu et al. (2010) reported a simple, efficient and green procedure for Knoevenagel condensation catalyzed by [$C_4$dabco][$BF_4$] ionic liquid in water.

*(XIII) COUPLING REACTION*

Chiappe et al. (2004) investigated the Stille cross-coupling reaction in 10 different ionic liquids. The physico-chemical properties of ionic liquids affect the transfer of vinyl and alkyl groups. Palladium acetate was immobilized in ionic liquid layers on the mesopore wall of hierarchical MFI zeolite, and tested as a catalyst for Suzuki coupling reaction in water (Jin et al., 2009).

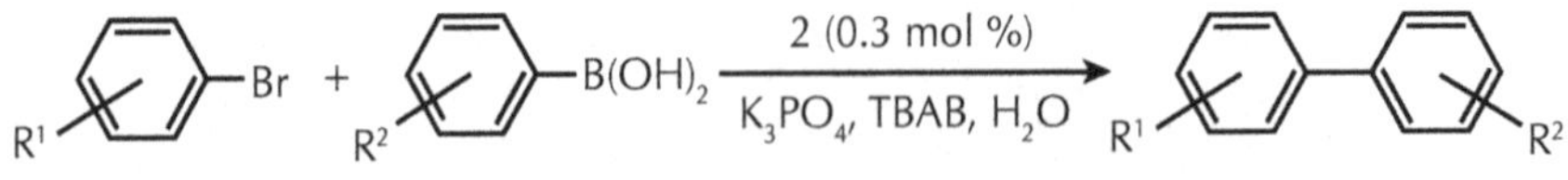

## *(XIV) ESTERIFICATION REACTION*

The application of ionic liquids as solvents for the enantioselective esterification of (R,S)-2-chloropropanoic acid with butan-1-ol using *Candida rugosa* lipase was reported by Gubicza et al. (2003). Liu et al. (2005) observed that the use of carbohydrates as renewable feedstocks is greatly hampered by their low solubility in any solvent but water. Ionic liquids that contain the dicyanamide ion (dca) dissolve approximately 200g $L^{-1}$ of glucose, sucrose and cyclodextrin. *Candida antarctica* lipase B (CaLB) mediated the esterification of sucrose with dodecanoic acid in [bmIm][dca].

*Candida antarctica* lipase B maintained transesterification activity upon dissolution in the ionic liquid $[Et_3MeN][MeSO_4]$, but not in other ionic liquids that dissolved it, such as [bmIm][[dca]. Cross-linked enzyme aggregates of CaLB, in contrast, remained active in this latter ionic liquid (Rantwijk et al., 2006). Esterification reaction of alcohol and the acetic acid in the presence of ionic liquid has been reported using 1-butylpyridinium chloride-aluminium chloride ionic liquid (Deng et al., 2001). Liu et al. (2011) used double $SO_3H$-functionalized ionic liquids for esterification of glycerol with acetic acid.

## *(XV) CYCLIZATION*

The utility of ionic liquids as a safe recyclable reaction media at 200°C in the presence of anhydrous scandium trifluoromethanesulfonate for a sequential reaction involving a Claisen rearrangement and cyclizations was reported by Zulfiqar et al. (2000). An Rh (I) catalyzed cycloisomerization of dienes with alkenes using ionic liquids as reaction media was investigated by Oonishi et al. (2009). It was found that the structure of ionic liquids strongly affected the recyclability of the catalyst. In this cycloisomerization, $[BDMI]^+$-based ionic liquid was more effective than a $[BMI]^+$-based one.

## (XVI) BIOCHEMICAL REACTION

Docherty and Kulpa (2005) examined the toxicity and antimicrobial activity of imidazolium and pyridinium ionic liquids. It provides information for future cross-disciplinary studies to test toxicity of ionic liquids, develop models and create green solvents. Chiappe et al. (2007) investigated effect of ionic liquids on epoxide hydrolase-catalyzed synthesis of chiral 1,2-diols while Cho et al. (2008) reported that toxicity of fluoride containing ionic liquid [BMIM][$BF_4$] to algal growth can increase due to hydrolysis of anions. The effects of ionic liquids on two freshwater algae were reported by Kulacki and Lamberti (2008). They examined *Scenedesmus quadricauda* and *Chlamydomonas reinhardtii*, under high and low nutrient test conditions (2008). Li et al. (2008) developed an efficient system for hydrolysis of lignocellulosic materials in ionic liquids with improved sugar yields at 100°C under atmospheric pressure.

## (XVII) BIODEGRADATION

With the goal of carbon free production of hydrogen from fossil fuels, a supported ionic liquid membrane for separating $CO_2$ from $H_2$ has been suggested for use in a separation enhanced reactor (Raeissi and Peters, 2009). Docherty et al. (2010) reported that pyridinium-based ionic liquids are biodegraded via different pathways depending upon the length of the substituted alkyl chain. Biodegradation products are non toxic to aquatic test organisms.

## (XVIII) SYNTHESIS

The complete synthesis of the pharmaceutical drug, pravadoline, in ionic liquid was reported by Earle et al. (1998). The best results were obtained [bmim][$PF_6$ ] at 150°C. The studies of carbonylation reactions of nitrogen containing compounds using $CO_2$ are relatively fewer in comparison of using CO most probably due to its chemical inert nature. Good yield (60–90%) of desired products were achieved by large amounts of dehydrating agents such as $PCl_5$, $POCl_3$, dicyclohexylcarbodiimide, and so on, but the worse thing was that the large amounts of by products were formed such as $Et_3NHCl$, $Et_3NHPOCl$, HCl, and so on. This indicates that using $CO_2$ as a carbonyl source to substitute phosgene in the synthesis of isocyanate and its derivatives were not only economically less favorable but it is also less eco-friendly.

Use of ionic liquid affords new opportunity to solve this problem because of the special solubility of carbon dioxide in ionic liquids. A series of aliphatic amines and even aromatic amines could react with $CO_2$ to afford corresponding urea with moderate to high yields in the presence of CsOH/ionic liquid catalyst system. Both cations and anions of the ionic liquids had a strong impact on the formation of urea when cyclohexylamine was used as substrate. The conversion was high enough, the desired product would precipitate. When water was added into the resulting liquid mixture after reaction because urea is insoluble in water, CsOH/ionic liquid still remain dissolved into the ionic liquid. Therefore, desired solid product with good yield could be recovered. An isolated yield of 98% was achieved after filtration and dryness, when bmimCl ionic liquid containing CsOH was employed.

$$C_6H_{11}\text{–}NH_2 \xrightarrow[\text{CsOH / Ionic liquid}]{CO_2} C_6H_{11}\text{–}NH\text{–}C(=O)\text{–}NH\text{–}C_6H_{11}$$

## *(XIX) MISCELLANEOUS*

The syntheses and reactions of alkynyl zinc reagents and Reformatsky-type reactions in ionic liquids (as a safe recyclable reaction medium) were reported by Kitazume and Kasai (2001). A simple method for the mono-N-alkylation of primary amines in ionic liquids was developed by Chiappe and Pieraccini (2003).

$$R-NH_2 + R'R''CH-X \xrightarrow{[bmim][PF_6]} R'R''CH-NH-R + (R'R''CH)_2N-R$$

Sarca and Laali (2004) carried out transacylation and deacylation of sterically crowded acetophenones in various imidazolium ionic liquids using triflic acid as catalyst. $BF_4^-$ based ionic liquids have potential as extractants in recovery of some amino acids from fermentation broth. Hydrophobicity of the amino acid, pH of the aqueous phase, and water solubility in the ionic liquid phase are key factors affecting this extraction (Wang et al., 2005). Weng et al. (2006) used novel quaternary ammonium ionic liquids as acidic catalysts for the synthesis of cinnamic acid.

$$RR'C(Cl)-CH(R'')-CCl_3 + 2\,H_2O \xrightarrow[120^\circ C,\ 8\ h]{\text{Ionic liquid}} RR'C{=}C(R'')COOH + 4\,HCl$$

A successful ionic liquid-based aqueous biphasic system was proposed for lipase extraction with recovery values around 80%. This will allow use of ionic liquids as withdrawal solvents and media for catalytic applications (Deive et al., 2011). Diego et al. (2011) reported efficient production of biodiesel in hydrophobic ionic liquids using immobilized lipase with the straightforward extraction of the products using the properties of appropriate ionic liquids.

Ressmann et al. (2012) extracted pharmaceutically active triterpene betulin from biomass by utilizing ionic liquids with significantly improved extraction yield and purity. A new functionalized ionic liquid [eimCH-

$_2$CONHBu]NTf$_2$ has been designed and synthesized by Huaxi et al. (2012) with high extraction efficiency and special selectivity for tryptophan. Duarte et al. (2012) investigated ionic liquids as foaming agents of semicrystalline natural-based polymers. Tetrahalogenoaurate anions are removed from water by precipitation with water-soluble ionic liquids or by extraction using hydrophobic ionic liquids (Papaiconomou et al., 2012).

Arce et al. (2007) presented rigorous thermodynamic analyses to evaluate the ability of the ionic liquid 1-ethyl-3-methylimidazolium ethylsulfate as an extracting solvent or a distillation entrainer for the separation of ethanol and ETBE. They also reported that ionic liquid 1-ethyl-3-methylimidazolium bis(trifluoromethyl) can selectively remove benzene from its mixtures with hexane. The organic product methyl-(Z)-$\alpha$-acetamido cinnamate can be crystallized from the ionic liquid 1-butyl-3-methylimidazolium tetrafluoroborate by either a shift to lower temperatures or a shift to higher carbon dioxide concentrations (Kroon et al., 2008). Shen et al. (2008) developed an efficient method for the synthesis of optically active O-acetyl cyanohydrins via one-pot lipase catalyzed kinetic resolution of *in situ* generated cyanohydrins and O-acetyl cyanohydrins in ionic liquid.

OAc
R CN
22-46% yield
60-92% cc
(i) TMSCN, [omim] $PF_6$
*(ii) Pscudomonas cepacia* lipase vinyl acclate
O
R H
(i) TMSCN, [omim] $PF_6$
*(ii) Candids antbactria lipase* *n*-BuOH/$Ac_2O$
OAc
R CN
34-57% yield
73-98% cc

Chiappe et al. (2010) achieved high concentrations of metal ions by dissolving metal salts in ionic liquids with common anions. Stahlberg et al. (2010) investigated the conversion of glucose to 5-(hydroxymethyl) furfural (HMF) in alkylimidazolium based ionic liquids together with lanthanide catalysts, which speeded up by the higher hydrophobicity of the imidazolium cation. Gas-liquid acetylene hydrochlorination proceed efficiently under catalysis of non mercuric metal chlorides by using ionic liquids as reaction media (Qin et al., 2011). The direct nucleophilic substitution reactions of alcohols can be promoted and modulated by a clean and recyclable reaction medium (zinc-based ionic liquid) [CHCl][$ZnCl_2$]$_2$ (Zhu et al., 2011).

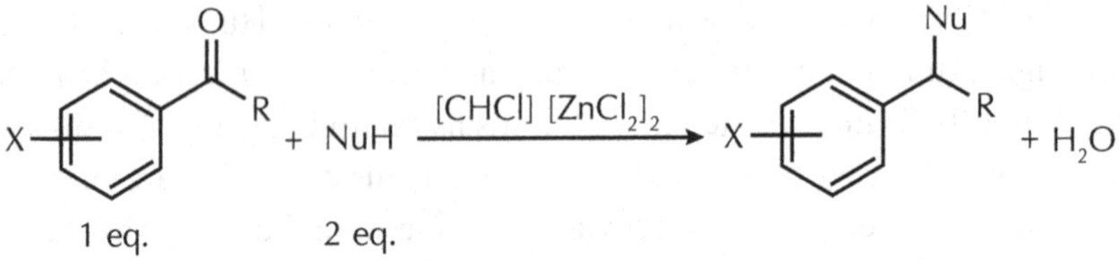

O
R
X
+ NuH
[CHCl] [ZnCl2]2
Nu
R
X
+ H2O
1 eq.
2 eq.

NuH = Amine, sulfonamide, amide, 1,3-dicarbonyl compound

Berger et al. (2011) used a simple ionic liquid-based system for the selective decomposition of formic acid to hydrogen and carbon dioxide. Gao et al. (2011) prepared Fe (III)-derived Lewis acid ionic liquid as an efficient and recyclable catalyst for the selective benzylation of arenes.

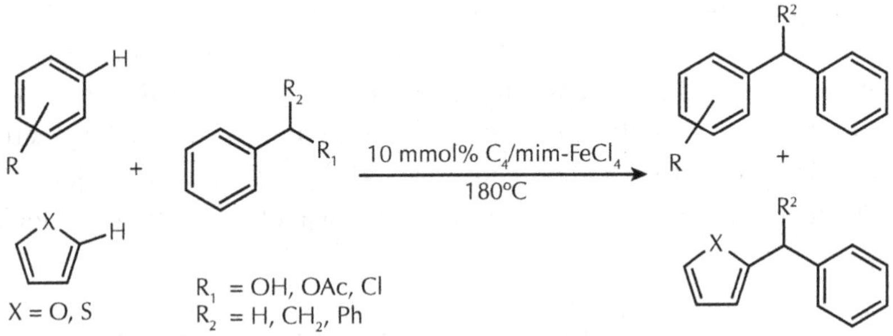

H
R
+
R2
R1
10 mmol% C4/mim-FeCl4
180°C
X
H
X = O, S
R1 = OH, OAc, Cl
R2 = H, CH2, Ph
R2
R
+
R2
X

Natrajan and Wen (2011) reported ionic liquids as unique reaction media for the efficient introduction of N-sulfopropyl groups in the synthesis of hydrophilic, chemiluminescent acridinium esters using reduced quantities of the carcinogenic reagent 1,3-propane sultone.

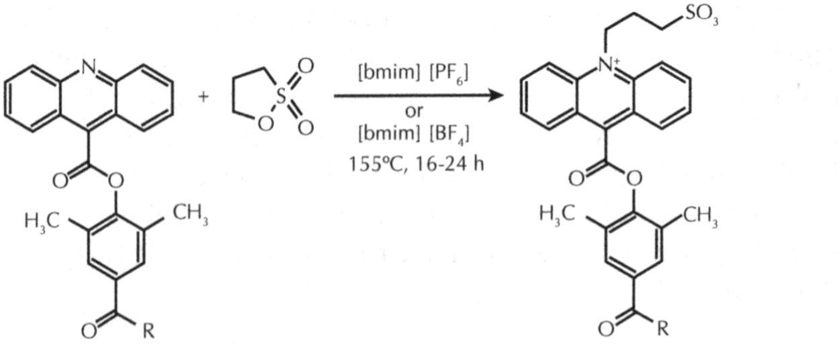

N
+
O
S
O
O
[bmim] [PF6]
or
[bmim] [BF4]
155°C, 16-24 h
SO3
N+
O
O
H3C
CH3
O
R

R = OMe, N-hydroxysuccinimidyl

> 80% Conversion minimal polysulfonation

A novel one-pot synthesis of 6-aminouracils via *in situ* generated ureas and cyanoacetylureas in the presence of ionic liquid 1,1,3,3-tetramethylguanidine acetate as a recyclable catalyst has been reported by Chavan and Degani (2012).

$R_1$–$NH_2$ + $R_3$–NCO + $R_3$–$CH_2$–CO–$CH_3$ → (product with $R_3$–N, O, N–$R_1$, $NH_2$)

[TMG] [Ac], 60°C
1.5 h and 90°C, 0.5-3.5 h

where, $R_1$ = Me, n-Pr, n-Bu, Ph, p-OMe-Ph, p-MePh, o-MePh, m-MePh, bn, p-OMe Bn, $PhCH_2CH_2$. p-n$Me_2$Ph, $R_3$-K

where, $R_1$ = Me, $R_3$ = Me and $R_1$ = n-Pr, $R_3$ = n-Pr

where, $R_1$ = Pb, Bn, $R_3$ = Allyl

PEG-functionalized basic ionic liquids proved to be efficient catalysts for dimethyl carbonate synthesis under atmospheric $CO_2$ through a "one-pot, two-stage" process (Yang et al., 2012). Sun et al. (2012) synthesized cyclic carbonates via cycloaddition of $CO_2$ to epoxides with bifunctional chitosan supported ionic liquid catalysts. Acidic ionic liquids functionalized, ordered and stable mesoporous polymers with excellent catalytic activities have been synthesized at high temperature (Liu et al., 2012).

Wellens et al. (2012) reported that cobalt can efficiently be separated from nickel by solvent extraction with the ionic liquid tri(hexyl)tetradecylphosphonium chloride. Li et al. (2012) reported hydrogen bonding-promoted dehydration of fructose to 5-hydroxymethylfurfural (HMF) in pure ionic liquid without any other additive or catalyst. Fehér et al. (2012) used Bronsted acidic ionic liquids, supported on silica gel in oligomerization of isobutene. The supported catalysts could be used several times without loss of activity or change in selectivity (2012).

Most of the chemical reactions require a solvent, and vapors of these solvents create environmental pollution. It is therefore necessary to find out a solvent with high boiling point; whereas most of the organic solvents are having a low boiling point. Ionic liquids enter the scene here. These are defined as designer solvents. Their properties can be so designed by selecting a proper cation and anion that they can serve our purpose.

## KEYWORDS


- **Beckmann rearrangement**
- **Diels–Alder reaction**
- **Fischer indole synthesis**
- **Friedel–Crafts reaction**
- **Michael addition reaction**
- **Wittig reaction**


## REFERENCES


1. Adams, C. J., Earle, M. J., Roberts, G. K., & Seddon, R. (1998). *Chemical Communications,* 2097–2098.
2. Arce, A., Earle, M. J., Rodríguez, H., & Seddon, K. R. (2007). *Green Chemistry, 9,* 70–74.
3. Arce, A., Rodríguez, H., & Soto, A. (2007). *Green Chemistry, 9,* 247–253.
4. Baj, S., Chrobok, A., & Słupska, R. (2009). *Green Chemistry, 11,* 279–282.
5. Baudequin, C., Bregon, D., Levillain, J., Guillen, F., Plaquevent, J. -C., & Gaumont, A. -C. (2005). *Tetrahedron: Asymmetry, 16,* 3921–3945.
6. Berger, M. E. M., Assenbaum, D., Taccardi, N., Spiecker, E., Wasserscheid, P. (2011). *Green Chemistry, 13,* 1411–1415.
7. Blasius E., Klein, W., & Schon, U. (1985). *Journal of Radioanalytical and Nuclear Chemistry, 89,* 389–398.
8. Bonhôte, P., Dias, A. -P., Papageorgiou, N., Kalyanasundaram, K., & Grätzel M. (1996) *Inorganic Chemistry, 35,* 1168–1178.
9. Boon, J. A., Levisky, J. A., Pflug, J. L., & Wilkes, J. S. (1986). *Journal of Organic Chemistry, 51,* 480–483.
10. Boulaire, V. L. & Gree, R. (2000). *Chemical Communications,* 2195–2196.
11. Chavan, S. S. & Degani, M. S. (2012). *Green Chem*istry, *14,* 296–299.

12. Chiappe, C., Imperato, G., Napolitano, E., & Pieraccini, D. (2004). *Green Chemistry, 6,* 33–36.
13. Chiappe, C., Leandri, E., Hammock, B. D., & Morisseau, C. (2007). *Green Chemistry, 9,* 162–168.
14. Chiappe, C., Malvaldi, M., Melai, B., Fantini, S., Bardi, U., & Caporali, S. (2010). *Green Chemistry, 12,* 77–80.
15. Chiappe, C. & Pieraccini, D. (2003). *Green Chemistry, 5,* 193–197.
16. Chiappe, C., Sanzone, A., & Dyson, P. J. (2011). *Green Chemistry, 13,* 1437–1441.
17. Cho, C., Pham, T. P. T., Jeon, Y., & Yun, Y. (2008). *Green Chemistry, 10,* 67–72.
18. Chum, H. L., Koch, V. R., Miller, L. L., & Osteryoung, R. A. (1975). *Journal of the American Chemical Society, 97,* 3264–3265.
19. Davis Jr., J. H. (2004). *Chemistry Letters, 33,* 1072–1077.
20. Deetlefs, M. & Seddon, K. R. (2003). *Green Chemistry, 5,* 181–186.
21. Deive, F. J., Rodríguez, A., Pereiro, A. B., Araújo, J. M. M., Longo, M. A., Coelho, M. A. Z. et al. (2011) *Green Chemistry, 13,* 390–396.
22. Deng, Y., Shi, F., Beng, J., & Qiao, K. (2001). *Journal of Molecular Catalysis A: Chemical, 165,* 33–41.
23. Diego, T. D., Manjón, A., Lozano, P., Vaultier, M., & Iborra, J. L. (2011). *Green Chemistry, 13,* 444–451.
24. Docherty, K. M., Joyce, M. V., Kulacki, K. J., & Kulpa, C. F. (2010). *Green Chemistry, 12,* 701–712.
25. Docherty, K. M., & Kulpa, C. F. (2005). *Green Chemistry, 7,* 185–189.
26. Duarte, A. R. C., Silva, S. S., Mano, J. F., & Reis, R. L. (2012). *Green Chemistry, 14,* 1949–1955.
27. Dyson, P. J., Grossel, M. C., Srivinasan N., Vine, T., Welton T., William A. J. P., & Zigars T. J. (1997). *Journal of Chemical Society Dalton Transactions,* 3465–3469.
28. Earle, M. J., McCormac, P. B., & Seddon, K. R. (1999). *Green Chemistry, 1,* 23–25.
29. Earle, M. J. & Seddon, K. R. (2000). *Pure and Applied Chemistry, 72,* 1391–1398.
30. Earle, M. L., McCormac, P. B., & Seddon, K. R. (1998). *Green Chemistry,* 2245–2246.
31. Fang, S. J., Hua Y., Ge, G., & Ru, G. X. (2005). *Chinese Chemical Letters, 16,* 321–324.
32. Fehér, C., Kriván, E., Hancsók, J., & Skoda-Földes, R. (2012). *Green Chemistry, 14,* 403–409.
33. Feng, S., Yanlong, G., Qinghua, Z., & Youquan, D., (2004). *Catalysis Surveys from Asia, 8,* 179–186.
34. Freemantle, M. (1998). *Chemical & Engineering News, 76,* 32–37.
35. Fu, S. K. & Liu, S. T. (2006). *Synthetic Communications, 36,* 2059–2067.
36. Gabriel, S. & Weiner, J. (1988). *Bit error rate, 21,* 2669–2679.
37. Gaillon, L. & Bedioui, F. (2001). *Chemical Communications,* 1458–1459.
38. Gao, J., Wang, J., Song, Q., & He, L. (2011). *Green Chemistry, 13,* 1182–1186.
39. Gloe, K., Muhl, P., Kholkin, A., Meerbote, M., & Beger J. (1982). *Isotopenpraxis, 18,* 170–175.
40. Gordon, C. M. & Ritchie, C. (2002). *Green Chemistry, 4,* 124–128.
41. Gubicza, L., Nemestóthy, N., Fráter, T., & Bélafi-Bakó, K. (2003). *Green Chemistry, 5,* 236–239.

42. Gui, J., Youquan, D., Zhide, H., & Zhaoin, S. (2004). *Tetrahedron Letters, 45,* 2681–2683.
43. Holbrey, J. & Seddon, K. R. (1999). *Clean Production Processes, 1,* 223–236.
44. Howarth, J., James, P., & Ryan, R. (2001). *Synthetic Communications, 31,* 2935–2938.
45. Huaxi, L., Zhuo, L., Jingmei, Y., Changping, L., Yansheng, C., Qingshan, L., Xiuling, Z., & Urs, W. (2012). *Green Chemistry, 14,* 1721–1727.
46. Jin, M., Taher, A., Kang, H., Choi, M., & Ryoo, R. (2009). *Green Chemistry, 11,* 309–313.
47. Kabalka, G. W. & Malladi, R. R. (2000). *Chemical Communications,* 2191.
48. Kalviri, H. A. & Kerton, F. M. (2011). *Green Chemistry, 13,* 681–686.
49. Kaufmann, D. E., Nouroozian, N. M., & Heuze, H. (1996). *Synlett,* 1091–1092.
50. Kim, Y. J. & Varma, R. S. (2005a). *Tetrahedron Letters, 46,* 1467–1469.
51. Kim, Y. J. & Varma, R. S. (2005b). *Journal of Organic Chemistry, 70,* 7882–7891.
52. Kitazume, T. & Kasai, K. (2001). *Green Chemistry, 3,* 30–32.
53. Kroon, M. C., Toussaint, V. A., Shariati, A., Florusse, L. J., Spronsen, J. V., Witkamp, G., & Peters, C. J. (2008). *Green Chemistry, 10,* 333–336.
54. Kulacki, K. J. & Lamberti, G. A. (2008). *Green Chemistry, 10,* 104–110.
55. Lee, C. W. (1999). *Tetrahadron Letter, 40,* 2461–2462.
56. Li, C., Wang, Q., & Zhao, Z. K. (2008). *Green Chemistry, 10,* 177–182.
57. Li, Y., Wang, J., He, L., Yang, Z., Liu, A., Yu, B., & Luan, C. (2012). *Green Chemistry, 14,* 2752–2758.
58. Liu, F., Zuo, S., Kong, W., & Qi, C. (2012). *Green Chemistry, 14,* 1342–1349.
59. Liu, Q., Janssen, M. H. A., Rantwijk, F. V., & Sheldon, R. A. (2005). *Green Chemistry, 7,* 39–42.
60. Liu, X., Ma, H., Wu, Y., Wang, C., Yang, M., Yan, P. et al. (2011). *Green Chem*istry, *13,* 697–701.
61. Moneiro, A. L., Zinn, F. K., de Souza, R. F., & Dupont, J. (1997). *Tetrahedron Asymmetry, 8,* 177–179.
62. Nara, S. J., Harjani, J. R., & Salunkhe, M. M. (2001). *Journal of Organic Chemistry, 66,* 8616–8620.
63. Natrajan, A. & Wen, D. (2011). *Green Chemistry, 13,* 913–921.
64. Ohno, H. & Fukumoto, K. (2007). *Accounts of Chemical Research, 40,* 1122–1129.
65. Oonishi, Y., Saito, A., & Sato, Y. (2009). *Green Chemistry, 11,* 330–333.
66. Papaiconomou, N., Vite, G., Goujon, N., Lévêque, J., & Billard, I. (2012). *Green Chemistry, 14,* 2050–2056.
67. Pegot, G. V. B. & Loupy, A. (2004). *European Journal of Organic Chemistry, 5,* 1112–1116.
68. Qin, G., Song, Y., Jin, R., Shi, J., Yu, Z., & Cao, S. (2011). *Green Chem*istry, *13,* 1495–1498.
69. Raeissi, S. & Peters, C. J. (2009). *Green Chemistry, 11,* 185–192.
70. Rantwijk, F. V., Secundo, F., & Sheldon, R. A. (2006). *Green Chemistry, 8,* 282–286.
71. Rebeiro, G. L. & Khadilkar, B. M. (2001). *Synthesis, 3,* 370–372.
72. Ressmann, A. K., Strassl, K., Gaertner, P., Zhao, B., Greiner, L., & Bica, K. (2012). *Green Chemistry, 14,* 940–944.
73. Ross, J. & Xiao, J. (2002). *Green Chemistry, 4,* 129–133.

74. Sahoo, S., Joseph, T., & Halligudi, S. B. (2006). *Journal of Molecular Catalysis A: Chemical, 242,* 179–182.
75. Sarca, V. D. & Laali, K. K. (2004). *Green Chemistry, 6,* 245–248.
76. Shen, Z., Cheong, H., Lai, Y., Loo, W., & Loh, T. (2012). *Green Chemistry, 14,* 2626–2630.
77. Shen, Z., Zhou, W., Liu, Y., Ji, S., & Loh, T. (2008). *Green Chem*istry, *10,* 283–286.
78. Song, C. E. & Roh, E. J. (2000). *Chemical Communications,* 837–838.
79. Ståhlberg, T., Sørensen, M. G., & Riisager, A. (2010). *Green Chemistry, 12,* 321–325.
80. Sun, J., Wang, J., Cheng, W., Zhang, J., Li, X., Zhang, S., & She, Y. (2012). *Green Chemistry, 14,* 654–660.
81. Thang, G. V., Pegot, B., & Loupy, A. (2004). *European Journal of Organic Chemistry, 5,* 1112–1116.
82. Walden, P. (1914). *Bulletin de l'Académie Impériale des Sciences de Saint Pétersbourg,* 405–422.
83. Wang, J., Pei, Y., Zhao, Y., & Hu, Z. (2005). *Green Chemistry, 7,* 196–202.
84. Wellens, S., Thijs, B., & Binnemans, K. (2012). *Green Chemistry, 14,* 1657–1665.
85. Welton, T. (1999). *Chemical Reviews, 99,* 2071–2083.
86. Weng, J., Wang, C., Li, H., & Wang, Y. (2006). *Green Chemistry, 8,* 96–99.
87. Wilkes, J. S., Levisky, J. A., Wilson, R. A., & Hussey, C. L. (1982). *Inorganic Chemistry, 21,* 1236–1264.
88. Wilkes, J. S. & Zaworotko, M. J. (1992). *Journal of the Chemical Society: Chemical Communications,* 965–967.
89. Xiao, Y. & Malothra, S. V. (2004). *Tetrahedron Letters, 45,* 8339–8342.
90. Xu, D., Liu, Y., Shi, S., & Wang, Y. (2010). *Green Chemistry, 12,* 514–517.
91. Xu, L., Chen, W., & Xiao, J. (2000). *Organometallics, 19,* 1123–1127.
92. Yang, Z., Zhao, Y., He, L., Gao, J., & Yin, Z. (2012). *Green Chemistry, 14,* 519–527.
93. Zakzeski, J., Bruijnincx, P. C. A., & Weckhuysen, B. M. (2011). *Green Chemistry, 13,* 671–680.
94. Zare, A., Hasaninejad, A., & Zare A. R. M. (2007). *Canadian Journal of Chemistry, 85,* 438–444.
95. Zhao, G., Jiang, T., Gao, H., Han, B., Huang, J., & Sun, D. (2004). *Green Chemistry, 6,* 75–77.
96. Zhu, A., Li, L., Wang, J., & Zhuo, K. (2011). *Green Chemistry, 13,* 1244–1250.
97. Zulfiqar, F. & Kitazume, T. (2000). *Green Chemistry, 2,* 296–297.

# CHAPTER 6

# SUPERCRITICAL FLUIDS

ABHILASHA JAIN, SHIKHA PANCHAL, SHWETA SHARMA, and RAMASHWAR AMETA

## CONTENTS

## 6.1 INTRODUCTION

A supercritical fluid is a substance at a temperature and pressure above its critical point, where distinct liquid and gas phases do not exist. It can effuse through solids like a gas and dissolve materials like a liquid. In addition, close to the critical point, small changes in pressure or temperature results in large changes in density allowing many properties of a supercritical fluid to be "fine-tuned". Supercritical fluids are suitable as a substitute for organic solvents in a range of industrial and laboratory processes. Carbon dioxide and water are the most commonly used supercritical fluids, and also being used for decaffeination and power generation, respectively.

## 6.2 SOME MAJOR SUPERCRITICAL FLUIDS

### *6.2.1 SUPERCRITICAL WATER (SCW)*

Water is described as superheated water, subcritical water or pressurized hot water between 100°C and its supercritical point at 370°C. SCW has been used in synthetic organic chemistry because of having some unique properties different from those of ambient water (Broli et al., 1999; Kajmoto, 1999; Savage, 1999; Siskin and Katrinzky, 2001). Water has similar properties to an organic solvent such as methanol. The near critical water (NCW) region is described as 250–300°C at 100–80bar (Raner et al., 1995; Stadler et al., 2003). It also has some unique properties and these characteristics are:

It has lower viscosity as compared to water that results in faster diffusion of the compound.

It has lower surface tension than water.

More solubility of two polar compounds due to less hydrogen bonding.

Its heat capacity is 2–6 times of liquid water, which improves transfer of heat.

Single homogeneous phase results in no interfacial mass transfer limitation.

Dramatic changes were observed in density and in ionic product of water near critical temperature.

All these properties results in an increased rate of reaction and extraction.

There are many types of reactions that occur in water at high temperature, for example,

(i) Polyester and other condensation polymers are cleaved to their starting materials at 573K.
(ii) Rubber tires are converted to oil with 44% yield at 673K.
(iii) Cellulose based waste can be converted to sugars.
(iv) It is a very effective medium for acid and base catalyzed reactions.
(v) Gasification of biomass can be done in it, which includes food waste, organic waste from industries, and waste from forestry or agriculture.
(vi) Oxidation reactions can be used for waste water treatment. The SCW oxidation (SCWO) technique is environmental friendly because no organic solvent or additives are required. In this technique, oxygen, air or hydrogen peroxide can be used as an oxidant. In SCWO, theoretically all organic compounds (pollutant) can be converted mainly into carbon dioxide and water. It is a cost effective technique.

The use of superheated water as suitable solvent for the extraction of non polar compounds from environmental sample was proposed by Hawthorne et al. in mid 1994. It is one of the best options as a solvent for chromatography because it is a green solvent which is cost effective, safe, non toxic, non inflammable, and recyclable. The reduction in viscosity with temperature enables high flow rates. Moreover, it can be used with wide range of detectors, where water has virtually no background signal. It is also compatible with flame ionization detection. So, it can be concluded that pressurized hot water and SCW is a great promise for the future in green chemical technology.

Organic reactions in supercritical water (SCW):

- Diel–Alder cycloaddition reactions have been carried out in SCW (Korzenski and Kolis, 1997).

- Heck coupling reaction was conducted in SCW (Diminnie et al., 1995; Reardon et al., 1995).
- Aldol condensation reaction could be conducted in SCW (An et al., 1997).
- Claisen, Rupe, Meyer-Schuster and pinacol-pinacolone rearrangements have also been carried out in SCW (Crittendon and Parsons, 1994; Kuhlmann et al., 1994).
- Dehydration of alcohol (Crittendon and Parsons, 1994; Kuhlmann et al, 1994; Xu and Antal, 1997) and decarboxylation of carboxylic acids (Carlsson et al., 1994; Yu and Savage, 1998) have been reported in SCW.
- In SCW, carbon–silicon bond cleavage of organo-silicon reaction is also reported (Itami et al., 2003). At 390°C, the reaction of arylsilanes in SCW causes C–Si bond cleavage.

R–$C_6H_4$–$SiMe_3$ —(SCW, 390°C; 27 M Pd, 300 mm)→ R–$C_6H_4$–H

(72-91% Yield)

Organic reaction in near critical water (NCW) region:

**(i) Diels–Alder cycloaddition**: This reaction was reported by heating with water in a microwave (MW) oven at 295°C for 20min, where 2,3-dimethylbutadiene and acrylonitrile react to give an addition product in good yield.

(Diene) + (Dienophile) CN —($H_2O$, 20 min; MW, 295°C)→ (Adduct)

**(ii) Claisen rearrangement**: It was reported in near critical water at 240°C in MW oven, which gives 84% yield (Raner et al., 1995).

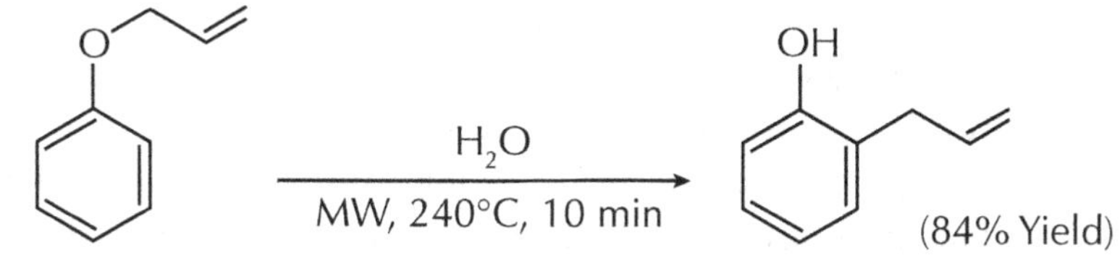

O

OH

$H_2O$

MW, 240°C, 10 min

(84% Yield)

Allyl phenyl ether Allyl phenyl ether

**(iii) Pinacol-Pinacolone rearrangement**: It has also been carried out in near critical water at 270°C (Kremsner and Kappe, 2005).

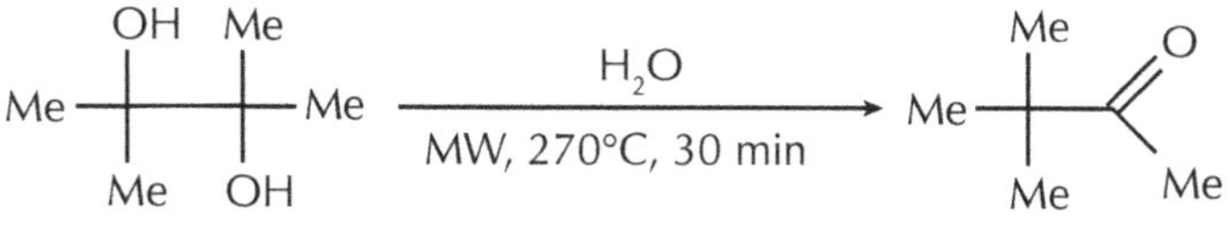

OH Me

Me Me

Me OH

$H_2O$

MW, 270°C, 30 min

Me O

Me

Me Me

Allyl phenyl ether Pinacolone

**(iv) Fischer indole synthesis**: When this reaction was carried out by heating with water in a MW oven at 270°C, 64% yield of indole derivative has been obtained (Strauss, 1999; Strauss and Trainor, 1995).

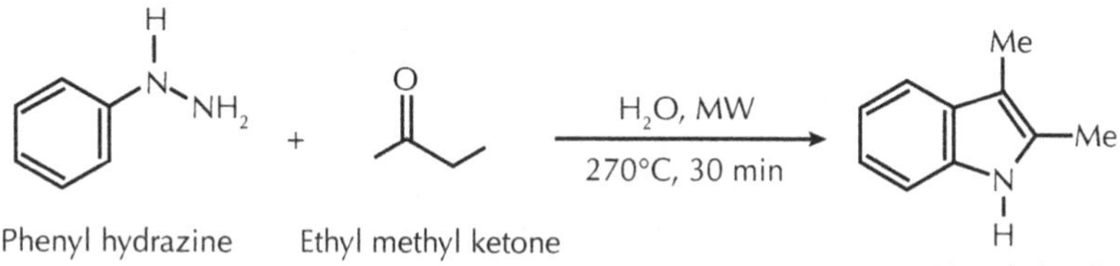

H

N

$NH_2$

\+

O

$H_2O$, MW

270°C, 30 min

Me

Me

N

H

Phenyl hydrazine Ethyl methyl ketone

2, 3-Dimethyl indone

**(v) Decarboxylation of carboxylic acids**: Decarboxylation of indole-2-carboxylic acid by heating with water at near critical region gives indole with 100% yield (An et al., 1997). It requires 255°C temperature and 20 min. for completion of reaction.

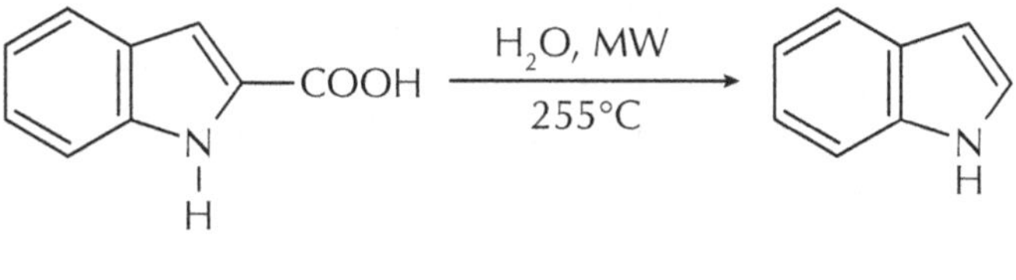

COOH

N

H

$H_2O$, MW

255°C

N

H

**(vi) Hydrolysis of amides and esters**: In NCW region, hydrolysis of amides and esters is not required the addition of acid/base or any other catalyst. Ethyl benzoate and benzamide are hydrolyzed to benzoic acid in MW at 295°C for 2–4hr, respectively, without addition of any catalyst (Kremsner and Kappe, 2005). It gives 92–95% yield of product.

$$C_6H_5\text{–N(=O)–y} \xrightarrow[295^\circ C]{H_2O,\ MW} C_6H_5\text{–C(=O)–OH}$$

$y = \text{-}OC_2H_5$ or $\text{-}NH_2$

**(vii) Some other reactions**: It was observed that the reactivity of some organic compounds in near critical region or at high temperature water is enhanced by some water soluble compounds, for example, mineral acid, ammonia, which are produced by hydrolysis of ester or amines and can act as acid or base catalyst (Akiya and Savage, 2002).

### *6.2.2 CARBON DIOXIDE*

Gore (1861) for the first time gave the process of preparing liquid CO. Carbon dioxide exists in three phases that is, solid, liquid and gas. Solid phase of $CO_2$ is called 'dry ice' and it is used for cooling. Gas phase is well known, and at atmospheric temperature and pressure the solid transforms to gas without liquification. Only in certain specified conditions, it can be liquefied. With the increasing pressure on gas or heating of solid $CO_2$, liquid phase can be achieved. The critical temperature of $CO_2$ is 31°C. At the temperature -56°C and 5.1atm, all three phases of carbon dioxide exists simultaneously. At 31°C and 73atm, it exists as supercritical fluid. At this condition, it has unique properties, that is, viscosity similar to the gas phase and density similar to the liquid phase.

Some advantages of supercritical carbon dioxide ($scCO_2$) are:

- A high diffusion rate offers potential for increased reaction rates.
- It has high compressibility. With relatively small change in pressure, large changes in solvent properties have been reported by which infinite range of solvent properties can be achieved. Small amount of co-solvent can further modify solvent properties.
- Due to the high solubility of light gases, some catalysts and substrates brings all compounds together in single homogeneous phase so that it has a potential for homogeneous catalytic processes.
- It is an excellent medium for oxidation and reduction reactions.
- Thus, $scCO_2$ can be used as a solvent in various ways, for example, for dry cleaning, extracting natural products, and solvent for organic reactions.
- In dry cleaning, the most commonly used solvent is PERC (perchloroethylene), which is a carcinogenic compound and is also one of the factor responsible for ozone layer depletion. This problem can be overcome by using $CO_2$ as a solvent for dry cleaning. It dissolves non polar substances, and with the addition of some surfactants the solubility of oils and greases in carbon dioxide can be increased. Nowadays, the micelle technologies have produced dry cleaning machines using liquid $CO_2$ and a surfactant to dry clean clothes.

$ScCO_2$ can be used as a solvent for extraction because of:

- No hydrolysis (which generally occurs in steam distillation).
- No loss of volatile components.
- No thermal degradation products.
- Free of inorganic salts or heavy metals.
- High concentration of valuable ingredients and high extraction yield.
- Free of any microbial life.
- Environmentally benign solvent.
- High solubility toward hydrocarbons, ethers, esters, and so on, whereas polar compounds, for example, sugars, tannins, glycosides, and so on are insoluble.

As $scCO_2$ has all these advantages; hence, it has numerous applications and is used for extraction of natural products:

- Essential oils from turmeric, coriander, ginger, ajowan, and so on have been extracted with $scCO_2$.
- It is extensively used for natural coffee decaffeination.

- Various food products, for example, pepper, cumin, cardamom, cloves, and so on can be isolated.

Vegetable's tannin materials have been isolated using sc$CO_2$.

The sc$CO_2$ can be used as a solvent for organic synthesis because it is environmentally benign and non toxic solvent (Anatas et al., 1998). Since the characteristics of sc$CO_2$ are intermediate between that of liquid and gas, and with the manipulation of temperature and pressure its solvent properties can be changed dramatically (Johnston, 1989; Mchugh and Krukonis, 1986). Some applications of sc$CO_2$ in organic synthesis are:

**(i) Diels–Alder reaction**: Reaction between 2-t-butyl-1,3-butadiene and methyl acrylate gives addition product with 54% yield, when carried out in sc$CO_2$ (Renslo et al., 1997).

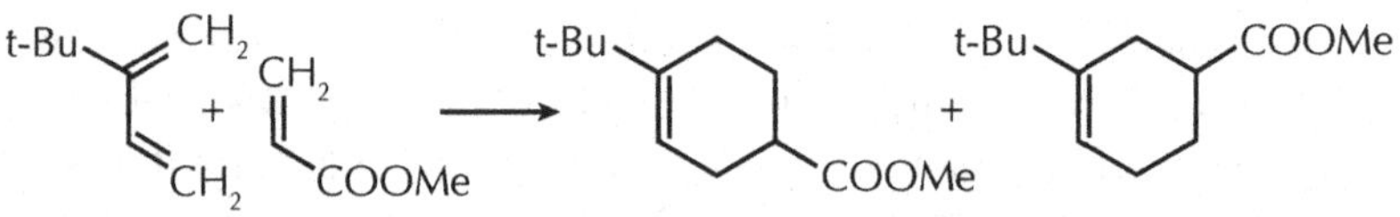

t-Bu, $CH_2$, $CH_2$ + $CH_2$, COOMe → t-Bu, COOMe + t-Bu, COOMe

An aza–Diels–Alder reaction of Danishefsky's diene with an imine in presence of scandium perfluoroalkane sulfonate in sc$CO_2$ gives the adduct with 99% yield (Matsuo et al., 2000).

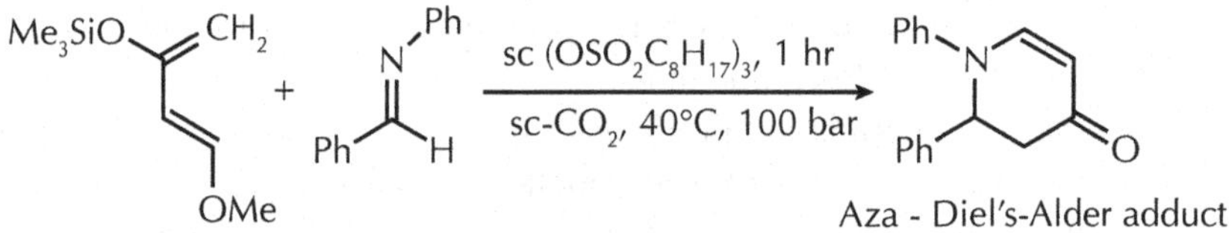

$Me_3SiO$, $CH_2$, OMe + Ph, N, Ph, H $\xrightarrow[\text{sc-CO}_2\text{, 40°C, 100 bar}]{\text{sc (OSO}_2\text{C}_8\text{H}_{17}\text{)}_3\text{, 1 hr}}$ Ph, N, Ph, O

Aza - Diel's-Alder adduct

Cott et al. (2005) carried out Diels–Alder reaction between maleic anhydride and furan derivatives in sc$CO_2$.

**(ii) Freidel–Crafts reaction:** The following transformation has been achieved by using Freidel–Crafts reaction in sc$CO_2$ (Burk et al., 1995).

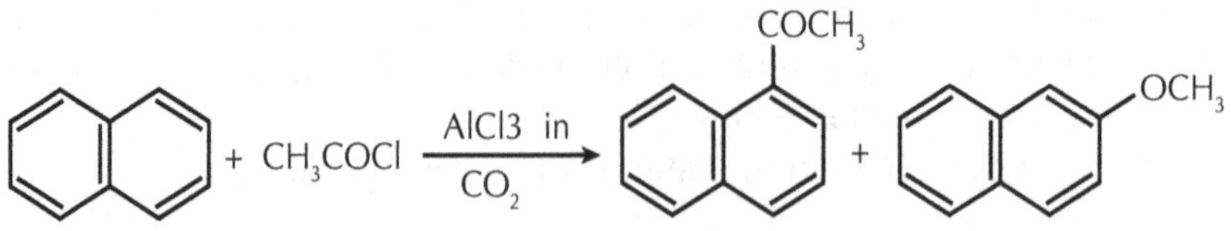

+ $CH_3COCl$ $\xrightarrow[\text{CO}_2]{\text{AlCl3 in}}$ $COCH_3$ + $OCH_3$

Continuous Friedal–Crafts alkylation in $scCO_2$ using solid acid catalysts is described by Amandi et al. (2005).

**(iii) Supercritical polymerization:** $scCO_2$ has been used for the polymerization, where it is mixed with a surfactant in order to enhance the solubility of hydrocarbon based molecules. Different types of polymers have been synthesized in $scCO_2$ (Cooper, 2000).

$$\text{epoxide (O, R)} + CO_2 \xrightarrow{\text{Catalyst}} H_3C\left[\text{C(=O)}-O-\text{CH(R)}-\text{CH}_2-O\right]_n CH_3 + \text{cyclic carbonate (O, O, O, R)}$$

It has been reported that use of water in $scCO_2$ enhances the solubility of organics (Johnston et al., 1996). The following polymerization reaction has also been accomplished using $scCO_{2.}$

$$CH_2{=}CH{-}C({=}O){-}O{-}CH_2(CF_2)_6{-}CF_3 \xrightarrow[\text{207 bar, 48 hrs., 59.4°C}]{\text{AIBN in } CO_2} (CH{-}CH)\,{-}C({=}O){-}O{-}CH_2(CF_2)_6{-}CF_3$$

Lee et al. (2012) used $scCO_2$ as an alternative solvent for the synthesis and purification of poly (L-lactic acid). This benign one pot process replaced the use of toxic organic solvents and produced polymers with tunable molecular weight. Yang et al. (2013) fabricated well controlled porous foams of graphene oxide modified poly (propylene carbonate) using $scCO_2$. Poly (propylene carbonate) is new amorphous, biodegradable, and biocompatible aliphatic polyester. It has a potentially wide range of applications like packing and biomedical materials. The fabricated porous materials were non cytotoxic and therefore, these are promising materials for tissue engineering applications.

Polymerization of 3-undecylbithiophene and preparation of poly(3-undecylbithiophene)/polystyrene composites in supercritical carbon was studied by Abbett et al. (2003). Poly(methyl methacrylate-ran-perfluoroalkyl ethyl methacrylate) copolymers having varying perfluoroalkyl ethyl methacrylate ester (Zonyl-TM) contents were synthesized by Cengiz et al. (2011) in $scCO_2$.

**(iv) Oxidation reactions:** Diastereoselective sulfoxidation of cystein derivative in $scCO_2$ was achieved giving the anti- diastereomer as the sole product of this reaction (Oakes et al., 1999).

TBHP Amberlyst
Sc-$CO_2$,40°C, 50-350 bar

It has been observed that oxidation of alcohols to carbonyl compounds can be carried out with high selectivity as well as at a high rate in $scCO_2$ (Jerner et al., 2000). Mello et al. carried out Baeyer–Villiger oxidation of ketones in $scCO_2$ at 250bar and 40°C under flow conditions with a silica-supported peracid. The reagent can be recycled by treatment with acidic 70% hydrogen peroxide at 0°C (2009).

$scCO_2$ $SiO_2$ $scCO_2$

Chapman et al. (2010) reported that primary and secondary alcohols can be oxidised selectively with $O_2$ in continuous flow in $scCO_2$. Selective oxidation of styrene to acetophenone has been investigated over Pd–Au catalysts with $H_2O_2$ in $scCO_2$ medium (Wang et al., 2007).

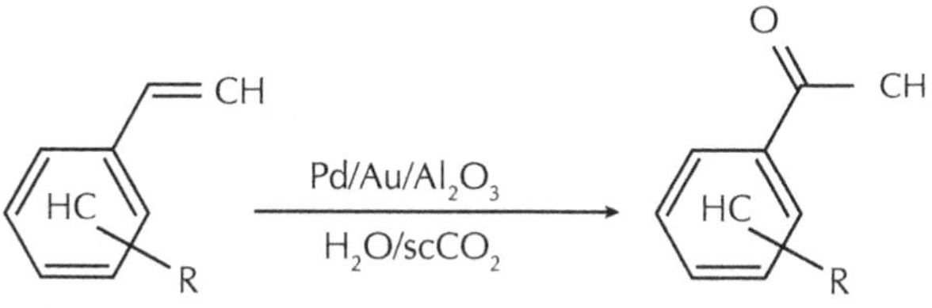

O
CH
CH
HC
HC
R
R
$Pd/Au/Al_2O_3$
$H_2O/scCO_2$

$R = H, 4\text{-}Me, 3\text{-}Cl, 3\text{-}NO_2$

A continuous stream of pentachlorophenol (PCP, 10–20mg $min^{-1}$) in $scCO_2$ was dechlorinated efficiently by a heated column (25 × 1cm diameter) of a zero-valent silver–iron ($Ag^0/Fe^0$) bimetallic mixture (Kabir and Marshall, 2001). Mello et al. (2012) carried out epoxidation of olefins with a silica-supported peracid in $scCO_2$ under Flow conditions.

**(v) Hydrogenation reactions**: The hydrogenation of alkene in $scCO_2$ has immense importance in industry (Baiker, 1999; Jessop et al., 1999), because hydrogen is soluble in $scCO_2$ and it is able to bring together substrates and catalysts in a single homogeneous reaction. γ-Butyolactone, which is one of the most valuable alternative of chlorinated solvent can be synthesized by hydrogenation of maleic anhydride with hydrogen in presence of $Pd/Al_2O_3$ using $scCO_2$ at 200°C (Pillai and Sahle-Demessi, 2002).

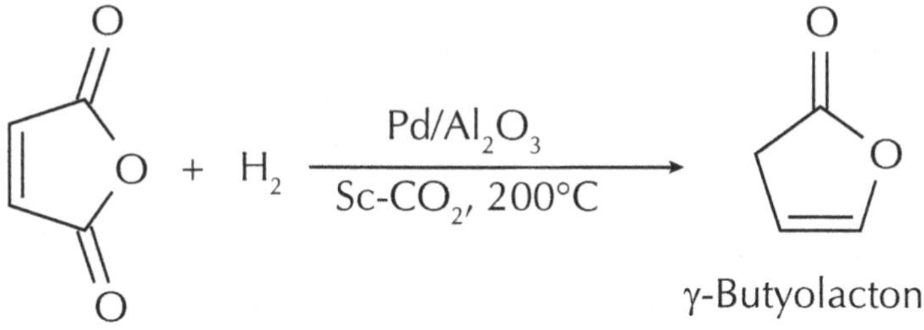

O
O
O + $H_2$
$Pd/Al_2O_3$
$Sc\text{-}CO_2$, 200°C
O
O

γ-Butyolactone

Various enantioselective hydrogenations have been achieved using $scCO_2$ as solvent (Burk et al., 1995). Ni (II) catalyzed hydrogenation of citral in $scCO_2$ results in the selective hydrogenation of C = O over C = C (Chatterjee et al., 2006).

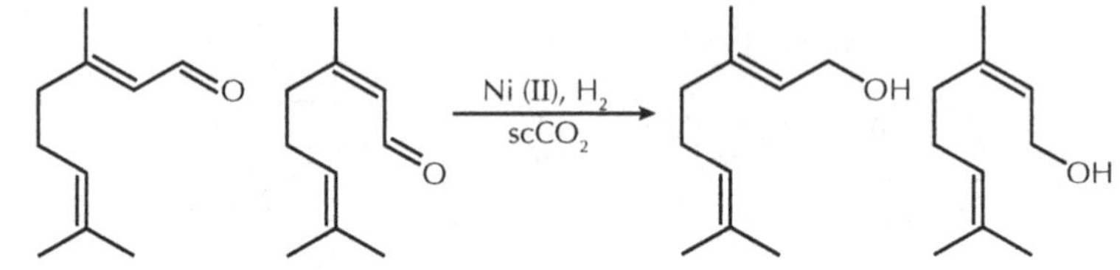

Ni (II), $H_2$
$scCO_2$

Trans-Citral cis-Cistral Genraniol Nctrol

Reduction of fatty acid methyl esters (FAME) to fatty alcohol mixtures in two different types of supercritical media ($H_2/CO_2$ and $H_2/C_3H_8$) was compared using two different hydrogenation catalysts (Andersson et al., 2000). Li et al. (2001) reported that Zn–$H_2$O–$CO_2$ is an excellent reducing reagent for the reduction of aldehydes in $scCO_2$. Selective hydrogenation of $\alpha,\beta$-unsaturated aldehydes to saturated aldehydes has been achieved using Pd/C catalyst in $scCO_2$ medium by Zhao et al. (2003). Hydrogenation of 2-butyne-1,4-diol to butane-1,4-diol has also been successfully conducted in $scCO_2$ by Zhao et al. (2003) at 323K with a high selectivity of 84% for butane-1,4-diol at 100% conversion.

The hydrogenation of benzonitrile to benzylamine with high conversion (90.2%) and selectivity (90.9%) was achieved in $scCO_2$ without using any additive over Pd/MCM-41 catalyst (Chatterjee et al., 2010). Advantageous heterogeneously catalyzed hydrogenation of carvone with green solvent $scCO_2$ was reported by Melo et al. (2011). Two Pd (II) complexes with the tridentate ligands (SOO/SOS donor atom) have been used as catalyst in homogeneous hydrogenation of olefins with molecular hydrogen in super critical carbon dioxide (Yilmaz et al., 2010).

**(vi) Heck reactions**: It has been reported that the reaction between iodobenezene and methyl acrylate is catalyzed by $Pd(OAc)_2$ in presence of fluorous ligand using $scCO_2$ giving 92% yield of methyl cinnamate.

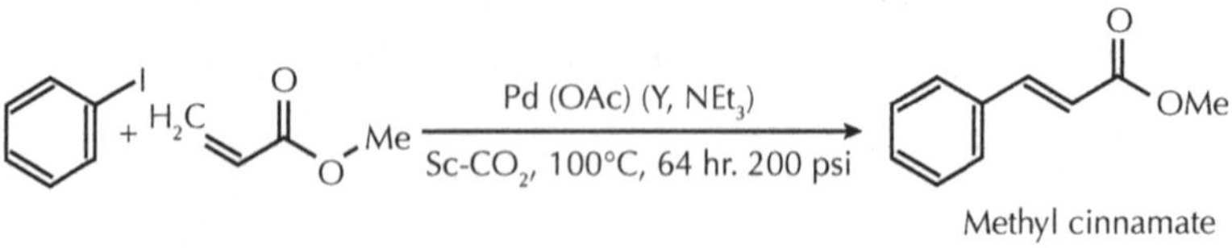

Pd (OAc) (Y, $NEt_3$)
Sc-$CO_2$, 100°C, 64 hr. 200 psi

Methyl cinnamate

Moreover, Heck reactions have been carried out, by using water soluble catalysts in $scCO_2$/water emitted biphasic system (Bhanage et al., 1999).

**(vii) Coupling reaction**: Pd catalyzed biaryl formation by homocoupling of iodobenezene in $scCO_2$ has been carried out (Shezad et al., 2002).

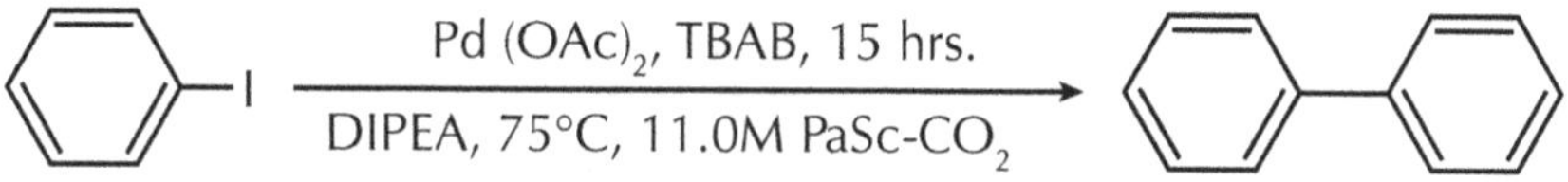

Pd $(OAc)_2$, TBAB, 15 hrs.
DIPEA, 75°C, 11.0M PaSc-$CO_2$

The homocoupling of iodoarenes catalyzed by $Pd(OCOCF_3)_2$/P(2-furyl)$_3$ also occurs best in $scCO_2$ and under solventless reaction conditions (Shezad et al., 2002).

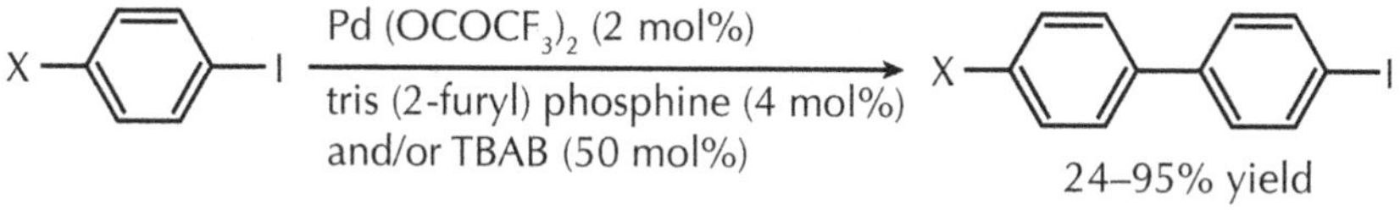

Pd $(OCOCF_3)_2$ (2 mol%)
tris (2-furyl) phosphine (4 mol%)
and/or TBAB (50 mol%)
24–95% yield

**(viii) Photochemical reaction**: Synthesis of 2-acyl-1,4-hydroquimone by photoinduced addition of aldehydes to *α, β*-unsaturated carboxyl compounds is an environmentally benign method, which can be further improved by use of $scCO_2$ in place of benzene as a solvent (Pacut et al., 2001).

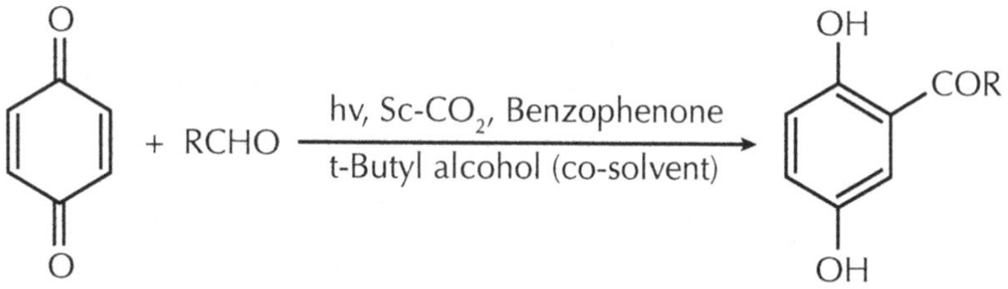

+ RCHO
hv, Sc-$CO_2$, Benzophenone
t-Butyl alcohol (co-solvent)

**(ix) Alkylation/Acylation**: The acetalization of terminal olefins with electron withdrawing groups was carried out smoothly in sc-$CO_2$under oxygen atmosphere using polystyrene supported benzoquinone as cocatalyst with palladium chloride (Wang et al., 2005). Amandi et al. (2007) used $scCO_2$ as a solvent for continuous alkylation of phenol using a solid acid catalyst, $\gamma$-$Al_2O_3$, with either cyclohexene or cyclohexanol as alkylating agents.

$$\text{Cyclohexanol} \xrightarrow[scCO_2]{Al_2O_3} \text{Cyclohexene} + H_2O$$

$$\text{Cyclohexene} + \text{Phenol} \xrightarrow[scCO_2]{Al_2O_3} \text{2-Cyclohexylphenol}$$

(x) **Esterification**: Esterification of 2-ethyl-1-hexanol with 2-ethylhexanoic acid to produce 2-ethylhexyl-2-ethylhexanoate has been investigated in $scCO_2$ by Ghaziaskar et al. (2006).

$$RCH_2OH + RCOOH \rightleftharpoons RC(=O)OCH_2R + H_2O$$

$$R = CH_3(CH_2)_3CH(C_2H_5)-$$

(xi) **Extraction**: The $scCO_2$ extraction of phenolic compounds from *Zostera marina* residues was optimized by developing a mathematical model based on mass transfer balances (Pilavtepe et al., 2012). Zarena et al. (2012) used supercritical fluid technology to extract the active constituents from *Mangosteen pericarp*. Xanthons were extracted and their antioxidant activity was measured by 2,2-diphenyl-1-picrylhydrazyl (DPPH) radical scavenging assay. The $scCO_2$ extraction of oleic sunflower seeds was shown to be effective and it yielded a product very similar to pure refined oil (Kiriamiti et al., 2002).

Vedaraman et al. (2004) have investigated the conventional methods of cholesterol extraction from cattle brain using $scCO_2$ and with different cosolvents, while carotenoids have been extracted from rosehip fruit using supercritical $CO_2$ at various extraction conditions by Machmudah et al. (2008). A two step process was used for extraction of alkannin derivatives from *Alkanna tinctoria* with supercritical $CO_2$ followed by alkaline hydrolysis of alkannin derivatives. The highest total alkannins (1.47%) was extracted with $scCO_2$, which was higher than conventional hexane extraction (1.24%) providing a solvent free alternative for industrial production (Akgun et al., 2011).

Soh and Zimmerman (2011) reported lipid extraction from wet algae using $scCO_2$, which was optimized for biodiesel production potential in terms of FAME yield and selectivity. Roop et al. (1989) determined the method of extraction of phenol from water with sc carbon dioxide, where $scCO_2$ has been found to be a good solvent for the extraction of flavor from milk fat (Haan et al., 1990).

**(xii) Catalysis**: McCarthy et al. (2002) reported a new procedure for multiphase catalysis, where the organometallic catalyst is immobilized in $scCO_2$ phase and water is used as the mobile phase for polar substrates and products. In the catalyst Ph-SBA-15-$PPh_3$-Pd, the Pd species were anchored inside the mesoporous material, and act as nanoreactors for Suzuki reaction in $scCO_2$ (Feng et al., 2010). Ciftci et al. (2012) studied enzymatic synthesis of phenolic lipids from flaxseed oil and ferulic acid in $scCO_2$ media using immobilized lipase from *Candida Antarctica.*

Enzymatic synthesis of phenolic lipids from flaxseed oil and ferulic acid was also studied in $scCO_2$ media using immobilized lipase from *Candida Antarctica* (Ghoreishi and Heidari, 2012). Enantioselective enzymatic hydrolysis of benzoyl benzoin catalyzed by *Candida cylindracea* (CCL) lipase was carried out in $scCO_2$ (Celebi et al., 2007) while the lipase catalyzed butanolysis of triolein was carried out in an ionic liquid and selective extraction of product was done using $scCO_2$ by Miyawaki and Tatsuno (2008).

**(xiii) Solubility**: Kazemi et al. (2012) measured the solubilities of ferrocene and acetylferrocene in $scCO_2$ using an analytical method in a quasi-flow apparatus, while the solubilities of benzene derivatives in $scCO_2$ was determined by the saturation method over the pressure range (9.5–14.5)MPa (Reddy and Madrass, 2011). Marceneiro et al. (2011) measured the solid solubilities of 1,4-naphthoquinone and 5-hydroxy-2-methyl-1,4-naphthoquinone (also known as plumbagin) in $scCO_2$ using a static analytical method at 308.2, 318.2, and 328.2K, and pressures between 9.1 and 24.3MPa.

Rajasekhar et al. (2010) used saturation method for the determination of the solubility of a drug, *n*-(4-ethoxyphenyl)ethanamide (phenacetin), in $scCO_2$ at 308, 318, and 328K, and 9–19MPa. Solid drug's solubility in supercritical fluids (SCFs) was also experimentally determined by Hojjati et al. (2007). Lee et al. (1999) measured the solubility of disperse dyes in $scCO_2$. An enhancement in solubility of poly(vinyl ester) in $scCO_2$ can be achieved by decreasing the strength of the polymer–polymer interactions (Girard et al., 2012). Solubility measurements of zopiclone and nimodipine in $scCO_2$ were also reported by Medina and Bueno (2001).

**(xiv) Synthesis**: Rohr et al. (2001) reported the solvent free synthesis of *β*-[(diethylcarbamoyl)oxy]styrene from phenylacetylene and diethylamine in $scCO_2$ is greatly accelerated for a series of ruthenium catalysts compared to the same reaction in toluene. $O_2$ and the mixed solvent (MeOH–$scCO_2$) can control the chemoselectivity of the palladium catalyzed carbonylation of amines. Methyl *N*-*n*-butylcarbamate or oxalbutyline could be obtained in high yields upon varying the conditions (Li et al., 2001). Du et al. (2005) reported organic solvent free process for the synthesis of propylene carbonate from $scCO_2$ and propylene oxide catalyzed by insoluble ion exchange resins. The process represents a simple, ecologically safer, and a cost effective route to cyclic carbonates with high product quality, as well as easy product recovery and catalyst recycling.

$$\text{propylene oxide} + CO_2 \xrightarrow[100^{\circ}C,\ S\ MPa]{\text{Ion exchange resin}} \text{propylene carbonate}$$

**(xv) Miscellaneous reactions**: Beside these reactions, there are number of reactions, which have been reported in $scCO_2$ with good yield and faster speed. For example, PausomKhand reaction (Jeong et al., 1997), Baylis–Hillman reaction (Rose et al., 2002), hydroboration of styrene (Carter et al., 2000), carbamate synthesis (Yoshida et al., 2000), 2-pyrone synthesis (Jessop and Leitner, 1999; Reetz et al., 1993), and polymerization reactions (Beckman, 2004).

It can now be concluded that $scCO_2$ is an environmentally benign medium. It has been used in many reactions, for example, hydrogenation, hydroformylation, coupling reaction, oxidation reaction, photochemical reaction as well as for stereochemical control reaction (Luzzio, 2001).

$\alpha$-Olefins undergo highly selective self-metathesis catalyzed by supported Re-oxide in the presence of $scCO_2$ (Selva et al., 2009).

$$2\ \text{(α-olefin)},\ n = 0\text{-}2 \underset{scCO_2,\ 80\text{-}150\ bar,\ 35^{\circ}C}{\overset{Re_2O_7/\gamma\text{-}Al_2O_3}{\rightleftharpoons}} \text{(internal olefin)},\ n = 0\text{-}2 + C_2H_4$$

Conversion:65-70%
Selectivity: >95%

The oxybromination of aniline and phenol derivatives was performed by Ganchegui and Leitner (2007) with high conversions, and selectivities in $H_2O/scCO_2$ exploiting the intrinsic reactivity of the biphasic system.

$$\text{Ar(X)(R)} \xrightarrow[\text{H}_2\text{O/scCO}_2,\ 40^\circ\text{C}]{\text{NaHCO}_3\text{(cat.), NaBr, H}_2\text{O}_2} \text{Ar(X)(R)Br}$$

X = OH, $NH_2$ or NHMe
R = H, Me, iBu...

Sulfur-cured natural rubber was devulcanized by using $scCO_2$ as swelling solvent (Kojima et al., 2004). Wacker reaction was carried out smoothly in $scCO_2$ or ROH/$scCO_2$. Results show that both $scCO_2$ and co-solvent can remarkably affect the selectivity toward methyl ketone and the presence of ROH accelerates this reaction (Jiang et al., 2000). Viguera et al. (2013) have used $scCO_2$ for the removal of lubricating oils from metallic contacts, while Weinstein et al. (2005) explored the use of liquid and $scCO_2$ for blending of poly(vinyl acetate) and citric acid.

Methanol-wetted zinc borates produced either from borax decahydrate and zinc nitrate hexahydrate ($2ZnO{\cdot}3B_2O_3{\cdot}7H_2O$) or from zinc oxide and boric acid ($2ZnO{\cdot}3B_2O_3{\cdot}3H_2O$) which were dried by both conventional and $scCO_2$ drying methods (Gonen et al., 2009). Pizarro et al. (2009) have determined the binary diffusion coefficients of 2-ethyltoluene, 3-ethyltoluene, and 4-ethyltoluene in $scCO_2$ in a pressure range of 15.0–35.0MPa and temperatures 313, 323, and 333K by means of the Taylor–Aris dispersion technique. Moderately thermophilic bacterial strain CC-HSB-11$^T$ (*Muricauda lutaonensis*) has been identified to produce zeaxanthin as a predominant xanthophyll by liquid chromatography–tandem mass spectrometry (LC–MS/MS). Micronization of zeaxanthin was achieved through $scCO_2$ antisolvent precipitation method (Hameed et al., 2011).

Xie et al. (2012) have reported a new method of post modifying the particle size and morphology of $LiFePO_4$ via $scCO_2$. Enantiodifferentiating photocyclization of 5-hydroxy-1,1-diphenyl-1-pentene sensitized by bis(1,2,4,5-di-*O*-isopropylidene-*α*-fructopyranosyl) 1,4-naphthalenedicarboxylate was performed in near critical and $scCO_2$ media containing organic entrainers to obtain a chiral tetrahydrofuran derivative in enantiomeric excess (ee) (Nishiyama et al., 2012).

Yuan et al. (2003) reported detoxification of aryl-organochlorine compounds by catalytic reduction in $scCO_2$. An automated continuous flow reactor has been built, and is used to study acid catalyzed etherification reactions in sc$CO_2$ (Walsh et al., 2005). Cid et al. (2005) carried out an excellent dye fixation on cotton in sc$CO_2$ using fluorotriazine reactive dyes. As a result, water usage and dye containing waste are eliminated and energy consumption can also be reduced. Catalytic and selective hydroxylation of phenol was carried out by Baldi et al. (2010) under environmentally benign reaction conditions in the presence of a Fe (III)-EPS (EPS = exopolysaccharide) catalyst. A novel method was developed by Long et al. (2012) for the dyeing of cotton fabric with a vinylsulfone reactive disperses dye in sc$CO_2$.

Zinc sulfide nanoparticles have been deposited on carbon nanotubes (CNTs) from a single source diethyldithiocarbamate (Casciato et al., 2012), while nanoporous silica materials have been prepared using an activated carbon as a mold and sc$CO_2$ as a solvent by Wakayama and Fukushima (2000). Kuo et al. (2011) investigated anti-inflammatory effects of sc$CO_2$ extract and its isolated carnosic acid from *Rosmarinus officinalis* leaves. Colloidal carbon spheres were synthesized by the carbonization of squalane, a non volatile hydrocarbon solvent in sc$CO_2$ (Barrett et al., 2011). Household steel wool was used as a reducing agent for the reductive dechlorination of polychlorinated biphenyls (PCBs) in sc$CO_2$ (Chen et al., 2012).

Solvents were always a problematic issue for organic synthesis and there was a regular search for an alternate solvent. Water and carbon dioxide can serve as solvents, but in critical conditions only. They are called supercritical fluids. Many organic syntheses have been already worked out in these solvents and many more are feasible in future.

## KEYWORDS


- **Supercritical carbon dioxide**
- **Supercritical fluid**
- **Supercritical polymerization**
- **Supercritical water**

## REFERENCES


1. Abbett, K. F., Teja, A. S., Kowalik, J., & Tolbert, L. (2003). *Macromolecules, 36,* 3015–3019.
2. Akgun, I. H., Erkucuk, A., Pilavtepe, M., & Celiktas, O. Y. (2011). *Journal of Supercritical Fluids, 57,* 31–37.
3. Akiya, N. & Savage, P. E. (2002). *Chemical Review, 107,* 2725–2750.
4. Amandi, R., Hyde, J. R., Ross, S. K., Lotz, T. J., & Poliakoff, M. (2005). *Green Chemistry, 7,* 288–293.
5. Amandi, R., Scovell, K., Licence, P., Lotz, T. J., & Poliakoff, M. (2007). *Green Chemistry, 9,* 797–801.
6. An, J., Bagnell, L., Cablewski, T., Strauss, C. R., & Trainor, R. W. (1997). *Journal of Organic Chemistry, 62,* 2505–2511.
7. An, T., Bagnell, L., Cablewski, T., Strauss, C. R., & Trainer, R. W. (1997). *Journal of Organic Chemistry, 62,* 2505–2511.
8. Anatas, P. T., Warner, P. T., & Warner, J. C. (1998). *Green chemistry: Theory and practice.* Oxford: Oxford University Press.
9. Andersson, M. B. O., King, J. W., & Blomberg, L. G. (2000). *Green Chemistry, 2,* 230–234.
10. Baiker, A. (1999). *Chemical Review, 99,* 453–473.
11. Baldi, F., Marchetto, D., Zanchettin, D., Sartorato, E., Paganelli, S., & Piccolo, O. (2010). *Green Chemistry, 12,* 1405–1409.
12. Barrett, C. A., Singh, A., Murphy, J. A., O'Sullivan, C., Buckley, D. N., & Ryan, K. M. (2011). *Langmuir, 27,* 11166–11173.
13. Beckman, E. J. (2004). *Journal of Supercritical Fluids, 28,* 121–191.
14. Bhanage, B. M., Ikushima, Y., Shirai, M., & Arai, M. (1999). *Tetrahedron Letters, 40,* 6427–6430.
15. Broli, D., Kaul, C., Kramer, A., Krammer, P., Richter, T., Jung, M., Vogel, H., & Zehner, P. (1999). *Angewandte Chemie International Edition, 38,* 2998–3014.
16. Burk, M. J., Feng, S., Gross, M. F., & Tumas, W. (1995). *Journal of the American Chemical Society, 117,* 8277–8278.
17. Burk, M. J., Feng, S., Gross, M. G., & Tumas, W. (1995). *Journal of the American Chemical Society, 117,* 8277.
18. Carlsson, M., Habenicht, C., Kam, L. C., Antal, M., Bian, N., Cunninghom, R. J., & Jones, M. (1994). *Industrial Engineering Chemistry Research, 33,* 1989–1996.
19. Carter, C. A. G., Baker, R. T., Nolan, S. P., & Tumas, W. (2000). *Chemical Communications, 5,* 347–348.
20. Casciato, M. J., Levitin, G., Hess, D. W., & Grover, M. A. (2012). *Industrial Engineering Chemistry Research, 51,* 11710–11716.
21. Celebi, N., Yildiz, N., Demir, A. S., & Calimli, A. (2007). *Journal of Supercritical Fluids, 4,* 386–390.
22. Cengiz, U., Gengec, N. A., Kaya, N. U., Erbil, H. Y., & Sarac, A. S. (2011). *Journal of Fluorine Chemistry, 132,* 348–355.
23. Chapman, A. O., Akien, G. R., Arrowsmith, N. J., Licence, P., & Poliakoff, M. (2010). *Green Chemistry, 12,* 310–315.

24. Chatterjee, M., Chatterjee, A., Raveendran, P., & Ikushima, Y. (2006). *Green Chemistry, 8,* 445–449.
25. Chatterjee, M., Kawanami, H., Sato, M., Ishizaka, T., Yokoyama, T., & Suzuki, T. (2010). *Green Chemistry, 12,* 87–93.
26. Chen, Y., Liao, W., & Yak, H. (2012). *Industrial Engineering Chemistry Research, 51,* 6625–6630.
27. Cid, M. V. F., Spronsen, J. V., Kraan, M. V. D., Veugelers, W. J. T., Woerlee, G. F., & Witkamp, G. J. (2005). *Green Chemistry, 7,* 609–616.
28. Ciftci, D. & Saldaña, M. D. A. (2012). *Journal of Supercritical Fluids, 72,* 255–262.
29. Cooper, A. L. (2000). *Journal of Materials Chemistry, 10,* 207–234.
30. Cott, D. J., Ziegler, K. J., Owens, V. P., Glennon, J. D., Graham, A. E., & Holmes, J. D. (2005). *Green Chemistry, 7,* 105–110.
31. Crittendon, R. C. & Parsons, E. J. (1994). *Organometallics, 13,* 2587–2591.
32. Diminnie, J., Metts, S., & Parsons, E. J. (1995). *Organometallics, 14,* 4023–4025.
33. Du, Y., Cai, F., Kong, D., & He, L. (2005). *Green Chemistry, 7,* 518–523.
34. Feng, X., Yan, M., Zhang, T., Liu, Y., & Bao, M. (2010). *Green Chemistry, 12,* 1758–1766.
35. G. Li., Jiang, H., & Li, J. (2001). *Green Chemistry, 3,* 250–251.
36. Ganchegui, B. & Leitner, W. (2007). *Green Chemistry, 9,* 26–29.
37. Ghaziaskar, H. S., & Daneshfar, A., & Calvo, L. (2006). *Green Chemistry, 8,* 576–581.
38. Ghoreishi, S. M. & Heidari, E. (2012). *Journal of Supercritical Fluids, 72,* 36–45.
39. Girard, E., Tassaing, T., Camy, S., Condoret, J., Marty, J., & Destarac, M. (2012). *Journal of the American Chemical Society, 134,* 11920–11923.
40. Gönen, M., Balköse, D., Gupta, R. B., & Ülkü S. (2009). *Industrial Engineering Chemistry Research, 48,* 6869–6876.
41. Gore, G. (1861). *Proceedings of the Royal Society of London, 11,* 85–86.
42. Haan, A. B. de., Graauw, J. de., Schaap, J. E., & Badings, H. T. (1990). *Journal of Supercritical Fluids, 3,* 15–19.
43. Hameed, A., Arun, A. B., Ho, H.-P., Chang, C.-M. J., Rekha, P. D., Maw-Lee, R. et al. (2011). *Journal of Agricultural and Food Chemistry, 59,* 4119–4124.
44. Hawthrone, S. B., Yang, Y., & Miller, D. J. (1994). *Analytical Chemistry, 66,* 2912–2920.
45. Hojjati, M., Yamini, Y., Khajeh, M., & Vatanara, A. (2007). *Journal of Supercritical Fluids, 41,* 187–194.
46. Itami, K., Terakawa, K., Yoshida, J. I., & Kajimoto, O. (2003). *Journal of the American Chemical Society, 125,* 6058–6059.
47. Jeong, N., Hwang, S. H., Lee, Y. W., & Lim, J. S. (1997). *Journal of the American Chemical Society, 119,* 10549–10550.
48. Jerner, G., Sueur, D., Mallat, T., & Baiker, A. (2000). *Chemical Communications,* 2247–2248.
49. Jessop, P. G., Ikaviya, T., & Noyoro, R. (1999). *Chemical Reviews, 99,* 475–493.
50. Jessop, P. G. & Leitner, W. (1999). *Chemical synthesis using supercritical fluids.* Washington: Wiley-VCH.
51. Jiang, H., Jia, L., & Li, J. (2000). *Green Chemistry, 2,* 161–164.

52. Johnston, K. P. (1989). In K. P. Johnston & J. M. L. Penninger (Eds.), *Supercritical fluid science and technology* (pp. 1–12). Washington D.C: American Chemical Society.
53. Johnston, K. P., Harrison, K. I., Clarke, M. J., Howdle, S. M., Heitz, M. P., Bright, F. V., Carlier, C., & Randolph, T. W. (1996). *Science, 271,* 624–626.
54. Kabir, A. & Marshall, W. D. (2001). *Green Chemistry, 3,* 47–51.
55. Kajmoto, O. (1999). *Chemical Review, 99,* 355–389.
56. Kazemi, S., Belandria, V., Janssen, N., Richon, D., Peters, C. J., & Kroon, M. C. (2012). *Journal of Supercritical Fluids, 72,* 320–325.
57. Kiriamiti, H. K., Rascol, E., Marty, A., & Condoret, J. S. (2002). *Chemical Engineering Processing: Process Intensification, 41,* 711–718.
58. Kojima, M., Tosaka, M., & Ikeda, Y. (2004). *Green Chemistry, 6,* 84–89.
59. Korzenski, M. B. & Kolis, J. W. (1997). *Tetrahedron Letters, 38,* 5611–5614.
60. Kremsner, J. M. & Kappe, C. O. (2005). *European Journal of Organic Chemistry,* 3672–3679.
61. Kuhlmann, B., Arnett, E. M., & Siskin, M. (1994). *Journal of Organic Chemistry, 59,* 5377–5380.
62. Kuhlmann, B., Arnett, E. M., & Siskin, M. (1994). *Journal of Organic Chemistry, 59,* 3098–3101.
63. Kuo, C. -F., Su, J. -D., Chiu, C. -H., Peng, C. -C., Chang, C. -H., & Sung, T. -Y. et al. (2011). *Journal of Agricultural and Food Chemistry, 59,* 3674–3685.
64. Lee, J. W., Min, J. M., & Bae, H. K. (1999). *Journal of the Chemical & Engineering Data, 44,* 684–687.
65. Lee, S. Y., Valtchev, P., & Dehghani, F. (2012). *Green Chemistry, 14,* 1357–1366.
66. Li, J., Jiang, H., & Chen, M. (2001). *Green Chemistry, 3,* 137–139.
67. Long, J., Xiao, G., Xu, H., Wang, L., Cui, C., Liu, J. et al. (2012). *Journal Supercritical Fluids, 69,* 13–20.
68. Luzzio, F. (2001). *Tetrahedron, 57,* 915–945.
69. Machmudah, S., Kawahito, Y., Sasaki, M., & Goto, M. (2008). *Journal of Supercritical Fluids, 44,* 308–314.
70. Marceneiro, S., Braga, M. E. M., Dias, A. M. A., & Sousa, H. C. D. (2011). *Journal of Chemical of Engineering Data, 56,* 4173–4182.
71. Matsuo, J., Tsuchige, T., Odashima, K., & Kobayashi, S. (2000). *Chemistry Letters, 2,* 178–179.
72. McCarthy, M., Stemmer, H., & Leitner, W. (2002). *Green Chemistry, 4,* 501–504.
73. Mchugh, M. & Krukonis, V. (1986). *Supercritical fluid extraction: Principles and practice* (pp. 1–11). Boston: Butterworths.
74. Medina, I. & Bueno, J. L. (2001). *Journal of Chemical and Engineering Data, 46,* 1211–1214.
75. Mello, R., Alcalde-Aragonés, A., Olmos, A., González-Núñez, M. E., & Asensio, G. (2012). *Journal of Organic Chemistry, 77,* 4706–4710.
76. Mello, R., Olmos, A., Parra-Carbonell, J., González-Núñez, M. E., & Asensio, G. (2009). *Green Chemistry, 11,* 994–999.

77. Melo, C. I., Bogel-Łukasik, R., Silva, M. G. D., & Bogel-Łukasik, E. (2011). *Green Chemistry, 13,* 2825–2830.
78. Miyawaki, O. & Tatsuno, M. (2008). *Journal of Bioscience and Bioengineering, 105,* 61–64.
79. Nishiyama, Y., Wada, T., Kakiuchi, K., & Inoue, Y. (2012). *Journal of Organic Chemistry, 77,* 5681–5686.
80. Oakes, R. S., Clifford, A. A., Bartle, K. D., Thornton-Pett, M., & Rayner, C. M. (1999). *Chemical Communications,* 247–248.
81. Pacut, R., Grimn, M. L., Kraus, G. A., & Tanko, J. M. (2001). *Tetrahedron Letters, 42,* 1415–1418.
82. Pilavtepe, M., Yucel, M., Helvaci, S. S., Demircioglu, M., & Celiktas, O. Y. (2012). *Journal of Supercritical Fluids, 68,* 87–93.
83. Pillai, U. R. & Sahle-Demessi, E. (2002). *Chemical Communications,* 422–423.
84. Pizarro, C., Iglesias, O. S., Medina, I., & Bueno, J. L. (2009). *Journal of Chemical and Engineering Data, 54,* 1467–1471.
85. Rajasekhar, Ch., Garlapati, C., & Madras, G. (2010). *Journal of Chemical and Engineering Data, 55,* 1437–1440.
86. Raner, K. D., Strauss, C. R., & Trainer, R. W. (1995). *Journal of Organic Chemistry, 60,* 2456–2460.
87. Reardon, P., Metts, S., Crittendon, C., Daugherity, P., & Parson, E. J. (1995). *Organometrallics, 14,* 3810–3816.
88. Reddy, S. N. & Madras, G. (2011). *Journal of Chemical and Engineering Data, 56,* 1695–1699.
89. Reetz, M. T., Konen, W., & Strack, T. (1993). *Chimica, 47,* 493.
90. Renslo, A. R., Weinstien, R. D., Tester, J. W., & Danherical, R. L. (1997). *Journal of Organic Chemistry, 62,* 4530–4533.
91. Rohr, M., Geyer, C., Wandeler, R., Schneider, M. S., Murphy, E. F., & Baiker, A. (2001). *Green Chemistry, 3,* 123–125.
92. Roop, R. K., Akgerman, A., Dexter, B. J., & Irvin, T. R. (1989). *Journal of Supercritical Fluids, 2,* 51–56.
93. Rose, P. M., Clifford, A. A., & Rayner, C. M. (2002). *Chemical Communications,* 968–969.
94. Savage, P. E. (1999). *Chemical Review, 99,* 603–622.
95. Selva, M., Perosa, A., Fabris, M., & Canton, P. (2009). *Green Chemistry, 11,* 229–238.
96. Shezad, N., Clifford, A. A., & Rayner, C. M. (2002). *Green Chemistry, 4,* 64–67.
97. Shezad, N., Clifford, A. A., & Rayner, C. M. (2002). *Green Chemistry, 4,* 64–67.
98. Siskin, M. & Katrinzky, M. (2001). *Chemical Reviews, 101,* 825–835.
99. Soh, L. & Zimmerman, J. (2011). *Green Chemistry, 13,* 1422–1429.
100. Stadler, A., Yousti, B. H., Dallinger, D., Walla, P., Vander Eycken, E., Kaval, N., & Kappe, C. O. (2003). *Organic Process Research and Development, 7,* 707–716.
101. Strauss, C. R. (1999). *Australian Journal of Chemistry, 52,* 83–96.
102. Strauss, C. R. & Trainor, K. W. (1995). *Australian Journal of Chemistry, 48,* 1665–1692.
103. Vedaraman, N., Brunner G., Kannan, C. S., Muralidharan, C., Rao, P.G., & Raghavan, K.V. (2004). *Journal of Supercritical Fluids, 32,* 231–242.

104. Viguera, M., Gómez-Salazar, J. M., Barrena, M. I., & Calvo, L. (2013). *Journal of Supercritical Fluids, 73,* 51–56.
105. Wakayama, H. & Fukushima, Y. (2000). *Chemistry of Materials, 12,* 756–761.
106. Walsh, B., Hyde, J. R., Licence, P., & Poliakoff, M. (2005). *Green Chemistry, 7,* 456–463.
107. Wang, X., Venkataramanan, N. S., Kawanami, H., & Ikushima, Y. (2007). *Green Chemistry, 9,* 1352–1355.
108. Wang, Z., Jiang, H., Qi, C., Wang, Y., Dong, Y., & Liu, H. (2005). *Green Chemistry, 7,* 582–585.
109. Weinstein, R. D., Gribbin., J. J., & Najjar, D., D. (2005). *Industrial and Engineering Chemistry Research, 44,* 3480–3484.
110. Xie, M., Zhang, X., Laakso, J., Wang, H., & Levänen, E. (2012). *Crystal Growth and Design, 12,* 2166–2168.
111. Xu, X. & Antal, M. J. (1997). *Industrial and Engineering Chemistry Research, 36,* 23–41.
112. Yang, G., Su, J., Gao, J., Hu, X., Geng, C., & Fu, Q. (2013). *Journal of Supercritical Fluids, 73,* 1–9.
113. Yılmaz, F., Mutlu, A., Ünver, H., Kurtça, M., & Kani, İ. (2010). *Journal of Supercritical Fluids, 54,* 202–209.
114. Yoshida, M., Hava, N., & Okuyama, S. (2000). *Chemical Communications,* 151–152.
115. Yu, J. & Savage, P. E. (1998). *Industrial and Engineering Chemistry Research, 37,* 2–10.
116. Yuan, T., Majid, A., & Marshall, W. D. (2003). *Green Chemistry, 5,* 25–29.
117. Zarena, A. S., Sachindra, N. M., & Sankar, K. U. (2012). *Food Chemistry, 130,* 203–208.
118. Zhao, F., Ikushima, Y., Chatterjee, M., Shirai, M., & Arai, M. (2003). *Green Chemistry, 5,* 76–79.

# CHAPTER 7

# OTHER GREEN SOLVENTS

ABHILASHA JAIN, RITU VYAS, AARTI AMETA, and P. B. PUNJABI

## CONTENTS

## 7.1 INTRODUCTION

Organic solvents are commonly used in organic syntheses and these are the major sources of generated wastes. These solvents are volatile, flammable and hazardous to human beings and the environment. There are two main approaches to solve the problems caused due to these solvents. In the first approach, the solvent is not used altogether. Many reactions can be carried out under the 'neat' conditions. But in solvent free techniques, many reactions are not possible and many reagents and intermediates are not stable outside the solution.

The second approach is the replacement of organic solvents with greener solvents. Now a days, ionic liquids and supercritical fluids have been used quite commonly. There are some more compounds, which can serve as green solvents in organic syntheses. The most popular among these green solvents are water, polyethylene glycol, glycerol, cyclopentylmethyl ether, 2-methyltetrahydrofuran, ethyl lactate, perfluorinated solvent, and so on.

## 7.2 SOME MAJOR GREEN SOLVENTS

### *7.2.1 WATER*

Water is one of the green solvent. It is readily available, least expensive, safe and non-hazardous to environment. Water is a "universal solvent" in nature. Living cells represent the most complex chemical reactions (termed as biochemical reaction) and all such reactions occur in environment with >90% water. Inorganic reactions are also carried out using water as a solvent.

Water is the best solvent among all the green solvent because it has many advantages like-

- Environmental benefits
- Cost/Economic factors
- Safety
- Synthetic efficiency

- Simple operation and
- Have great potential for new synthetic methodologies.

## CHARACTERISTICS PROPERTIES OF WATER

It exists in three forms—solid (ice), liquid and vapor.

(i) **Boiling point and freezing point**: Boiling point of water is 373.15K, whereas freezing point is 273.15K at 1 atm pressure.

(ii) **Viscosity:** The viscosity of water is 1.01mm$^2$ s$^{-1}$ at 20°C. Its viscosity affects the movement of solute in water as well as sedimentation rate of suspended particles.

(iii) **Surface tension and density:** The surface tension of water is 72.75mN/m$^2$, whereas density is 1.0000kg m$^{-3}$ at 4°C.

(iv) **Heat capacity**: Heat capacity of water is 4,182J kg$^{-1}$ at 20°C. Due to high heat capacity, the rapid exchange in temperature results in low changes in water temperature. So in large scale processes, temperature of endo- and exothermic reactions can be controlled.

(v) **Dielectric constant:** The value of dielectric constant of water is 80. Due to its high value, polar compounds are readily soluble in water.

(vi) **Hydrogen bonding:** One molecule of water can form four hydrogen bonds with other water molecules. Hydrogen bonding in water endows a unique characteristic, which is known as dual activator property. Water can activate as both, nucleophiles and elecrophiles and accelerates polar reactions, which involve polar transition state or intermediates.

Water is a versatile solvent in nature and it is used in synthetic organic chemistry. Even with less solubility of most of the organic compounds in it, the entire scenario has changed by the pioneer discovery of Rideout and Breslow (1980). Grieco et al. (1983) used water as a solvent in organic synthesis for the first time. Water has many physicochemical characteristics, which makes it a unique solvent and these characteristics can be modified as per requirement of the reaction by using various catalysts and additives like surfactants (Lindstrom, 2002).

Most of the important reactions in organic synthesis have been tried using water as a solvent or one of the components in the solvent mixture; of course, with some modifications in the conventional methodologies.

## *APPLICATIONS*

(i) **Diels–Alder reaction**—It is a [4 + 2] cycloaddition reaction between diene and a dienophile.

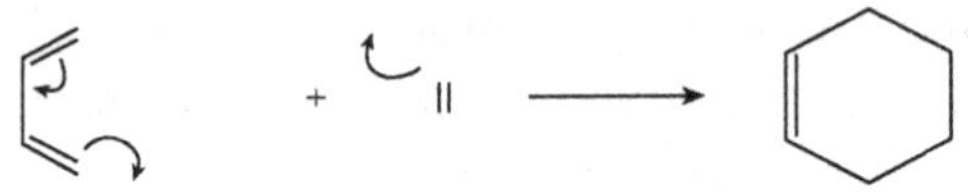

Butadiene (Diene) + Ethene (Dienophile) ⟶ Cyclohexane (Adduct)

In the beginning of 19th century, this reaction was carried out in aqueous medium, for the first time for example, furan reacted with maleic anhydride in hot water to give the adduct; where maleic anhydride gives maleic acid, which reacts with furan (Diels and Alder, 1931).

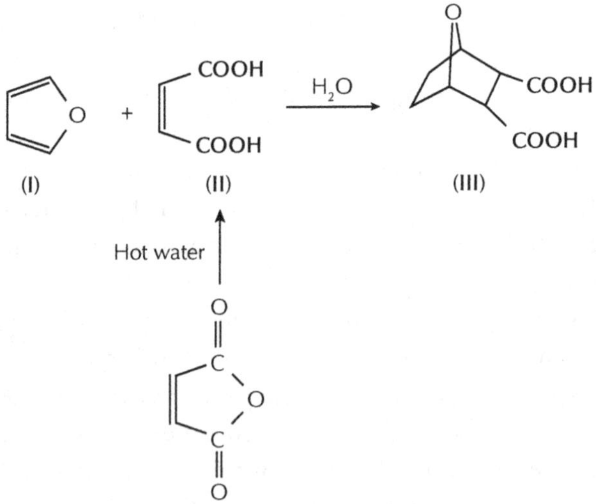

(I) + (II) —$H_2O$→ (III)

Hot water

The solvent affects the stereoselectivity of some reaction (Breslow et al., 1983; Breslow and Maitra, 1984). For example, cyclopentadiene reacts with butanone to give a product, where the ratio of endo/exo products was 21.4, when these are stirred at 0.15M concentration in water, whereas the ratio is only 8.5 in ethanol. The stereochemical changes could be explained by the need to minimize the transition state surface area in water solution favoring the more compact endo-structure (Berson et al., 1962; Samil et al., 1985).

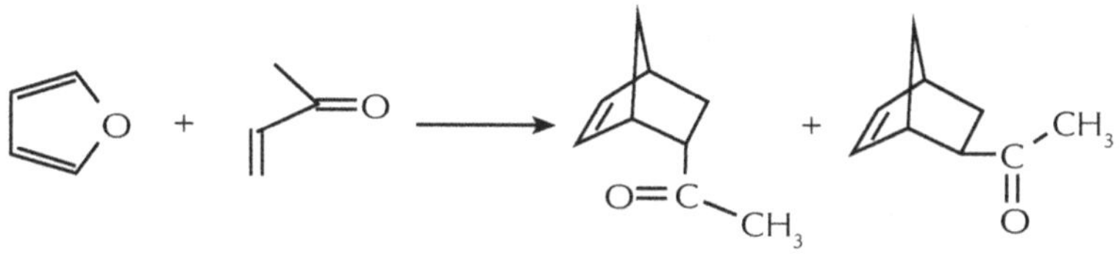

*HETERO DIELS–ALDER REACTION*

Hetero Diels-Alder reaction was reported in aqueous medium for the first time by Larsen and Grieco (1985). In this reaction, simple eiminium salts generated *in situ* under Mannich like conditions, reacted with dienes in water give aza-Diels-Alder reaction products. Retro Diels-Alder reaction and Aza–Diels–Alder reactions also occurred readily in water (Grieco et al., 1987). It has also been reported that rate of given Diels–Alder reaction increases more than 700 times when carried out in water.

(ii) **Claisen rearrangement**

This reaction involves [3, 3]—sigmatropic shift.

In Claisen rearrangement of chorismic acid, pure water is used to promote the reaction.

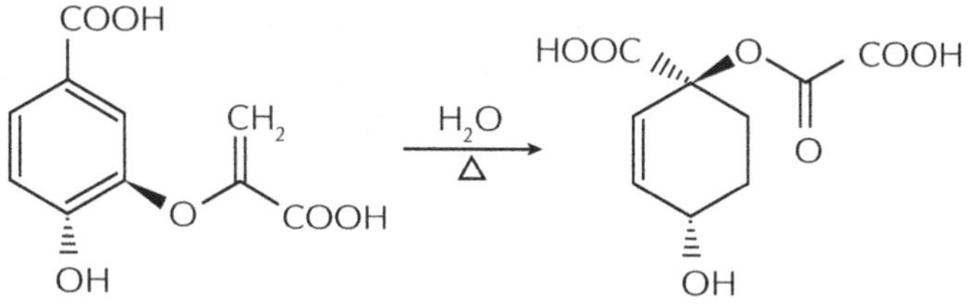

Chorismic acid

It was observed that the rate of this reaction is increased with the use of water as a polar solvent compared to non-polar solvent (Ponaras, 1983; White and Wolfarth, 1970). It has been reported that Claisen rearrangement of allyl vinyl ether in water gives the aldehyde with 82% yield (Grieco et al., 1989).

This rearrangement of naphthyl ether derivative gives maximum yield in water (100%) as compared to toluene (16%) (Narayan et al., 2005, Table 7.1).

120 hrs.
Room temp.

**TABLE 7.1** Comparative yields in various solvents.

| Solvent | Yield (%) |
|---|---|
| Toluene | 16 |
| Methanol | 56 |
| Neat | 73 |
| Water | 100 |

**(iii) $2\sigma + 2\sigma + 2\pi$ Cycloaddition**

Quadricyclane with dimethyl acetylenedicarboxylate DMAD gives an addition product. In water, the rate of reaction was found maximum (Narayan et al., 2005, Table 7.2).

COOMe
MeOOC
Room temp.
COOMe
COOMe

**TABLE 7.2** Time taken in different solvents.

| Solvent | Time (hrs) |
|---|---|
| Toluene | >120 |
| Acetonitrile | 84 |
| Methanol | 18 |
| Neat | 48 |
| Water | 1/6 (10min) |

**(iv) Aldol condensation**

The aldol condensation involves self addition of aldehydes containing α-hydrogen atom, giving β-hydroxy aldehyde, which undergoes dehydration resulting into α, β-unsaturated aldehyde. Aldol reaction of silyl enol ethers with aldehydes was reported, which was promoted by water (Lubineau, 1986; Lubineau and Meyer, 1988).

**(v) Benzoin condensation**

The reaction of aromatic aldehydes in presence of sodium or potassium cyanide in aqueous ethanolic solution gives α-hydroxy ketones (Buck, 1948; Lapworth, 1903; Lapworth, 1904).

$$2\,\mathrm{Ph}-\overset{\displaystyle H}{\underset{}{C}}=O \xrightarrow[\mathrm{EtOH,\ H_2O}]{\mathrm{CN^{\ominus}},\ \Delta} \mathrm{Ph}-\overset{\displaystyle O}{\overset{\|}{C}}-\underset{\displaystyle H}{\overset{\displaystyle OH}{C}}-\mathrm{Ph}$$

The benzoin condensation in aqueous medium using inorganic salts is about 200 times faster than in ethanol (Koal and Breslow, 1988).

(vi) **Claisen–Schmidt condensation**

This reaction involves the condensation of aromatic aldehydes, which do not have α-hydrogen, with an aliphatic aldehyde or ketone having α-hydrogen in presence of a strong base to form α, β-unsaturated aldehyde or ketone (Claisen and Claparede, 1981; Schmidt, 1881).

$$C_6H_5CHO + CH_3CHO \xrightarrow{NaOH} HCH_5C_6 \quad CH.CHO$$

Cinnamaldehyde

The reaction of cyclohexanone with benzaldehyde in water gives high yield.

**(vii) Heck reaction**

It involves the coupling of an alkene with a halide in presence of Pd (0) catalyst to form a new alkene.

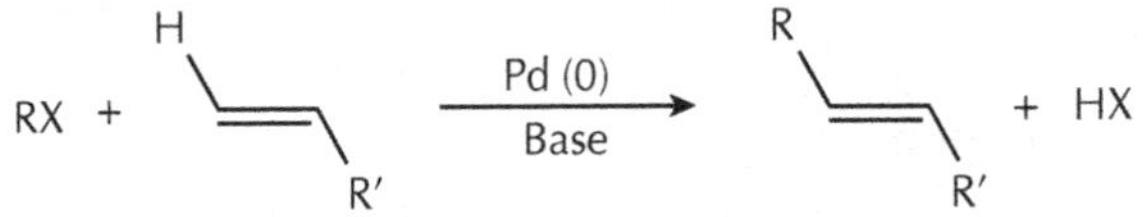

$$RX + H\text{-}CH{=}CH\text{-}R' \xrightarrow[\text{Base}]{\text{Pd (0)}} R\text{-}CH{=}CH\text{-}R' + HX$$

Heck reaction can proceed very well in water. Water plays an important role in the transformation of catalyst precursor into Pd (o) species and to generate zero valent Pd species. It can be performed under mild conditions in presence water and acetate ion.

**(viii) Knoevenagel reaction**

This is a condensation reaction of an aldehyde or a ketone with the compound having active $-CH_2-$group in presence of a weak base (Johnson, 1942; Knoevenagel, 1898).

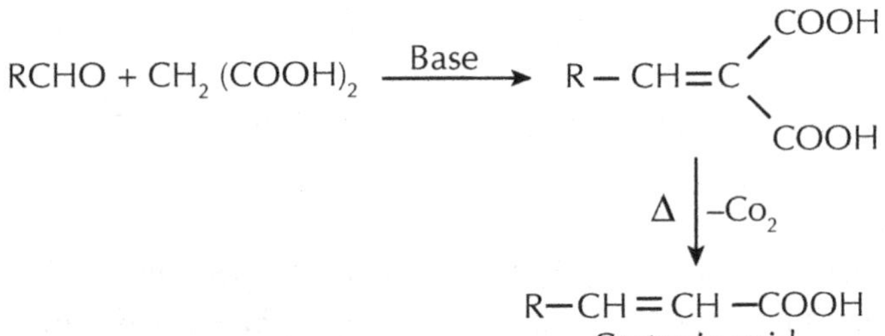

$$RCHO + CH_2\,(COOH)_2 \xrightarrow{Base} R-CH{=}C(COOH)_2$$

$$\xrightarrow{\Delta,\ -Co_2} R-CH=CH-COOH$$

Crotonic acid

This reaction has also been carried out between aldehydes and acetonitrile in water.

Knoevenagel type addition product can be obtained by the reaction of acryclic derivative in presence of water and DABCO [1,4-diazabicyclo [2,2.2] octane with 90–98% yield (Auge et al., 1994).

(ix) **Oxidation reactions**

Although many of the oxidation reactions are carried out in aqueous medium, now a days some more innovative oxidation reactions have been reported in aqueous medium.

*Oxidation of alkenes and alkynes*—Oxidation of alkene using aqueous solution of $KMnO_4$ and in presence of a PTC or 18-crown-6 gives carboxylic acid with good yield.

$$\underset{\text{Alkene}}{R{-}(CH_2)_n{-}CH{=}CH_2} \xrightarrow[\text{PTC or 18-Crown-6}]{\text{aq. } KMnO_4} \underset{\text{Carboxylic acid}}{R{-}(CH_2)_n{-}CH_2{-}COOH}$$

Similarly, alkynes can also be oxidized and they give a mixture of carbonic acids.

$$R{-}C{\equiv}C{-}R' + 4\,[O] \xrightarrow[\text{aq. medium}]{KMnO_4} R\,COOH + R'\,COOH$$

*Oxidation of aldehydes and ketones*—Oxidation of aromatic aldehydes by aqueous performic acid has been reported at low temperature (0–4°C). Baeyer-Villiger oxidation of ketone was also carried out in aqueous heterogeneous medium using m-chlordroxybenzole acid MCPBA at room temperature (Fringuelli et al., 1989).

$$\text{4-R-cyclohexanone} \xrightarrow[\text{20°C, 1 hr.}]{\text{MCPBA, } H_2O} \text{5-R-oxepan-2-one (lactone)}$$

*Oxidation of amines*—Oxidation of aromatic amines having –OH or –COOH group to nitro compound by oxone in 20–50% aqueous acetone gives 73–84% yield (Webb and Seneviratrie, 1995). By using oxone in water under neutral or basic condition, it is possible to synthesise N-oxide from amino pyridine directly (without protection of amino group) in good yield (Robke and Behrman, 1993).

*Oxidation of sulphides*—Oxidation of sulphides to sulphone can be carried out by using sodium perbonate (SPB) in aqueous methanolic sodium hydroxide (Mckillop and Tarbin, 1987) whereas oxidation of sulphides to sulphoxides can be done in 70% aqueous tetrabutyl phosphonium hydroxide (TBPH) in water in heterogeneous phase at low temperature (20–70°C) (Fringuelli et al., 1993).

(x) **Reduction reactions**

Reduction of many compounds in aqueous medium has been reported with high yield.

*Reduction of alkene and alkyne*—The reduction of carbon-carbon double bond of α, β-unsaturated carbonyl compounds by using $Zn/NiCl_2$ in methoxyethanol-water system has been reported (Petrier and Luche, 1987).

O= COOEt, $CH_3$ $\xrightarrow[\text{Methoxyethanol, }-H_2O]{Zn/NiCl_2,\ 30°C,\ 2\ hr.}$ O= COOEt, $CH_3$

Reduction of alkynes can be carried out with water soluble monosulphonated and trisulphonated triphenylphosphine (Larpent and Meignan, 1993).

*Reduction of aldehydes and ketones*—The reduction of carbonyl compound in aqueous media has been carried out by sodium borohydride at room temperature.

Samarium iodide in aqueous THF and cadmium chloride-magnesium in the $H_2O$-THF system can be employed for the reduction of carbonyl compounds (Hasegawa and Curran, 1993; Mordolol, 1993).

*Reduction of aromatic ring*—The heterocyclic compounds can be reduced by $SmI_2$–$H_2O$ system with good yield at 0°C (Hanessian and Girand, 1994). Aromatic compound can also be reduced in aqueous medium with ruthenium chloride/trioctylamine (TOA) at room temperature (Fache et al., 1995).

O R $\xrightarrow[H_2,\ MeOH\text{-}H_2O,\ R.T.]{RuCl_3/TOA}$ O R

(xi) **Photochemical reactions**

Photodimerisation of thymine and uracil has been reported in aqueous medium (Ramamurthy, 1986).

Organic compounds like stilbenes, alkyl cinnamates, coumarin and so on. can also be dimerized in water with good yields.

Some more examples of photochemical reactions in aqueous medium are:

(a) $OCH_3$ / F $\xrightarrow[KCN,\ H_2O]{hv}$ $OCH_3$ / CN (Major) + $OCH_3$ / OH (Minor)

(b) $OCH_3$ / F $\xrightarrow[KCN,\ H_2O]{hv}$ $OCH_3$ / CN / CN (Major) + $OCH_3$ / OH / OH (Minor)

(xii) **Reaction of carbanion equivalents**

Reactions of carbonyl compounds and imines with allyl halides allow use of water as a solvent along with no requirement of low temperature unlike the organolithium and organomagnesium reagent where metal such as indium, tin and zinc are used as metal mediators (Keh et al., 2003).

(xiii) **Barbier–Grignard type carbonyl alkylation**

*Barbier type allylation*—This reaction is highly regio- and stereoselective. In this reaction, the use of water as a solvent has very high selectivity for α-adduct as a product (100%) as compared to (dimethyl sulphide) (DMS) with 65% yield. Here, water is required for the formation of oxonium ion intermediate, which furnishes α-adduct (Tan et al., 2003).

(xiv) **Reaction of radicals**

*Formation of lactone*—It has been reported that yield of lactone formation is 78% in water while it is 0% in benzene and hexane (Yorimitsu et al., 2000).

$Et_3B$/Trace $O_2$ → Slovent, 3hr.

(xv) **Dehalogenation**

Organic halides are treated with phosphinic acid in the presence of radical initiator azo-bis-isobutyronitrile (AIBN) and a base ($NaHCO_3$), in aqueous ethanol to give corresponding reduced products in high yields.

O–C₆H₄–I → aq. $H_3PO_4$, $NaHCO_3$ / AIBN, aq. Ethanol → RO–C₆H₄–H

(xvi) **Multicomponent reactions**

*Passerini reaction:* The rate of this reaction is increased 300 times, when it is carried out in water (Pirrung and Sarma, 2004).

*Ugi reaction:* There is an acceleration in the rate of this reaction as well as high yields are obtained, when water is used as a solvent (Pirrung and Sarma, 2004).

(xii) **Miscellaneous**

A highly efficient, one-pot and three component reaction of amine and carbon disulfide with alkyl vinyl ether via Markovnikov addition reaction was carried out in water under a mild and green procedure with excellent yield and complete regiospecificity (Halimehjani et al., 2010).

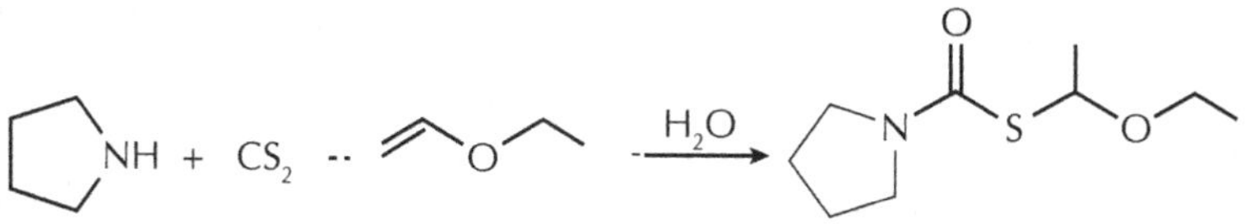

Water was used as a promoter for the synthesis of 2-aminobenzothiazoles or 2-aminobenzoxazoles in tandem reaction by Zhang et al. (2011).

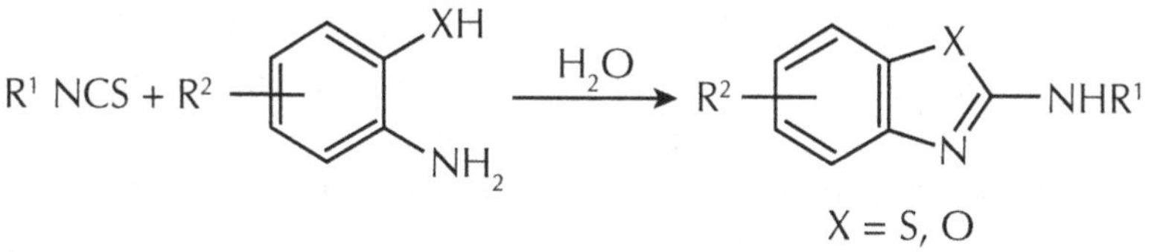

Peng et al. (2010) used stilbazo as a promoter for the synthesis of biaryl compounds by a ligand-free Suzuki–Miyaura reaction in water at room temperature.

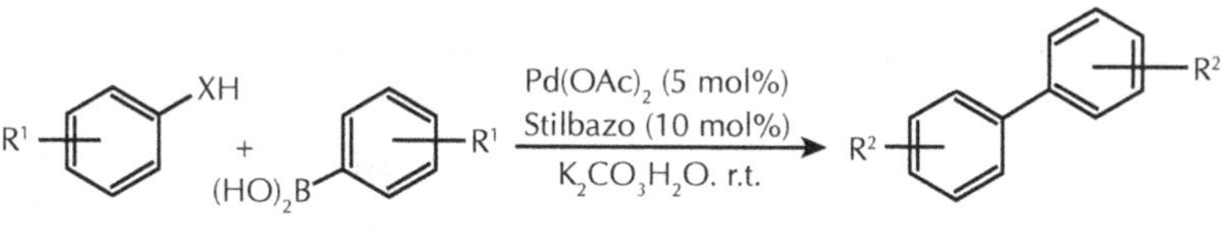

Addition of ethylene to aniline to generate N-ethylaniline was carried out by Dub et al. (2010) in the presence of a catalyst $PtBr_2/Br^-$ in aqueous medium. It was found to be an atom economical addition.

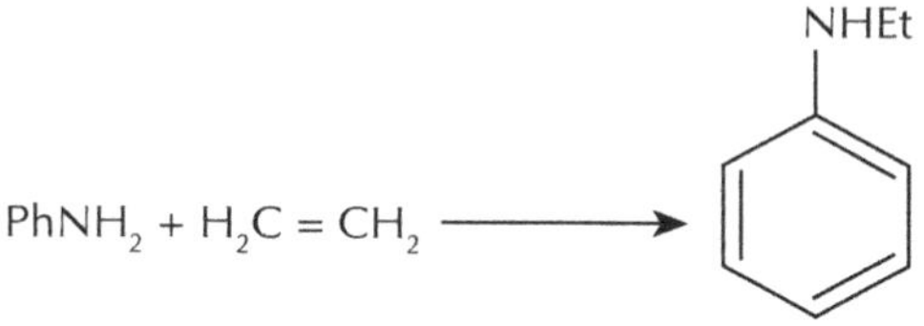

## 7.2.2 POLYETHYLENE GLYCOL (PEG)

Polyethylene glycol is a linear polymer formed from the polymerization of ethylene oxide. It is available in a variety of molecular weights. The numerical designations of PEG indicates the average molecular weight for example, PEG-200, PEG-400, PEG-2000 and so on. Low molecular weight PEGs are liquid and completely miscible in water whereas PEGs having high molecular weight are waxy white solids and highly soluble in water.

The PEG is inexpensive, recoverable, biologically compatible, non-toxic, thermally stable and biodegradable (Harris and Zolipsky, 1997). In addition to this, PEG and its monomethyl ethers have low vapour pressure and are nonflammable so that these compounds present simple workup procedure and can be recovered as well as recycled. Thus, it can be considered as an environmentally benign solvent.

The PEG is employed as a support for the various transformations (Dickerson et al., 2002). It is a biologically acceptable polymer, which has immense importance in drug delivery, bioconjugates (Harris, 1992) and bioseparations (Albertsson, 1986). According to US FDA, it is considered safe and approved for internal consumption (Herold et al., 1989; Peglation, 2002). It can also be used as an efficient medium for PTC. Now a days, it is used as a solvent for many organic reactions. The PEG is commercially available and it is low cost. Generally, low molecular weight (<2,000) PEGs have been used because of low melting point or their existence as liquid at the room temperature. The PEGs are stable to acid, base and high temperatures (Chen et al., 2004a, b; Guo et al., 2002; Naik and Doraiswamy, 1998). These are not affected by oxygen, hydrogen peroxide or other oxidation systems (Haimov and Neumann, 2002).

## *APPLICATIONS*

### *OXIDATION REACTION*

In the oxidation dihydroxy compounds of olefins, PEG-400 is used as a solvent and $OsO_4$ acts as a catalyst for the reaction (Chandrasekhar et al., 2003).

$$H_5C_6\text{–CH–CH–}C_6H_5 \xrightarrow[\text{PEG - 400}]{O_sO_4,\ \text{NMO}} H_5C_6\text{–}\underset{OH}{CH}\text{–}\underset{OH}{CH}\text{–}C_6H_5$$

After, the extraction of the product, PEG—400 can be reused.

$K_2Cr_2O_7$ has been used (which is soluble in PEG—400) for the oxidation of benzyl bromide to benzeldehyde with good yield (Santaniello et al., 1979).

## REDUCTION REACTION

It has been reported that the reduction of alkyl and acyl esters to the corresponding alcohols by sodium borohydride was enhanced in PEG-400.

$$R'COOR \xrightarrow[\text{PEG - 400}]{NaBH_4} R'\text{–}CH_2\text{–}OH$$

Carbonyl compounds can be reduced by $NaBH_4$ more easily and efficiently, when PEG-400 is used as solvent rather than THF. The reduction of halides as well as acyl chloride can also be accomplished by $NaBH_4$ in PEG-400 conveniently (Santaniello et al., 1983).

## SUBSTITUTION REACTION

In PEG-300, the reaction of tert-butyl chloride with water gives corresponding alcohol (Leininger et al., 2002).

$$(CH_3)_3C\text{–Cl} + H_2O \xrightarrow{\text{PEG-400}} (CH_3)_3CH\text{–OH}$$

The reaction of alkyl halides ($RCH_2Br$) with acetate, iodide and cyanide gives the corresponding substituted products in PEG-400 (Santaniello, 1984). Potassium thioacetate in PEG-400 has been used as a nucleophilic reagent for the alkyl halides (R—CHXR'), which gives 92–98% yield of product (Ferravoski et al., 1987).

## DIELS–ALDER REACTION

Diels-Alder reaction of 2,3-dimethyl-1,3-butadiene with acrolein in PEG-300 give the addition product in good yield.

$$Me{-}C(=CH_2){-}C(=CH_2){-}Me + CH_2{=}CH-CHO \xrightarrow{\text{PEG - 300}} \text{3,4-dimethylcyclohex-3-ene-1-carbaldehyde (Me, Me, CHO)}$$

## HECK COUPLING REACTION

It has been reported that Heck coupling reaction can be carried out in molten liquid PEG-2000 at 80°C with good yield (Chandrasekhar et al., 2002). After the extraction of product, PEG-2000 and $Pd(OAc)_2$ could be reused.

$$R - C_6H_4{-}Br + CH_2 {-}CH{-}X \xrightarrow[\text{Pd (OAc)}_2]{\text{PEG - 400}} R - C_6H_4{-}CH{=}CH{-}X$$

$$X = -COOEt;\ C_6H_5$$

## HECK REACTION

The use of PEG-2000 as an efficient reaction medium for the Pd catalysed C–C bond formation reaction (named as Heck reaction) has been reported. This transformation is more rapid, high yielding and more regio- and stereoselective in PEG. Besides being an effective solvent for this reaction; PEG also acts as a phase transfer catalyst for the C–C- bond formation.

$$PhBr \xrightarrow[\text{TEA, PEG-2000}]{\text{Pd (OAc)}_2\text{, 80°C}} \begin{cases} \xrightarrow{CH_2=CH-Ph} PhCH{=}CHPh\ (E)\ 93\% + PhCH{=}CHPh\ (Z)\ 7\% \\ \xrightarrow{CH_2=CH-COOEt} PhCH{=}CHCOOEt\ (E)\ 90\% + PhCH{=}CHCOOEt\ (Z)\ 10\% \end{cases}$$

## *BAYLIS-HILLMAN REACTION*

Unreactive aldehydes and activated olefins react, when PEG-400 is used as a recyclable solvent for Baylis–Hillman reaction (Chandrasekhar et al., 2004).

$$C_6H_5CHO + CH_2=CH\text{-}CO\text{-}OEt \xrightarrow[\text{PEG-400, 2hr, R.T.}]{\text{DABCO (20 mol\%)}} C_6H_5CH(OH)C(=CH_2)COOEt \; (92\%)$$

## *SUZUKI CROSS-COUPLING REACTION*

Suzuki cross-coupling reaction is a Pd catalysed C–C coupling reaction of organoboron compounds with aromatic aldehydes in presence of a base using PEG-400 as a reaction medium. The yield of product biaryl was found to be 55–81% (Namboodiri and Varma, 2001).

$$C_6H_5Br + (HO)_2B\text{-}C_6H_4\text{-}CH_3 \xrightarrow[\text{PEG - 400, MW}]{PdCl_2,\ \text{KF, 50 sec}} C_6H_5\text{-}C_6H_4\text{-}CH_3$$

## *DECARBOXYLATION OF CINNAMIC ACID*

The decarboxylation of substituted α-phenyl cinnamic acid derivative has been achieved by using catalytic amount of methylimidazole and aqueous. $NaHCO_3$ in PEG under microwave irradiation.

## *SYNTHESIS OF 2-AMINO-2 CHROMONES*

The condensation reaction of aldehyde, malononitrile, and α-napthnol in PEG-$H_2O$ gave 2-amino-2-chromones in high yield at room temperature. The reaction was catalysed by nano-sized MgO (Kumar et al., 2007).

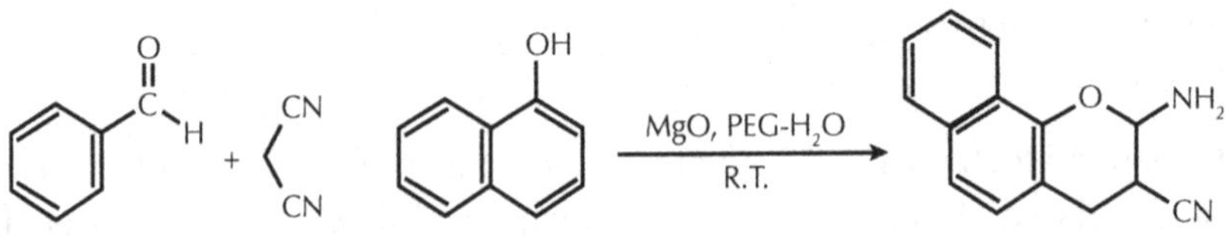

$\xrightarrow[\text{R.T.}]{\text{MgO, PEG-}H_2O}$

This reaction has very simple experimental procedure, cost effectiveness while recyclability of catalyst and use of environmental friendly solvent are another beneficial features of this reaction.

On the basis of these applications and its characteristics, it can be concluded that PEG and its aqueous solution can be one of the best alternative of organic solvents. It can be used as phase transfer catalyst also. The use of PEG in many enzytmatic transformations has also been reported (Kondo et al., 1994; Mandenius et al., 1988; Persson et al., 1992; Tjerneld et al., 1985)

### *7.2.3 GLYCEROL*

Glycerol is **not only a green solvent** but it can also serve as an **effective replacement for many other solvents.**

#### *ADVANTAGES OF USING GLYCEROL AS A SOLVENT*

- **Polarity:** Glycerol, as a polar molecule, is **capable of dissolving many polar organic** compounds as well as hydrophobic substrates such as ethers and hydrocarbons.
- **Volatility and boiling point:** Boiling point of glycerol is 290°C and it has nonvolatile behavior so **it could easily be separated** from a solute using distillation. Its high boiling point enables glycerol to be used at higher temperatures.
- **High yields**: In certain reactions, use of glycerol results in **high yields** due to possibility of hydrogen bonding by the glycerol molecule. The reaction selectivity is more in glycerol for various reasons including its polarity, structure and solubility properties resulting in higher yields.

- **Microwave heating:** It has been reported that glycerol can tolerate the heating, even when microwaves are used. This is used in many microwave assisted organic synthesis and results in a cleaner reaction and less reaction time.

Apart from these, there are some key properties of glycerol that make it a green solvent. These are low toxicity, low vapor pressure, low environmental impact, availability, easy handling and storage. It can also be used with catalysts.

## *APPLICATIONS*

### *ORGANIC SYNTHESIS*

It has been reported that some organic synthesis reactions for example, Heck reactions, Suzuki reaction and hydrogenation reactions are carried out with high yields in glycerol. Aza-Michael reaction between p-anisidine and n-butyl acrylate can proceed smoothly under catalyst free conditions in glycerol as a solvent, whereas many other solvents were found to be ineffective (Gu and Jerome, 2010).

$NH_2$ OMe + OBu$^n$ O — Catalyst - Free solvent, 100°C, 20 hrs. → NH OBu$^n$ O OMe

Similarly, Michael addition of indole to nitrostyrene was found to give 80% yield of the desired product under catalyst free conditions in glycerol.

N H + $NO_2$ — Catalyst - Free solvent, 90°C, 20 hrs. → $NO_2$ N H

## ENHANCING REACTION SELECTIVITY

The selectivity of certain reactions has been improved by using glycerol as a solvent, in order to produce higher product yields, for example, ring opening reaction of p-anisidine with styrene oxide can be performed without any catalyst, if glycerol is used as a solvent. This reaction demonstrates a better regioselectivity. It was also reported that reaction involving styrene, paraformaldehyde and dimedone shows an extremely higher yield in glycerol. The rate of the hetero-Diels-Alder reaction was also increased in glycerol because it is a polar protic solvent. This property allows the solvation of reactant intermediates, where the intermediate energy decreases relative to the starting material, resulting into the rate enhancement.

Overall, the structure of glycerol, its polarity and the intermolecular forces can make a reaction more selective. Thus, the product yield obtained depends on the way, in which a solvent interacts with the starting material.

Moreover, glycerol has been used uniquely for its high selectivity to give desired products in one-pot two step reactions. By using a single reactor, the yield of the product increases on one hand, while separation processes and purification of intermediate compounds are avoided on the other, making this process green for example, (i) reaction involving arylhydrazines, β-ketone esters, formaldehyde and styrenes (ii) reaction involving indoles, arylhydrazine, β-ketone and esters and (iii) one-pot sequential reaction involving phenylhydrazine, ethyl 4-methoxybenzoylacetate, α-methylstyrene and paraformaldehdye in glycerol.

## SOLVENT FOR BIOCATALYSIS

Glycerol can be used as a solvent giving high yield with the use of natural catalysts due to its low toxicity and high affinity for hydrophilic compounds for example, Baker's yeast catalyzed reduction of ketones and bioreductions of 2'-chloroacetophenone using glycerol as a co-solvent to obtain high yields.

## *CATALYST DESIGN AND RECYCLING*

In the process of homogeneous catalyst recovery, solvents are used to immobilize and recycle the catalyst. It reduces waste and recycles the catalyst. Glycerol has a better solubility with the homogeneous catalyst with ionic compounds and hence, it can be used for catalyst recycling. It has been observed that in the catalytic formation of bisindolylmethane in glycerol over $CeCl_3$/Lewis acid, the reaction products can be selectively extracted from the glycerol/$CeCl_3$ mixture by liquid phase extraction with ethyl acetate and therefore, allowing a convenient recycling of both $CeCl_3$ and glycerol.

Glycerol can also dissolve organometallics complexes; thus, allow non-ionic catalysts to be recovered for example, the hydrogenation reaction catalyzed by the $[Ru(p\text{-cumene})Cl_2]_2$ complex using glycerol. This Ru complex catalyst is not ionized in glycerol, and is recycled after extraction of the reaction products with diethyl ether (Gu and Jerome, 2010).

## *SOLVENT FOR SEPARATION*

Bioethanol is one of the important biofuel and it is a green alternative to gasoline. The purification of bioethanol by extractive distillation was achieved by glycerol. It was found that the separation process of ethanol was more effective, when glycerol was used. Here the ethanol, water and glycerol were all recovered with more than 99% purity.

## *USE IN MATERIALSYCHEMISTRY*

Glycerol shows promising properties in the preparation of materials. Its high boiling point and low vapor pressure allow it to be used for the reactions at high temperatures. Good solubility of inorganic and organic compounds in glycerol is another added advantage.

One simple method to prepare metal particles is by heating a metallic salt and then reducing it. Glycerol has a high boiling point and it can also act as a reducing reagent. Some copper particles were successfully obtained with a purity greater than 99% when $Cu(OH)_2$, CuO and $Cu(OAc)_2$ are reduced under atmospheric conditions in glycerol at a temperature below 240°C, (Gu and Jerome, 2010).

Similarly, in an another process, silver nitrate produces silver particles with a high yield and uniformity using glycerol as a solvent as well as reducing reagent. The size of the silver particles can be changed by varying the amount of silver nitrate.

### 7.2.4 CYCLOPENTYL METHYL ETHER (CPME)

Cyclopentyl methyl ether (CPME) is new hydrophobic ether solvent. Unlike other common ether solvents, CPME has some unique properties. It has many other properties that make it greener, easy to use and a more cost effective process solvent for many types of synthesis. Watanabe et al. reported cyclopentyl methyl ether as a new and alternative process solvent. The CPME is an alternative to other ethereal solvents such as tetrahydrofuran (THF), 2-methyl tetrahydrofuran (2-MeTHF), dioxane (carcinogenic), and 1,2-dimethoxyethane (DME) (2007).

Some properties of CPME are listed below:

- **High hydrophobicity:** It is easily separated and recovered from water, reduces emissions and wastewater. It is widely applicabile as a reaction, extraction and crystallization solvent, in simple and one-pot syntheses.
- **Wide liquidity range**: It has wide range of applications from lower to higher temperatures for accelerating reaction rate.
- **Low heat of vaporization:** The heat of vaporization of CPME is 69.2Kcal/kg and due to this low value, it saves energy for distillation and recovery.
- **Resist peroxide formation:** Ethereal solvents have some explosive nature arising from peroxide generation. The ether radical is more stable and more peroxide is accumulated during the storage. The ether radical of CPME is unstable compared to other ethereal solvents. Thus, it is concluded that the peroxide formation from CPME is very slow as compared to other ethers.
- **Relatively stable to acids and bases:** It has good stability under acidic and basic conditions.

It has limited solubility in water, easy drying, narrow explosion area and high boiling point. Formation of azeotropes with water coupled with a narrow explosion range make.

These unique properties of CPME give high recovery rate (>90%). All these properties collectively contribute to greener chemistry through a reduction in the total amount of solvents used, waste water and waste solvent created and carbon dioxide emissions. The CPME also contributes to process innovation, where it can save process time, facilitates by shortening work-up time and simplifying the total process, which means fixed costs are reduced.

The melting point of CPME is 140°C whereas boiling point is 106°C. The viscosity of CPME is found to be 0.55cP and its surface tension is 25.17mN/m. It has low specific heat value (0.4346Kcal/kg K) and its dielectric constant is very less compared to water, (4.76). It has low density (0.86g/cm$^3$).

## *APPLICATIONS*

The CPME is better alternative for other ethereal solvents. As a solvent, it can be used in following processes

- Extraction
- Crystallization
- Polymerization
- Coating

(i) **Reactions with alkylating agents**
Methylation of alcohol with methyl triflate (Me-OTf) (which is a powerful methylating agent) has been carried out in CPME (Watanabe et al., 2007).

(ii) **Friedel-Crafts type reactions**
In Friedel–Crafts reaction, Lewis acid catalyst and halogenated solvents are required. As CPME can attain anhydrous conditions, it is useful to keep anhydrous reaction media for metal triflates or prevent the decomposition. But CPME cannot be compatible with a strong Lewis-acid catalyst ($AlCl_3$). Due to this, some Ti Lewis acids are recommended with CPME (Watanabe et al., 2007).

## *GRIGNARD TYPES REACTION*

The CPME is a preferred solvent for Grignard type reactions because it can maintain anhydrous conditions without any particular precautions. Magnesium turnings were placed on a flask for the preparation of Grignard reagent. Then it was covered with CPME and a small piece of iodine was added. A solution of alkyl bromide was also added while heating. After completion of the addition, the mixture was heated for a while. A small amount of magnesium still remained in the flask. The Grignard reagent thus prepared was cooled to 0°C. The solution became cloudy with the precipitation of the Grignard reagent. After that a solution of carbonyl compound was added to the Grignard reagent and allowed it to attain room temperature. The reaction achieved completion at this stage (Watanabe et al., 2007).

## *REACTIONS WITH TRANSITION METAL CATALYSTS*

Transformations with transition metal catalysts (Pd, Ni, Rh, Ir), ethereal solvents are used. High reaction temperature is preferable and anhydrous conditions are useful for multicomponent couplings for such catalytic reactions. Like other ethers and toluene, CPME can also take part in the Pd-catalyzed asymmetric allylic alkylation. Due to high boiling point, CPME is successfully employed as the solvent for such coupling reactions for completion of reaction within short time. High boiling points and easy workup made the overall reaction sequence very effective and convenient (Watanabe et al., 2007).

Recently, Kawatsura et al. (2010) reported a Suzuki-Miyaura coupling reaction catalyzed by a ruthenium complex. In this reaction, substituted halobenzene were coupled with $ArB(OH)_2$ to give diaryl compounds. Such an arylation is also reported with 5- member heterocyclic rings using aryl halide in presence of Pd-base and CPME (Beydoun and Doucet, 2011).

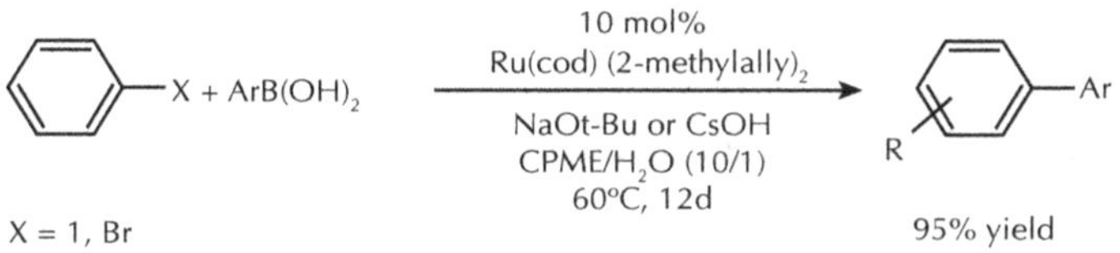

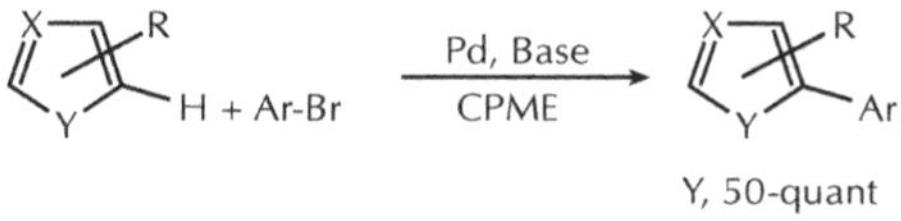

X + $ArB(OH)_2$ → (10 mol% Ru(cod) (2-methylally)$_2$; NaOt-Bu or CsOH; CPME/$H_2O$ (10/1); 60°C, 12d) → R–Ar

X = 1, Br 95% yield

H + Ar-Br → (Pd, Base; CPME) → Ar

Y, 50-quant

## *PALLADIUM CATALYZED DIRECT ARYLATION OF HETEROAROMATICS*

Cyclopentyl methyl ether can be employed for the palladium catalyzed direct arylation of heteroaromatics. The direct 5-arylation of thiazoles, thiophenes or furans was carried out by using aryl bromides (a coupling partners) in the presence of CPME and 0.5–1mol% of palladium catalysts with moderate to high yields (Watanabe et al., 2007).

## *CONDENSATION REACTIONS*

Aldol and Claisen condensations require basic media; thus, it is compatible with CPME. Claisen-Schmidt condensation with a pyridine derivative is feasible in CPME with the proper base catalysis. These base catalyzed reactions indicate the usefulness of CPME for general condensation reactions. In addition to this advantage, the desired product can be isolated by the simple extraction from CPME, leaving the polar starting materials in aqueous solution. Thus, use of CPME makes the workup very easy (Watanabe et al., 2007).

## *ENOLATE CHEMISTRY*

An interesting solvent effect of CPME emerged during enolate formation. An asymmetric methylation (chirality transfer reaction), which might be manifested through the formation of a rigid enolate from *O*-methyl mandelic acid, in the medium of CPME was reported by Kawabata's group. Aggregated enolate formation is controlled by the interference of CPME with base or enolate itself.

## *TRANSFORMATIONS*

Classical transformations, which have been utilized in the pharmaceutical industries, are renewed by the use of CPME as a solvent. It demonstrates the application of CPME in the manipulation of the heterocyclic intermediates. It provides a green solution for improving chemical process by minimizing the solvent waste stream and improving laboratory safety due to CPME's unique composition, which resists the formation of peroxides. It is stable than THF and 2-MeTHF (stabilizer required) and resists peroxide formation. Therefore, the frequency of peroxide testing is reduced.

It is novel hydrophobic ether solvent and used in many organometallic reactions, where it provides better yield and higher selectivity than THF. It forms an azeotrope rich with water and can be more easily dried as compared to THF and 2-MeTHF. It has limited miscibility in water (1.1g/100g at 23°C), so easily separated and recovered from water reducing the waste stream. The CPME has higher boiling point (106°C) as compared to THF and 2-MeTHF and higher reaction temperature reduces overall reaction time. It has low heat of vaporization and therefore, less solvent is lost during reflux.

## *REDUCTION*

Cho et al. (2012) studied ring-expansion reactions of heterocyclic ketoximes and carbocyclic ketoximes with several reductants such as $AlHCl_2$, $AlH_3$ (alane), $LiAlH_4$, $LiAlH(OtBu)_3$, and $(MeOCH_2CH_2O)_2AlH_2Na$ (Red-Al). Among reductants, $AlHCl_2$ ($LiAlH_4$:$AlCl_3$ = 1:3) in cyclopentyl methyl ether (CPME) has been found to be a suitable reagent for the reaction, and excellent yields were obtained of the rearranged cyclic secondary amines.

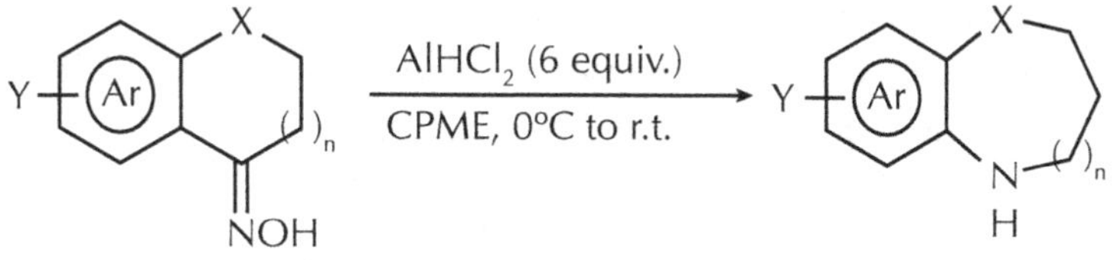

Y–(Ar) X ( )$_n$ NOH $\xrightarrow[\text{CPME, 0°C to r.t.}]{AlHCl_2 \text{ (6 equiv.)}}$ Y–(Ar) X N ( )$_n$ H

$X = CH_2$, O. S
Y = H, OMe n = 0, 1, 2

A system offers a facile excess to optically active vicinal diamines, when reduction using aryl $LiAlH_4$ in cyclopentyl methyl ether was carried out. It also provides a practical rout to α- amino alcohols and α- amino ketones with desired stereochemistry (Ooi et al., 2005).

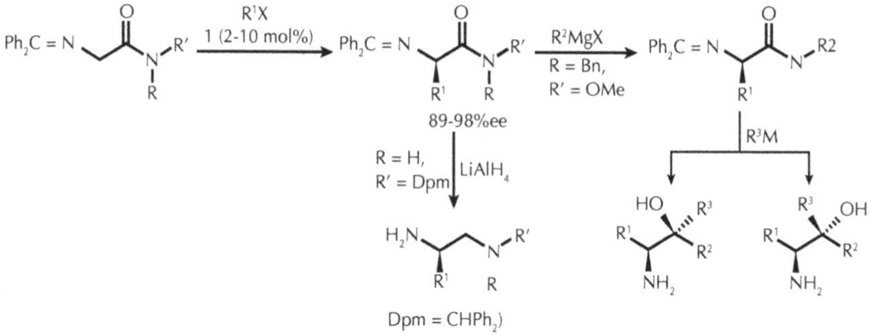

$Ph_2C = N$ O N R′ R $\xrightarrow{R^1X,\ 1\ (2\text{-}10\ mol\%)}$ $Ph_2C = N$ O N R′ $R^1$ R (89-98%ee) $\xrightarrow[R = Bn,\ R' = OMe]{R^2MgX}$ $Ph_2C = N$ O N R2 $R^1$

$R^3M$: HO $R^3$ $R^1$ $R^2$ $NH_2$; $R^3$ OH $R^1$ $R^2$ $NH_2$

R = H, R′ = Dpm, $LiAlH_4$: $H_2N$ N R′ $R^1$ R

Dpm = $CHPh_2$)

Hydrostannation, hydrosilylation, hydrothiolation, and tributyltin hydride mediated reductions were successfully carried out in CPME (Kobayashi et al., 2013).

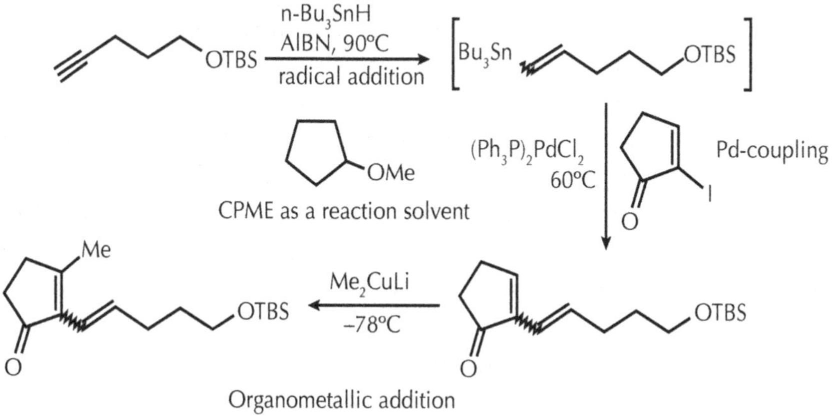

OTBS $\xrightarrow[\text{radical addition}]{n\text{-}Bu_3SnH,\ AIBN,\ 90°C}$ [ $Bu_3Sn$ OTBS ]

OMe
CPME as a reaction solvent

$(Ph_3P)_2PdCl_2$, 60°C; O I; Pd-coupling

Me, O, OTBS $\xleftarrow[-78°C]{Me_2CuLi}$ O, OTBS

Organometallic addition

(x) **Reaction in presence of acid**

Watanabe et al. (2009) observed Pinner reaction in CPME, where cyanide is converted to methoxy imine.

$$R-CN \xrightarrow[\text{Solvent, 0°C, 1 h, then 0.5°C, 48 h}]{\text{4M HCl solution, } CH_3OH} R-C(=NH)-OCH_3 \cdot HCl$$

*NANOPARTICLES*

Nanoparticles of gold were obtained by Sugie et al. (2008) by treating $HAuCl_4$ with a thiol.

$$HAuCl_4-4H_2O + HS-nC_{12}H_{25} \xrightarrow[\text{CPME}]{Et_3SiH} \text{Gold nanoparticles}$$

### *7.2.5 2-METHYLTETRAHYDROFURAN (2-METHF)*

2-MeTHF is derived from renewable resources such as corn cobs and sugar cane bagasse. 2-MeTHF offers both; economical and environmentally friendly advantage over tetrahydrofuran, when used as solvent. It is a aprotic solvent, which resembles toluene in its physical properties. Bromo and iodo Grignard reagents tend to be more soluble in 2-MeTHF, while chloro Grignard reagent tend to be less soluble in it. It forms an azeotrope rich with water, which can be more easily dried than dichloromethane or tetrahydrofuran. It is a truly green alternative to dichloromethane and tetrahydrofuran. It has limited miscibility in water (14g/100g at 23°C). 2-MeTHF is easy to separate and recovering it from water reduces its quantity in the waste stream. Its boiling point is (80°C), which is higher as compared to tetrahydrofuran (66°C); thus, higher reaction temperature reduces overall reaction time. It has low heat of vaporization, so less solvent is lost during reflux and therefore, it saves energy during distillation and recovery of solvent.

It is used as an alternative to tetrahydrofuran for organometallic reactions.

- Grignard
- Reformatskii (Reformatsky)
- Lithiation
- Hydride reduction and
- Metal catalyzed coupling (Heck, still, Suzuki)

It is used as an alternative to dichloromethane for biphasic reactions for example, alkylation, amidation and nucleophilic substitution reactions.

2-MeTHF has an added advantage of having some beneficial physical and chemical properties. Corn cobs and sugar cane are renewable resources for furfural, which on hydrogenation yields 2-MeTHF while THF is obtained from 1,4-butandiol, which is an oil derived substance. 2-MeTHF is the only aprotic solvent similar to THF derived from renewable resources and it is industrially available.

The incineration of solvents in most of the chemical industries adds to green house gases causing green house effect, but incineration of 2-MeTHF does not contribute to it as it returns the carbon dioxide back to the atmosphere, which was captured by previous years crop.

2-MeTHF is not miscible with water and it is comparable or even better than THF in terms of its chemical properties. However, 2-MeTHF resembles toluene in term of its physical properties. It provides an easy and clean phase separation during work up. These advantages make the process simpler and more robust, translating into higher through-put and reduced cost.

2-MeTHF reduces the solvent and energy variable costs. Thus, it has better extractive properties than the classic THF/toluene mixture. This means that the number of extraction steps can be reduced by using it and simultaneously, the recovery of the product is increased. 2-MeTHF solution of crude product can be dried through a simple distillation at atmospheric pressure. The water rich 2-MeTHF azeotrope will create rapidly an anhydrous solution providing the option to add a new reagent without product isolation for example, the classic reaction sequence carbonyl to alcohol followed by alcohol to ester is particularly adequate by cutting almost 50% and it is improved by avoiding isolation of the intermediate alcohol. It is much easier to recycle and dry as compared to THF.

2-MeTHF requires only simple distillation at atmospheric pressure whereas THF is recycled and dried using swing distillation. The recycling and drying of 2-MeTHF is cost effective (Comanita, 2006a, b). 2-MeTHF is a versatile solvent covering a wide range of applications (Comanita and Aycock, 2005) and it has comparatively more stability than THF even in acidic reactions (Aycock, 2007). The hydrolysis of 2-MeTHF is also much slower because of its immiscibility with water. Lithium aluminum hydride ($LiAlH_4$) is more soluble in 2-MeTHF (about 10%). This can reduce aldehydes, esters and acids in 2-MeTHF and the products are similar as obtained in THF.

A one-pot dehydrobromination of a 2-bromoacrylate ester with lithium 2,2,6,6-tetramethylpiperidide (LTMP) and aldehydes in 2-MeTHF has been carried out by Pace et al. (2012).

$CO_2R'$ Br — LTMP (2 equiv.), RCHO / 2-MeTHF, 40°C / One pot procedure → R, OH, $Co_2R'$ — 16 examles 80-96% yields

R′ = Me, Et, Bu, Bn
R = Aryl, heteroaryl, tertiary-alkyl, α,β-unsaturated

Some esterifications were also carried out in 2-methyltetrahydrofuran (as an excellent substitute for THF) in biocatalysed processes in organic media. This application for this green solvent is a proof of opening a new field utilizing MeTHF in biotransformation (Simeó et al., 2009).

HO, O, Base, HO, $R_2$, $R_1$ — Lipase / Acyl donor / THF or MeTHF → $R_4O$, O, Base, $R_3O$, $R_2$, $R_1$

Milton and Clarke (2010) reported the cross-coupling of Grignard reagents at 5M in methyltetrahydrofuran with no added reaction solvents. 2-MeTHF was also used in an improved microwave accelerated synthesis of the $[Pd(L)Cl_2]$ pre-catalysts from sodium tetrachloropalladate in very high yield.

2-Methyltetrahydrofuran is a useful bio-based co-solvent for benzaldehyde catalyzed reactions (Shanmuganathan et al., 2010).

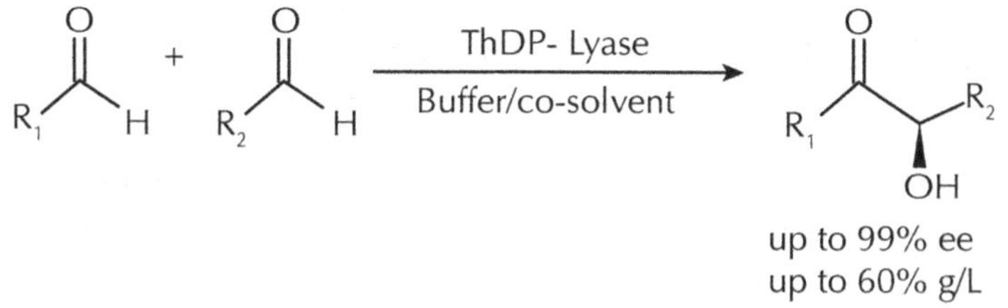

ThDP- Lyase
Buffer/co-solvent

up to 99% ee
up to 60% g/L

## *ETHYL LACTATE*

It is also known as lactic acid ethyl ester. Lactate esters are commonly used as solvents in the paint and coating industry. These have numerous attractive advantages like 100% biodegradablity, easy to recycle, non-corrosive and non-carcinogenic. Ethyl lactate has replaced solvents including NMP, toluene, acetone and xylene, which has resulted in making the work place relatively safer. Ethyl lactate has been used as a green solvent because it has low VOC, high solvency power for resin and polymer and high boiling point. It has almost eliminated the common use of chlorinated solvents.

Pereira et al. (2012) reviewed the field of use of ethyl lactate as green solvent because ethyl lactate is formed by the esterification of lactic acid by ethanol, which are biomass derived compounds (2011). Wan et al. developed a sustainable catalyst system consisting of $H_2O$/ethyl lactate, $Pd(OAc)_2$ and $K_2CO_3$. They used it for Suzuki-Miyaura reactions using various aryl bromides and iodides to incorporate arylboronic acids under ligand free conditions.

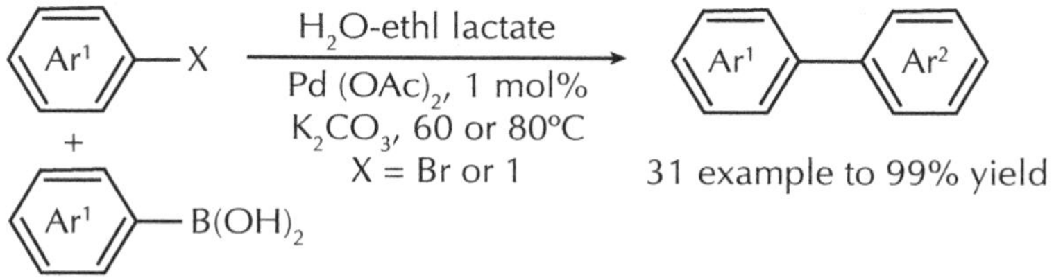

$H_2O$-ethl lactate
Pd $(OAc)_2$, 1 mol%
$K_2CO_3$, 60 or 80°C
X = Br or 1

31 example to 99% yield

Bennett et al. (2009) used ethyl lactate as a tunable solvent for the synthesis of aryl aldimines.

### *7.2.6 PERFLUORINATED (FLUOROUS) SOLVENTS*

Horvath and Rabai (1994) introduced this term for the first time, which has analogy with aqueous medium. These compounds were defined by Gladysz and Curran (2002) as being compounds that are highly fluorinated and based upon $sp^3$ hybridized carbon. Perfluorinated hydrocarbons are found to be unique solvents. These compounds are immiscible with water and most of the common organic solvents and form third liquid phase. These are chemically benign and environment friendly because these are non-toxic, nonflammable, thermally stable, recyclable and having high ability to dissolve oxygen. Fluorous fluids have high density, low intermolecular interaction, low surface tension, low dielectric constant and high stability

Perflurous liquids for example, perfluoroethers, perfluoroalkanes, perfluroamines and so on, exhibit unique characteristics, which make them suitable alternative to most of the common organic solvents. The boiling point of these liquids depend on their molar mass and it is lower than the corresponding alkanes. The density of perfluorous alkanes are higher than water and other organic moleules. Oxygen, carbon dioxide and hydrogen like gases are highly soluble in perfluorocarbons. Thus, these perfluorinated hydrocarbons permit some selective and efficient oxidation reaction under mild conditions.

Melting point and boiling point of some perfluorinated solvents are –

**TABLE 7** Melting and boiling points of perfluorinated solvents

| Compound | **Formula** | **M. P. (°C)** | **B.P. (°C)** |
|---|---|---|---|
| Perfluorohexane | $C_6F_{14}$ | -87°C | 75°C |
| Perfluoroheptane | $C_7F_{16}$ | -78°C | 82°C |
| Perfluorodecalin | $C_{10}F_{18}$ | -10°C | 142°C |
| Perfluoromethylcyclohexane | $C_7F_{14}$ | -45°C | 72°C |
| Perfluorotributyl amine | $C_{12}F_{27}N$ | -50°C | 173°C |

Perfluorocarbons are non-ozone depleting compounds. Newly introduced fluoroiodocarbons, which are non-flammable, non-corrosive and non-ozone depleting, are also emerging as possible replacements for CFC's.

Fluorous phase technique has a different green approach. Although these are solvents but not simply solvent replacements. Because of their extremely non-polar characteristics, these are not suitable for organic reactions and are used in conjunction with organic solvent to form biphasic system (Clarke et al., 2004). In this technique, reagents or catalysts, which are soluble in fluorous fluids, remain in fluorous phase whereas starting materials or substrates are dissolved in organic solvents or water, which are immiscible with fluorous fluids. These two distinct layers become homogeneous. On heating, reactants and substrate come in contact with each other and thus, the reaction takes place. These layers are separated again, when temperature is lowered down. As products remain in organic layer while unused reactant and catalyst remain in fluorous phase, which leads to easy separation of products as well as recycling of the catalysts. Thus, the use of organic solvent for extraction can be avoided with this technique.

## *APPLICATIONS*

### *OXIDATION REACTIONS*

In the oxidation of various aliphatic and benzylic alcohols, fluorous biphasic system (FBS) has been used by Nishimura et al. (2000). By using FBS, the formation of epoxides have been reported on the oxidation of alcohols and alkenols by Maayan et al. (2003).

### *ASYMMETRIC ALLYLIC ALKYLATION*

In the asymmetric palladium catalyzed alkylation of the allylic substrate, the use of fluorous chiral bisoxazolines was reported by Bayardon and Sinou (2003). The high yield of product has been obtained.

## CHLORINATION AND BROMINATION OF ALCOHOL

Chlorination and bromination of alcohol was achieved by Nakamura et al. (2003) by using phase vanishing reactions. A perfluorohexane layer regulates the rate of reagent transport in these processes.

## STILLE COUPLING

It has been reported that palladium catalyzed cross-coupling reactions of organotin reagents with electrophiles (Hoshino et al., 1997; Still, 1986).

## MISCELLANEOUS REACTIONS

Friedel–Craft acylation, Mitsunobu reaction and palladium–catalysed C–C cross coupling reactions such as Suzuki reaction, Heck reaction and so on, have been carried out under fluorous conditions (Barrett et al., 2000; Betzemeier and Knochel, 1997; Dandapani and Curran, 2002; Mikami et al., 2001; Monineau et al., 1999). Thus, it can be concluded that fluorous fluids are found to be greener solvent in organic synthesis.

Besides these compounds, there are some more compounds, which can be considered as green solvents in future. Some of these are N, N-Dimethylpropyleneurea (DMPU), 1, 3-propanediol, 1, 3-dioxolan, and so on.

Water is a universal as well as a green solvent. Cyclopentyl methyl ether, 2-methyl tetrahydrofuran, polyethylene glycol, eth yl lactate, glycerol and so on, are among other green solvents. Search is still on for solvents, which are greener in nature and can replace traditional toxic solvents.

## KEYWORDS

- **Cyclopentyl methyl ether**
- **Ethyl lactate**
- **Organic solvents**
- **2-Methyl tetrahydrofuran**
- **Polyethylene glycol**

## REFERENCES


1. Albertsson, P. A. (1986). *Partion of cell particales and macromolecules,* (3rd edn.). New York: Wiley.
2. Auge, A., Lubin, M., & Lubineau, A. (1994). *Tetrahedron Letters, 35,* 7947–7948.
3. Aycock, D. F. (2007). *Organic process research & development, 11,* 156–159.
4. Barrett, A. G. M., Braddock, D. C., Catterick, D., Chadwick, D., Henschke, K. P., & Mckinnell, R. M. (2000). *SYNLETT: Fluorous biphase catalytic Friedel-Crafts acylation: Ytterbium tris(perfluoroalkanesulfonyl) methide catalysts, 6,* 847–849.
5. Bayardon, J., & Sinou, D. (2003). *Tetrahedron Letters, 44,* 1449–1451.
6. Bennett, J. S., Charles, K. L., Miner, M. R., Heuberger, C. F., Spina, E. J., Bartels, M. F., & Foreman, T. (2009). *Green Chemistry, 11,* 166–168.
7. Berson, J. A., Hamlet, Z., & Muller, W. A. (1962). *Journal of American Chemical Society, 84,* 297–304.
8. Betzemeier, B., & Knochel, P. (1997). Angerwandte Chemie International Edition, *36,* 2623–2624.
9. Beydoun, K., & Doucet, H. (2011). *ChemSuschem, 4,* 526–534.
10. Breslow, R., & Maitra, U. (1984). *Tetrahedron Letters, 25,* 1239–1240.
11. Breslow, R. Maitra, U., & Rideout, D. (1983). *Tetrahedron Letters, 24,* 1901–1904.
12. Buck, J. S. (1948). *Organic Reactions, 4,* 269–304.
13. Chandrasekhar, S., Narsihmulu, Ch., Sultana, S. S., & Reddy, N. R. (2002) *Organic Letters, 4,* 4399–4401.
14. Chandrasekhar, S., Narsihmulu, Ch., Sultana, S. S., & Reddy, N. R. (2003). *Chemical. Communication,* 1716–1717.
15. Chandrasekhar, S., Narsihmulu, Ch., Saritna, B., & Sultana, S. S. (2004). *Tetrahedron Letters, 45,* 5865–5867.
16. Chen, J., Spear, S. K., Huddleston, J. G., Holbrey, J. H., & Rogers, R. D. (2004a). *Industrial & Engineering Chemistry Research, 43,* 5358–5364.
17. Chen, J., Spear, S. K., Huddleston, J. G., Holbrey, J. H., & Rogers, R. D. (2004b). *Journal of Chromatography B: Biomedical sciences and applications, 807,* 145–149.
18. Cho, H., Iwama, Y., Mitsuhashi, N., Sugimoto, K., Okano, K., & Tokuyama, H. (2012). *Molecules, 17,* 7348–7355.
19. Claisen, L., & Claparede, A. (1981). *Berichte der deutschen chemischen Gesellschaft, 14*(2), 2460–2468.
20. Clarke, D., Ali, M. A., Clifford, A. A., Parratt, A., Rose, P., Schwinn, D., Bannwarth, W. C., & Rayner, M. (2004). *Current Topics in Medicinal Chemistry, 4,* 729–771.
21. Comanita, B. (2006a). *Industrie Pharma Magazine,* October.
22. Comanita, B. (2006b). *Specialty Chemical Magazine,* October.
23. Comanita, B., & Aycock, D. (2005). *Industrie Pharmaceutical Magazine, 17,* 54–56.
24. Dandapani, S., & Curran, D. P. (2002). *Tetrahedron,* 58, 3855–3864.
25. Dickerson, T. J., Reed, N. N., & Janda, K. D. (2002). *Chemical Reviews, 102,* 3325–3344.
26. Diels, O., & Alder, K. (1931). *Liebigs Annalen der Chemie, 490,* 243–257.
27. Dub, P. A., Zubiri, M. R., Baudequin, C., & Poli, R. (2010). *Green Chemistry, 12,* 1392–1396.

28. Fache, F., Lehueds, S., & Lemaine, M. (1995). *Tetrahedron Letters, 36,* 885–888.
29. Ferravoski, P., Fiecchi, A., Grisenti, P., Santaniello, E., & Trave, S. (1987). *Synthetic Communications, 17,* 1569–1575.
30. Fringuelli, F., Germani, R., Pizzo, F., & Savelli, G. (1989). *Gazzetta Chimica Italiana, 119,* 249–249.
31. Fringuelli, F., Pellegrino, R., Piermattic, O., & Pizzo, F. (1993). *Synthetic Communication, 23,* 3157–3163.
32. Gladysz, J. A., & Curran, D. P. (2002). *Tetrahedron, 58,* 3823–3825.
33. Grieco, P. A., Rrandes, E. B., Mecann, S., & Clark, J. D. (1989). *The Journal of Organic Chemistry, 54,* 5849–5851.
34. Grieco, P. P., Garner, A., & He, Z. (1983). *The Journal of Organic Chemistry, 25,* 1807.
35. Grieco, P., Parker, D. T., Cornwell, M., & Ruckle, R. (1987). *Journal American Chemistry Society, 109,* 5859–5861.
36. Gu, Y., & Jerome, F. (2010). *Green Chemistry, 12,* 1127–1138.
37. Guo, Z., Li, M. Willauer, H. D., Huddleston, J. G., April, G. C., & Rogers, R. D. (2002). *Industrial & Engineering Chemistry Research, 41,* 2535–2542.
38. Haimov, A., & Neumann, R. (2002). *Chemistry Communication,* 867–876.
39. Halimehjani, A. Z., Marjani, K., & Ashouri, A. (2010). *Green Chemistry, 12,* 1306–1310.
40. Hanessian, S., & Girand, C. (1994). One step a-deoxygenation of unprotected aldonolactones using samarium dioxide—THF/H2O System—A New Synthesis of 2-Deoxy-D-Ribose. *Synlett,* 861–862.
41. Harris, J. M. (Ed.). (1992). *Polyethylene glycol chemistry. Biotechnological and biomedium applications.* New York: Plenum Press.
42. Harris, J. M., & Zalipsky, S. (Eds.). (1997). *Polyethylene (Ethylene Glycol): Chemistry and biological applications.* Washington DC: ACS Books.
43. Hasegawa, C., & Curran, D. P. (1993). *The Journal of Organic Chemistry, 58,* 5008–5010.
44. Herold, D. A., Keil, K., & Bruns, D. E. (1989). *Biochemical Pharmacology, 38,* 73–76.
45. Horvath, I. T., &. Rabai. J. (1994). Facile catalyst separation without water: Fluorous. biphase hydroformylation of olefins. *Science,* 266, 72–75.
46. Hoshino, M., Degenkolb, P., & Curran, D. P. (1997). *The Journal of Organic Chemistry, 62,* 8341–8349.
47. Johnson, J. R. (1942) *Organic Reactions, 1,* 210–265.
48. Kawatsura, M., Kamesaki, K., Yamamoto, M., Hayase, S., & Itoh, T. (2010). *Chemistry Letters, 39,* 1050–1051.
49. Keh, C. C. K., Wei, C., & Li, C. J. (2003). *Journal of American Chemical Society, 125,* 4062–4063.
50. Knoevenagel, F. (1898). *Berichte der deutschen chemischen Gesellschaft, 31,* 2596–2619.
51. Koal, E. T., & Breslow, R. (1988). *Journal of American Chemical Society, 110,* 1596–1597.
52. Kobayashi, S., Kuroda, H., Ohtsuka, Y., Kashihara, T., Masuyama, A., & Watanabe, K. (2013). *Tetrahedron, 69,* 2251–2259.

53. Kondo, A., Urabe, T., & Higashitani, K. J. (1994). *Journal of Fermentation and Bioengineering, 77,* 700–703.
54. Kumar, D., Reddy, V. B., Mishra, B. G., Rana, R. K., Nadagouda, M. N., & Varma, K. S. (2007). *Tetrahedron, 63,* 3093–3097.
55. Lapworth, A. J. (1903). *Journal of Chemical Society, 83,* 995–1005.
56. Lapworth, A. J. (1904). *Journal of Chemical Society, 85,* 1206–1214.
57. Larpent, C., & Meignan, G. (1993). *Tetrahedron Letters, 34,* 4331–4334.
58. Larsen, S. D., & Grieco, P. A. (1985). *Journal of American Chemical Society, 107,* 1968–1769.
59. Leininger, N. F., Clontz, R., Gainer, J. L., & Kirwan, D. V. (2002). In M. A. Abraham, L. Moens (Eds.). *Clean solvents, alternative media for chemical reactions and processing (ACS Sympossium Series)* (pp. 208–223, 819). Washington DC: American Chemical Society.
60. Lindstrom, U. M. (2002). *Chemical Reviews, 102,* 2750–2772.
61. Lubineau, A. (1986). *Journal of Organic Chemistry, 51,* 2142–2144.
62. Lubineau, A., & Meyer, E. (1988). *Tetrahedron, 44,* 6065–6070.
63. Maayan, G., Fish, R. H., & Neuman, R. (2003). *Organic Letters, 5,* 3547–3550.
64. Mandenius, C. F., Nilsson, B., Persson, I., & Tjerneld, F. (1988). *Biotechnology and Bioengineering, 31,* 203–207.
65. Mckillop, A., & Tarbin, J. A. (1987). *Tetrahedron, 43,* 1753–1758.
66. Mikami, K., Mikami, Y., Matsumota, Y., Nishikido, J., Yamamoto, F., & Nakajima, H. (2001). *Tetrahedran Letters, 42,* 289–292.
67. Milton, E. J., & Clarke, M. L. (2010). *Green Chemistry, 12,* 381–383.
68. Monineau, J., Pozzi, G., Quici, S., & Sinou, D. (1999). *Tetrahedron Letters, 40,* 7683–7686.
69. Mordolol, M. (1993). *Tetrahedron Letters, 34,* 1681–1684.
70. Narayan, S., Muldoon, J., Finn, M. G., Fokin, V. V., Kolb, H. C., & Sharpless, K. B. (2005). *Angewandte Chemie International Edition, 44,* 3275–3279.
71. Naik, S. D., & Doraiswamy, L. K. (1998). *AIChE Journal, 44,* 612.
72. Nakamura, H., Usui, T., Kudora, H., Ryu, I., Matsubara, H., Yasuda, S., & Curran, D. P. (2003). *Organic Letters, 5,* 1167–1169.
73. Namboodiri, V. V., & Varma, R. S. (2001). *Green Chemistry, 3,* 146–148.
74. Nishimura, T., Maeda, Y., Kakiuchi, N., & Uemura, S. (2000). *Journal of the Chemical Society, Perkin Transactions 1,* 4301–4305.
75. Ooi, T., Takeuchi, M., Kato, D., Uematsu, Y, Tayama, E., Sakai, D., & Maruoka, K. (2005). *Journal of the American Chemical Society,* 127, 5073–5083.
76. Pace, V., Castoldi, L., Alcántara, A. R., & Holzer, W. (2012). *Green Chemistry, 14,* 1859–1863.
77. Peglation, M. G. (2002). *Cancer Treatment Reviews, 28,* 13–16.
78. Peng,Y., Liu, J., Lei, X., & Yin, Z. (2010). *Green Chemistry, 12,* 1072–1075.
79. Pereira, C. S. M., Silva, V. M. T. M., & Rodrigues, A. E. (2011). *Green Chemistry, 13,* 2658–2671.
80. Persson, I., Stalbrand, H., Tjerneld, F., & Hahn-Hagerdal, B. (1992). *Applied Biochemistry and Biotechnology, 27,* 27–36.
81. Petrier, C., & Luche, J. L. (1987). *Tetrahedron Letters, 28,* 1234.

82. Pirrung, M. C., & Sarma, K. D. (2004). *Journal of American Chemical Society, 126,* 444–445.
83. Ponaras, A. A. (1983). *Journal of Organic Chemistry, 48,* 3866–3868.
84. Ramamurthy, R. (1986). *Tetrahedron Letters, 42,* 5753–5839.
85. Rideout, D., & Breslew, R. (1980). *Journal of. American Chemical Society, 102,* 7816–7817.
86. Robke, G. D., & Behrman, E. (1993). *Journal of Chemical Research, Synopses,* 412.
87. Samil, A. A. Z., Desavignac, A., Rico, I., & Latters, A. (1985). *Tetrahedron, 41,* 3683–3688.
88. Santaniello, E., Fiecchi, A., Manzocchi, A., & Ferraboschi, P. (1983). *Journal of Organic Chemistry, 48,* 3074–3077.
89. Santaniello, E., Manzochi, A., & Sozzani, P. (1979). *Tetrahedron Letters, 20,* 4581–4584.
90. Santanielto, E. (1984). In L. J. Mathias, C. E. Jr. Carrater (Eds.). *Crown ethers and phase transfer catalysis in polymer science* (p. 397). New York: Plenum.
91. Schmidt, J. G. (1881). *Berichte der deutschen chemischen Gesellschaft, 14,* 1459.
92. Shanmuganathan, S., Natalia, D., Wittenboer, A. V. D., Kohlmann,C. Greiner, L., & De María, P. D. (2010). *Green Chemistry, 12,* 2240–2245.
93. Simeó,Y., Sinisterra, J. V., & Alcántara, A. R. (2009). *Green Chemistry, 11,* 855–862.
94. Still, J. K. (1986). *Angewandte Chemie International Edition, 25,* 508–524.
95. Sugie, A., Somete, T., Kanie, K., Muramatsu, A., & Mori, A. (2008). *Chemical Communications,* 3882–3884.
96. Tan, K. T., Chng, S. S., Cheng, H. S., & Loh, T. P. (2003). *Journal of American Chemical Society, 125,* 2958–2963.
97. Tjerneld, T., Persson, I., Albertsson, P. A., & Hagerdal, B. H. (1985). *Biotechnology and Bioengineering, 27,* 1036–1043.
98. Wan, J., Wang, C. Zhou, R., & Liu, Y. (2012). *RSC Advances, 2,* 8789–8792.
99. Watanabe, K., Kogoshi, N., Miki, H., & Torisawa, Y. (2009). *Synthetic Communications, 39,* 2008–2013.
100. Watanabe, K., Yamagiwa, N., & Torisawa, Y. (**2007**). *Organic Process Research and Development, 11,* 251–258
101. Webb, K. S., & Seneviratrie, V. (1995). *Tetrahedron Letters, 36,* 2377–2378.
102. White, W. N., & Wolfarth, E. F. (1970). *Journal of Organic Chemistry, 35,* 2196–2199.
103. Yorimitsu, H., Nakamurat, T., Shinokubo, H., Oshime, K., Omoto, K., & Fujimoto, H. (2000). *Journal of American Chemical Society, 122,* 11041–11047.
104. Zhang, X., Jia, X., Wang, J., & Fan, X. (2011). *Green Chemistry, 13,* 413–418.

# CHAPTER 8

# PHOTOCATALYSIS: AN EMERGING TECHNOLOGY

INDU BHATI, PARAS TAK, H. S. SHARMA, and RAKSHIT AMETA

## CONTENTS

## 8.1 INTRODUCTION

Although 70% of the earth area is covered with water, inspite of that it is well known that only 1% of this water can be used for human consumption and other purposes. Men started industrialization of a number of products for the development. As a result, there are a huge number of small and big industries all over the world. A number of heavy metals, inorganic compounds, organic compounds like phenols, pesticides, fertilizers, surfactants, drugs, detergents, insecticides, dyes, and other chemical products are disposed directly into the water resources by these industries, without any effective treatment strategy. All these substances are toxic and harmful to human beings, animals, and plants. Many of these compounds can be readily absorbed through the skin and create health problems.

Innovative steps should be taken for purification and recycling of wastewater of various industries, as water has now become a key symbol of protest all around the world. Color removal from the textile wastewaters has become an issue of interest during the last few years because of the toxicity of the dyes and more often, the colored wastewater from the textile industries also decreases the visibility of the receiving water affecting the animal kingdom adversely.

One technique gaining popularity in recent years and is quite promising also for the treatment of resilient pollutants, is the photocatalysis.

## 8.2 BASICS AND APPLICATIONS

### *8.2.1 PHOTOCATALYSIS*

Photocatalysis includes such reactions, which utilize light to activate a substance (particularly a semiconductor), which modifies the rate of a chemical reaction without being involved itself. The definition of 'Photocatalysis' accepted by IUPAC after long debate is a catalytic reaction involving light absorption by a substrate (1988). Similarly, the substrate, which is a semiconductor, absorbs light, and acts as a catalyst for that chemical reaction, is known as photocatalyst. Extensive reviews of this field have been made from time to time by various researchers (Ameta et al., 1992, 2010; Gratzel, 1983; Kamat, 1993; Pelizzetti et al., 1986).

## *PHOTOCATALYTIC REACTIONS*

The photocatalytic reactions can be classified into two categories on the basis of physical state/appearance of reactants.

- Homogeneous photocatalysis
  (a) When the catalyst and reactant; both are in the same phases, that is, gas, solid or liquid, then the photocatalytic reaction is called homogeneous photocatalysis. Different dyes/organic substances and colored coordination compounds are best example of homogeneous photocatalysts.
- Heterogeneous photocatalysis
  (a) When the catalyst and reactant both are in different phases, then the photocatalytic reaction is called heterogeneous photocatalysis. The common example of this kind is a solid photocatalyst in contact with either liquid or a gas phase.

## *PHOTOCATALYSTS*

All the photocatalysts are normally semiconductors, but all semiconductors are not necessarily photocatalysts. Semiconductor is a substance, where the energy gap between conduction band (lowest unoccupied molecular orbital, LUMO) and valence band (highest occupied molecular orbital, HOMO), ranges from 1.5 to 3.0eV.

## *BAND GAP*

The energy difference between the valence band and the conduction band is known as the band gap ($E_g$). On the basis of this band gap, the materials are classified in three categories (i) $E_g < 1.0$eV, metal or conductor, (ii) $E_g > 5.0$eV, insulator or non conductor and (iii) $E_g \sim 1.5-3.0$eV, semiconductor. This classification is shown in Figure 8.1.

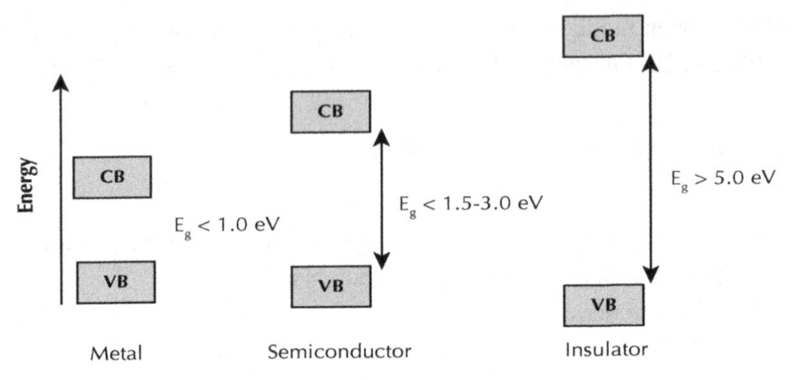


**FIGURE 8.1** Band gap and nature of materials.

The band gaps of various semiconductors are given in Table 8.1.

**TABLE 8.1** Band gaps of different semiconductors.

| Semiconductor | Band gap (eV) at 300K |
|---|---|
| ZnS (Wurtzite) | 3.91 |
| ZnS (Zinc blende) | 3.54 |
| $TiO_2$ | 3.20 |
| ZnO | 3.03 |
| $WO_3$ | 2.60 |
| CdS | 2.42 |
| CdSe | 1.70 |

When the sufficient energy (equal to or more than the band gap) is provided to a semiconductor, $e^-$ from the valence band absorbs this energy and jumps to the conduction band leaving behind a hole in valence band. Therefore, a semiconductor, which is an insulator at normal temperature and pressure, may conduct some electric current on exposure to light containing energy equal to or more than the corresponding band gap of semiconductor. Hence, semiconductors are capable of conducting electricity even at room temperature in presence of light and that is why, they work as photocatalyst.

### 8.2.2 MECHANISM OF PHOTOCATALYSIS

When the photocatalyst is illuminated by light, the energy of photons is utilized by the $e^-$ of valence band promoting it to the conduction band, while a hole ($h^+$) is created in valence band by removal of this electron. This process creates 'photo-excitation state'. Thus, a photocatalyst absorbs appropriate radiations (Vis/UV) from sunlight or illuminated light source (fluorescent lamps), producing a pair of electron ($e^-$) and hole ($h^+$). This excited $e^-$ can be used for reducing an acceptor substrate whereas the $h^+$ may be utilized for oxidation of donor molecules (Figure. 8.2).

In photogenerated catalysis, the photocatalytic activity depends on the ability of the catalyst to create $e^- – h^+$ pairs. But in absence of any other influence or process, the excited state may diffuse down the concentration gradient of $e^-–h^+$ pair, as there is an equal possibility for $e^-$ to lose the energy and get back its position in valence band, where hole was present.

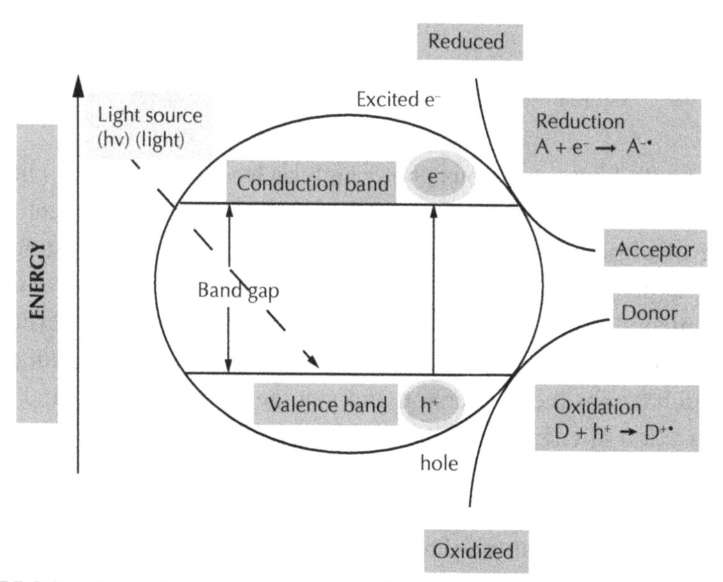


**FIGURE 8.2** Mechanism of photocatalysis ($e^-–h^+$ pair generation).

## *FATE OF EXCITED ELECTRON–HOLE PAIR*

The fate of excited electron and hole is decided by the relative positions of the conduction and valence bands of semiconductor and the redox levels of the substrate (Figure 8.3). The importance of photocatalysis lies in the fact that by choosing a photocatalyst (semiconductor) of the desired band gap, one can drive a reaction in a desired direction, as photocatalyst provides an oxidation as well as reduction environment, simultaneously.

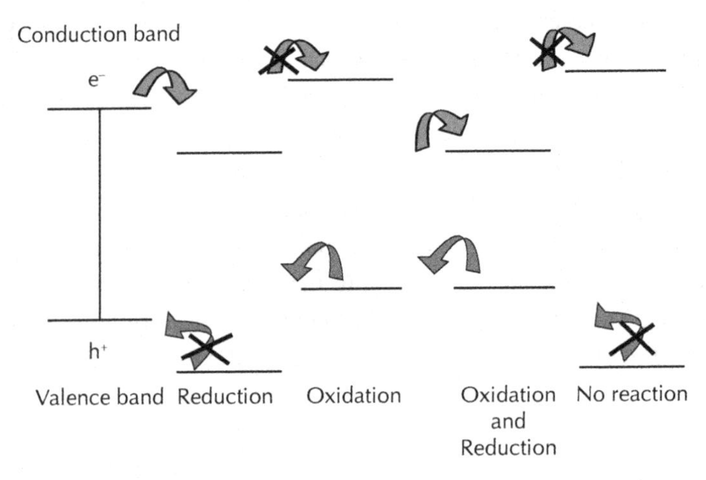


**FIGURE 8.3** Various possibilities of reactions.

The field of heterogeneous photocatalysis has already grown out of its infancy and now emerges as a major field of research. The recent development of ultrafine semiconductor particles with many interesting photocatalytic properties has added some newer dimensions. The ability of such semiconductors to carry out redox processes with greater efficiency and selectivity than in the homogeneous solutions has made them potential candidates for the conversion and storage of solar energy and in the mineralization of chemical pollutants. Photocatalysis has a wide scope of its uses in variety of applications, which includes waste water treatment, conservation and storage of energy, self cleaning, deodorization, sterilization, air purification, antifouling and antifogging, synthesis in nitration, sulphonation, halogenations, and so on.

### 8.2.3 WASTEWATER TREATMENT PROCESSES

A typical system for surface water treatment generally consists of pre-settling, coagulation/flocculation (Montgomery, 1985) (sediment removal), granular filtration (Faust and Aly, 1999) (sediment removal), corrosion control (USEPA, 1989) (pH adjustment or addition of corrosion inhibitors), and disinfection. These methods suffer from the one or other drawbacks. However, photocatalysis provides more cleaner and environment healthy/benign methods for waste water treatment (Ameta et al., 2003). Mainly, the responsible chemicals for water contamination are dyes, pesticides, surfactants, phenols, halo-organics, herbicides, organic compounds, pharmaceuticals, and so on.

#### DYES

Many of the dyeing, textile, and printing industries discharge their colored effluents in the nearby water resources without proper treatment, which pollute the water. This colored water cannot be used for any useful purpose due to its hazardous nature. Photocatalysis provides an eco-friendly method for removal of these toxic dye molecules from effluents. Various semiconductors have been used by researchers for this purpose. Tanaka et al. (2000) photocatalytically degraded some commercially used azo dyes in $TiO_2$ suspension and proposed that the disappearance of azo dyes proceeds through both oxidation and reduction processes, while $TiO_2$/UV photocatalytic degradation of methylene blue has been investigated in aqueous heterogeneous suspensions (Tatsuma et al., 1999; Houas et al., 2001).

Arabatzis et al. (2003) synthesized and characterized Au/$TiO_2$ thin films, which were used for degradation of azo dye, while Wei et al. (2007) used V doped $TiO_2$ for photocatalytic degradation of methyl orange. Cerium doped $TiO_2$ catalyst was prepared by controlled hydrolysis of titanium alkoxide based on esterification reaction by Tang et al. (2007).

Ameta et al. (2000) studied the photocatalytic degradation of orange-G over ZnO in presence of a surfactant whereas Pare et al. (2009) studied the photocatalytic degradation of lissamine fast yellow using ZnO in the aqueous suspension. ZnO was also used for photodegradation of methylene blue and eosin by Chakrabarti and Dutta (2004). Photocatalytic bleaching of rhodamine-B and rhodamine-6G over zinc oxide and lead oxide was studied by Mansoori et al. (2004) whereas photobleaching of amaranth dye in presence of ZnO photocatalyst was reported by Kothari et al. (2004). Ameta et al. (2002, 2006) have done an extensive work on photocatalytic bleaching of some dyes. Kothari (2007) observed the photocatalytic bleaching of Evans blue over zinc oxide and Vyas et al. (2005) investigated the photocatalytic bleaching of eosin using zinc oxide and effect of surface charge. Punjabi et al. (2005) reported the photoreduction of congo red by ascorbic acid and EDTA as reductant and cadmium sulphide as photocatalyst.

Photodegradation of rhodamine-B (RB) catalyzed by $TiO_2$ films was demonstrated by Ma and Yao. The mechanism of this reaction was proposed as,

$$RB \xrightarrow{h\acute{\imath}} RB^*$$

$$RB^* + TiO_2 \longrightarrow RB^{+\bullet} + TiO_2\,(e^-)$$

$$RB^* + O_2 \longrightarrow RB^+ + O_2^{-\bullet}$$

$$TiO_2\,(e^-) + O_2 \longrightarrow TiO_2 + O_2^{-\bullet}$$

$$RB^{+\bullet} \xrightarrow{O_2} \text{Products}$$

Doping of semiconductor with metals or non-metals enhances its activity. Vaya et al. (2008) reported the effect of transition metal ions doping on the photocatalytic activity of ZnS, whereas Li et al. (2008a) prepared ZnS polymer composites which exhibited high efficiency for degrading methyl orange, methylene blue, and eosin B. Puretedal et al. (2009) used nanoparticle zinc sulphide doped with manganese, nickel, and copper as a nanophotocatalyst for the degradation of organic dyes; methylene blue and safranin. The photocatalysts like $TiO_2$, ZnO, CdS, and so on can oxidize organic pollutants like dyes and other harmful wastes into non toxic and less harmful materials (Behnajady, 2006; Bilgi and Demir, 2005; Papic et al., 2006; Poulis et al., 2003).

A new method for doping metal ions on to the $TiO_2$ surface has been developed by Xu et al. (2004). It was observed that the surface doped $TiO_2$ exhibited higher photocatalytic activity than pure $TiO_2$ for the degradation of methyl orange in water. $Fe^{3+}$ doped $TiO_2$ catalyst was prepared by Tong et al. (2008) and used for degradation of methyl orange. Preparation and characterization and use of nanosilver doped mesoporous titania photocatalyst for dye degradation was investigated by Binitha et al. (2009).

Sol–gel synthesis of $TiO_2$ nanoparticles and photocatalytic degradation of methyl orange in aqueous $TiO_2$ suspensions was investigated by Yang et al. (2006). Tian et al. (2008) synthesized Au/$TiO_2$ catalyst from Au (I)–thiosulphate complex and observed its photocatalytic activity for the degradation of methyl orange, while Wang et al. (2009) characterized and reported the photocatalytic activity of poly (3–hexylthiophene) modified $TiO_2$ for degradation of methyl orange under visible light.

Li et al. (2008b) $TiO_2$ prepared immobilized nanoparticles of $TiO_2$ supported on natural porous mineral and used it for photocatalytic degradation of azo dyes. On the other hand, Magalhaes and Lago (2009) grafted $TiO_2$ on expanded polystyrene beads, which was used for the solar degradation of dyes. UV/titanium dioxide degradation of two xanthene dyes, erythrosine B and eosin Y, was studied by Pereira et al. (2013) in a photocatalytic reactor, while Mahadwad et al. (2011) carried out photocatalytic degradation of reactive black-5 (RB-5) dye using supported $TiO_2$ photocatalyst based adsorbent as a semiconductor photocatalyst in a batch reactor. Santhanalakshmi (2012) investigated kinetic studies for degradation of reactive yellow, acid blue, methyl orange, and acid green with $TiO_2$ in $H_2O_2$ aqueous solution.

Visible light-induced degradation of methylene blue in the presence of photocatalytic ZnS and CdS nanoparticles has been carried out by Soltani et al. (2012). Saravanan et al. (2011) studied photocatalytic degradation of methylene blue using nano ZnO. Ameta et al. (2011) used $Sb_2S_3$ semiconductor for photocatalytic degradation of naphthol green B. They proposed that the dye molecules were oxidized by hydroxyl radicals, which were generated due to the reaction between hydroxyl anion and hole of the semiconductor. The role of hydroxyl radicals as an active oxidizing species was confirmed by carrying out the reaction in presence of a hydroxyl radicals scavenger that is, 2-propanol, where the rate was drastically reduced. They have given the following mechanism for the degradation of the naphthol green B.

$$SC \xrightarrow{h\nu} h^+(VB) + e^-(CB)$$

$$h^+ + OH^- \rightarrow {}^{\bullet}OH$$

$${}^{\bullet}OH + Dye \rightarrow Products$$

Shanthi and Kuzhalosai (2012) studied photocatalytic degradation of acid red 27 azo dye in aqueous solution using nanosized ZnO, while Madhusudan et al. (2011) studied photocatalytic degradation of coralene dark red 2B azo dye using calcium zincate nanoparticle in presence of natural sunlight. Yang and Luan (2012) synthesized and characterized a composite polymer polyaniline/$Bi_2SnTiO_7$ and studied its photocatalytic activity using methylene blue system, while Kaur et al. (2011) synthesized, characterized, and studied the photocatalytic activity of $La_2CoO_4$ using azure B as the model system. Municipal waste water and brilliant blue dye was photocatalytically degraded using hydrothermally synthesized surface-modified silver-doped ZnO by Parvin et al. (2012). $Sm_2FeTaO_7$ photocatalyst was used for degradation of indigo carmine dye under solar light irradiation by Torres-Martínez et al. (2012). Ameta et al. (2010) used nanosized chromium doped $TiO_2$ supported on zeolite for methylene blue degradation, while Sacco et al. (2012) carried out photocatalytic degradation of methylene blue under visible light on N-doped $TiO_2$ photocatalysts.

### *PESTICIDES*

Pesticides and herbicides have their own importance to save crops from insects, but due to their bioaccumulation nature; these create problems.

These pollutants are often toxic and cause adverse effects on human and animal life, when present even in low concentrations. Organo chemicals such as dicofol, BHC, cypermethrin, and organophosphorous pesticides are commonly used pesticides and insecticides. Among all, photocatalytic oxidation is one of the best suitable technologies used for the elimination of these organic pollutants (Doong and Chang, 1997; Kerzhentsev et al., 1996). Photocatalytic oxidation processes have been successfully utilized for the removal/degradation of pesticides and herbicides from water.

Photocatalytic degradation of pesticide pirimiphos-methyl has been carried out by Hermann et al. (1999). Methoxychlor and p, p'-DDT have successfully been degraded by $TiO_2$ in aqueous suspension (Zaleska et al., 2000). A $TiO_2$ film developed by Choi et al. (2000) has been used to degrade polychlorinated dibenzo-p-dioxine under UV irradiation, while Muszkat et al. (2002) studied the photocatalytic degradation of pesticides and biomolecules in water. $TiO_2$ nanoparticles have been used for transformation of chlorinated volatile organic compounds by Liu et al. (2002).

Tamini et al. (2006) studied the degradation of pesticide methomyl in aqueous solution by ultra-violet irradiation in the presence of $TiO_2$, whereas Konstantinou and Albanis (2003) used photocatalytic process for the transformation of pesticides in aqueous titania suspension using artificial and solar light. Yu et al. (2007) carried out photocatalytic degradation of organochlorine pesticides on a nano-$TiO_2$ coated film, while Senthilnathan and Philip (2010a) degraded mixed pesticides. $Re^{3+}$ doped nano-$TiO_2$ was used by Zhang et al. (2010) for photocatalytic degradation of a carbofuran, whereas nitrogen doped titania with different nitrogen containing organic compounds was used under UV and visible light for degradation of lindane (Senthilnathan and Philip, 2010b).

Machuca-Martínez and Colina-Márquez (2011) studied the effect of pH and the catalyst concentration on $TiO_2$-based photocatalytic degradation of three commercial pesticides that is, 2,4-D, diuron, ametryne, while Verma et al. (2013) studied titanium dioxide mediated photocatalytic degradation of malathion in aqueous suspension. Zinc oxide has also been used for photocatalytic oxidation of organophosphorous pesticides by Fadaei and Dehghani (2012). Miguel et al. (2012) studied photocatalytic degradation of 44 organic pesticides with titanium dioxide in natural water.

## *SURFACTANTS*

Surfactants are extensively used in many fields of technology, research, pharmacy, cosmetics, textile, industry, agriculture, biotechnology and in daily life due to their favorable physicochemical properties. A large quantity of surfactant is released to the aquatic environment after its use. It causes serious threat to aquatic environment because of their high foaming and lower oxygenation potential, which leads to death of water borne organism. Surfactants cause short term as well as long term changes in the ecosystem. Many environmental and public health regulatory authorities have fixed limit of 0.5–1.0mg/L. Surfactants are classified into ionic and non ionic types according to their chemical nature.

Singhal et al. (1997) carried out the degradation of cetylpyridinium chloride, while Rao and Dubey (1996) studied photocatalytic degradation of binary and ternary mixture of three surfactants, viz., dodocylbenzene sulphonic acid, sodium salt (DBS), cetylpyridinium chloride (CPC), and triton X-100 (TX-100) using $TiO_2$ as a photocatalyst. Hidaka et al. (1990) degraded various kinds of cationic (*e.g.*, $C_{16}$-HTAB and $C_{12}$-BDDAC), anionic (*e.g.*, $C_{12}$-DBS, $C_{12}$-DS, $C_{12}$-LES-3, $C_{14}$-AOS, and $C_{12}$-DG) and non ionic (*e.g.*, NPE*n* (n = 7, 9, 17, 50), $C_{18}$-PEA-15, $C_{12–14}$-BHA, $C_{12–14}$-PAE-10, $C_{12–14}$-NOE, and $C_{14–16}$-NOE) surfactants by $TiO_2$ semiconductor particles under UV illumination. These surfactants are converted completely into $CO_2$.

Solar photodegradation of two commercial surfacatants, SDS and DBS has been studied by Amat et al. (2004) whereas Kimura et al. (2004) investigated photocatalytic degradation of non ionic surfactant polyoxyethylene alkyl ether in water using immobilized $TiO_2$ photocatalyst. The photocatalytic degradation of two industrial grade surfactants, sodium lauryl sulphate (SLS) and sodium dodecylbenzenesulphonate was achieved using $TiO_2$ immobilized on glass (Lizama et al., 2005) and photocatalytic system was also used for the degradation of a cationic surfactant that is, $C_{19}H_{49}ClN$ (Hassan et al., 2008). $TiO_2$–$CO_2$ composite oxide was used for photocatalytic degradation of dodecylbenzenesulfonate (DBS) under visible irradiation (Han et al., 2009). Bardos et al. (2011) have done titanium dioxide-mediated photocatalyzed degradation of benzenesulfonate.

Zhang et al. (2011) carried out photocatalytic degradation of phenanthrene (PHE) over $TiO_2$ in aqueous solution containing non ionic surfactant micelles, while Giahi et al. (2012) studied photocatalytic degradation of anionic surfactant using zinc oxide nanoparticles. Various industries do not give proper treatment to their effluents before throwing it in environment, which cause environment pollution. Different organic compounds present in industrial effluents create water pollution, even at ppm level. Organic pollutants have a wide variety which includes phenols; alcohols; halo, nitro and carbonyl compounds; hydrocarbons, and so on. Photocatalysis is a promising technology, which degrade various organic pollutants completely into harmless products.

## *HYDROCARBONS*

Platinum loaded $TiO_2$ has been used for photocatalytic hydroxylation of aromatic ring by using water as an oxidant (Park and Choi, 2005). Lair et al. (2007) studied photocatalytic degradation of naphthalene over $TiO_2$ in the presence of inorganic anions. Tang et al. (2008) carried out the synthesis of different sized cuprous oxide nano-crystallites and studied their photocatalytic activity, while Fuerte et al. (2002) synthesized nanosized Ti-W oxide and observed the effect of doping level in the photocatalytic degradation of toluene under sunlight excitation.

Fujihara et al. (1984) studied heterogeneous photocatalytic oxidation of aromatic compounds on $TiO_2$. They reported that there is photochemical production of $H_2O_2$ also at semiconductor during this photooxidation.

$$2\,TiO_2 \xrightarrow{h\nu} 2\,e^- + 2\,h^+ \xrightarrow{H_2O} \tfrac{1}{2}\,O_2 + 2\,H^+ \longrightarrow O_2 + 2\,H^+$$

$$H_2O_2 \xrightarrow{Fe^{2+}} {}^{\bullet}OH^- + Fe^{3+}$$

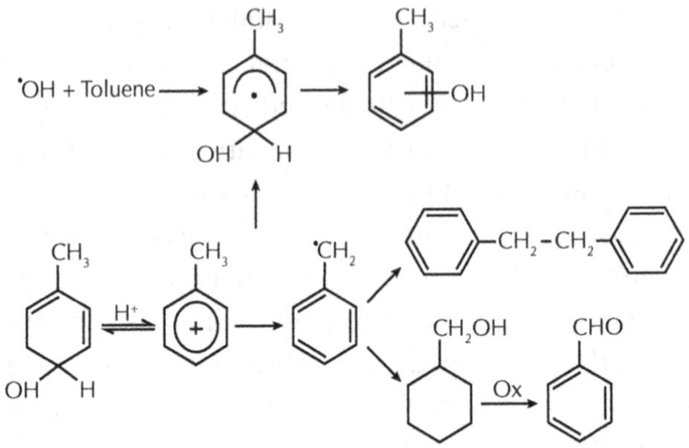

Marci et al. (2003) reported the photocatalytic oxidation of toluene on irradiated $TiO_2$, while Zhang et al. (2003) carried out a comparative study on decomposition of gaseous toluene by $O_3$/UV, $TiO_2$ /UV, and $O_3$/ $TiO_2$/UV. Photocatalytic degradation of 1,4–dioxane in $TiO_2$ suspension has been observed by Yamazaki et al. (2007), and Garcia et al. (2005) investigated oxidation of light alkanes over titania supported palladium/ vanadium catalyst.

## *ALCOHOLS*

V-doped $TiO_2$ photocatalyst was prepared by Klosek and Raftery (2001), which was used for photoxidation of ethanol in visible light. It is believed that under visible irradiation, the vanadium center donates an electron to the $TiO_2$ conduction band, which allows the oxidation of surface absorbed molecules. Mohamed et al. (2002) studied the photocatalytic oxidation of some selected aryl alcohols in acetonitrile whereas heterogeneous photocatalytic dehydrogenation of ethanol over $TiO_2$ has been reported by Kawai and Sakata (1980). Under flourescent visible light lamp Kirchnerova et al. (2005) performed oxidation of n-butanol over commercial $TiO_2$.

## *PHENOLS*

Phenolic compounds are widely used in industrial and daily life, but due to their stability and carcinogenic character, they threat the human health and water ecosystem. Hatipoglu et al. (2004) studied the photocatalytic degradation of m-cresol while Kang et al. (2000) reported the photocatalytic degradation of 4-chlorophenol in water over $TiO_2$ and $TiO_2$/CdS powder. San et al. (2001) investigated photodegradation of 3-aminophenol and Chen et al. (2002) examined the photooxidation of phenol and benzene using $TiO_2$. The influence of co-doping of Zn (II) + Fe (III) on the photocatalytic activity of $TiO_2$ for phenol degradation was studied by Yuan et al. (2002), whereas Sakthivel and Kisch (2003) reported that C/$TiO_2$ photocatalyst was active under artificial solar light and could efficiently decompose tetrachlorophenol.

Semiconductor promoted photooxidation of phenol was carried out by Okamoto et al. (1985). They proposed a mechanism for the reaction, which involved stepwise hydroxylation, via di-, tri, and tetra-hydroxybenzenes (mixtures of the various isomers), leading to formic acid, which was finally oxidized to carbon dioxide.

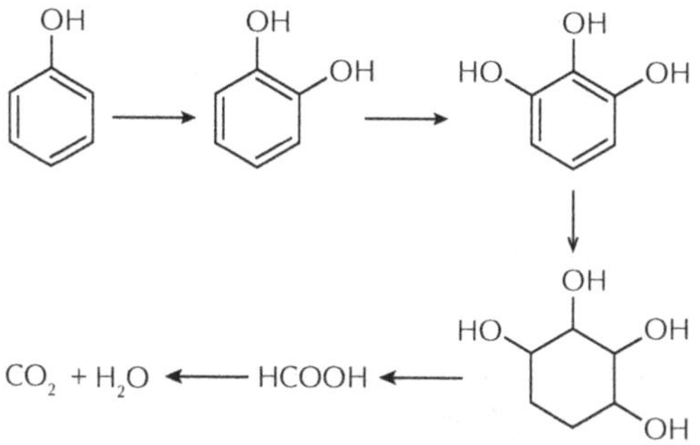

$$CO_2 + H_2O \leftarrow HCOOH \leftarrow$$

McManamon et al. (2011) improved photocatalytic degradation rates of phenol using novel porous $ZrO_2$-doped $TiO_2$ nano-particulate powders, while photocatalytic degradation of phenol in natural seawater using visible light active carbon modified (CM)-n-$TiO_2$) nanoparticles under UV light and natural sunlight illuminations was carried out by Shaban et al. (2012).

## *HALO COMPOUNDS*

The halo-organic compounds are stable compounds; thus, they show low chemical reactivity These properties make them long lasting and non biodegradable. The halo-organic compounds can be used as solvents, agro-chemicals, dyes, drugs and so on. These compounds represent a major class of environmental pollutants as these compounds passed to the aquatic plants and animals in different ways, which are consumed as food by human beings and other animals becoming a part of food chain.

Various methods like extraction, incineration, chemical degradation, electrochemical treatment, sonochemical destruction, and photochemical process are in use for the treatment of halo-organics pollutants. Among all these methods, photochemical processes have been found useful for degradation of halo-compounds into less toxic compounds in water effluents (Cesareo et al., 1986; Ollis, 1982; Ollis et al., 1991). The photocatalytic degradation of $CHCl_3$, $CHBr_3$, $CCl_4$ and $CCl_3CO_2^-$ has been investigated by Choi and Hoffmann (1997) in aqueous $TiO_2$ suspension.

Krysova et al. (1998) used $TiO_2$ for the photocatalytic degradation of diuron, whereas Jirkovsky et al. (1997) studied a direct photolysis of diuron under various conditions at 254nm. Bhatkhande et al. (2004) carried out photocatalytic degradation of chlorobenzene, while Pandiyan et al. (2002) used photochemical methods for the destruction and dehalogenation of chlorophenols.

## CARBONYL DERIVATIVES

Yanagida et al. (1989) reported the photoreduction of aldehyde and related derivatives by ZnS. Identification of intermediates and reaction mechanism of the photo-mineralization of polycarboxylic benzoic acid (trimellitic acid) in UV-irradiation aqueous suspensions of titania have been studied by Assabbane et al. (2000). The photo-assisted decomposition of salicylic acid on $TiO_2$ and Pd/$TiO_2$ films was reported by Sukhaser et al. (1995). A comparative study of the degradation of benzamide and acetic acid on $TiO_2$ was also made by Heintz et al. (2000).

## NITROGEN CONTAINING COMPOUNDS

Nitrogen containing compounds comprise different kinds of compounds such as amino acids, proteins, nitro, drugs, herbicides, pesticides, dyes, and many more. Some of them cause serious environmental threat because of their stability and toxicity. There is a pressing demand for such technologies for removal of such pollutants to get clean and pure water, which should be economic and environmentally benign.

The oxidation of dissolved nitrogen in natural water has been observed by Takeda and Fujiwara (1996) using titanium dioxide and platinized titanium dioxide. Ciping et al. (1993) investigated the free radical intermediates generated in photocatalytic oxidation of some organic compounds containing nitrogen atom, whereas photocatalytic degradation of a series of primary, secondary, tertiary amines, and other N containing organic compounds over UV-illuminated film of $TiO_2$ has been studied by Low et al. (1991). The photocatalytic mineralization of nitrobenzene, nitrosobenzene, phenylhydroxylamine, aniline and 4-nitrosophenol has been investigated by Piccinini et al. (1997) in the presence of $TiO_2$, while Tayade and Key (2010) carried out photocatalytic degradation of nitrobenzene with $TiO_2$ nanotubes.

Alberici et al. (2001) used $TiO_2$/UV–Vis photocatalytic process in the destruction of nitrogen-containing organic compounds. Pyridine ($C_5H_5N$), propylamine ($C_3H_7NH_2$), and diethylamine ($C_4H_{10}NH$) were used by them as the substrates for the photodegradation in the presence and absence of oxygen. Aqueous $TiO_2$ suspension has also been used for photocatalytic degradation of six membered heteroatomic compounds such as pyridazine, pyrimidine, and pyrazine. A complete mineralization has been achieved in this case.

The oxides of nitrogen [NO and $NO_2$ ($NO_x$)] have variety of negative impact on environment and human health. Photocatalysis has been proved to be a successful method for removal of $NO_x$ from the atmosphere (Devahasdin et al., 2003; Wang et al., 2007). During the past 15 years, Jing et al. (2011) reviewed this field with a focus on the progress of nitrogen-containing organic compounds removal in wastewater by $TiO_2$-mediated photocatalytic activity. They have summarized the important factors affecting the relevant photocatalytic activity. They have also proposed that the different rates of photocatalytic degradation of N-containing compounds are the result of a strong interaction among pollutants structure, $TiO_2$ properties and photocatalytic reaction conditions. Nitroaromatics are toxic compounds, which are insoluble in nature and therefore, these are not easily biodegraded. Dillert et al. (1995, 1996) studied photocatalytic degradation of trinitrotoluene and other nitro aromatic compounds, whereas Augugliaro et al. (1991) studied photocatalytic degradation of nitrophenols in aqueous titanium dioxide dispersion.

Nahen et al. (1997) observed the photocatalytic degradation of trinitrotoluene, trinitrobenzene and dinitrobenzene on exposure to UV light using titanium dioxide as catalyst and found that degradation follows oxidative and reductive pathways; both, while Schmelling and Gray (1995) compared the transformation and mineralization of TNT (Trinitrotoluene) under photocatalytic and direct photolytic reactions. 5-Nitro-1,2,4-triazol-3-one (NTO) is a powerful explosive present in industrial waste water, which was completely mineralized in 3hr with $TiO_2$ by Campion et al. (1999).

Wang et al. (1999) used photocatalytic technique to degrade 2-nitrophenol in aqueous solution in the presence of titanium dioxide, whereas the effect of nano-$TiO_2$ on photocatalytic degradation of nitrobenzene was investigated by Makarova et al. (2000) for the removal of nitro aromatic compounds from contaminated water streams. Vohra and Tanaka (2002) employed $TiO_2$ mediated photocatalytic degradation process to treat aqueous 2-, 3-, and 4-nitrotoluene pollutants. The mineralization products included $NH_4^+$, $NO_3^-$, and $CO_2$.

## *OTHERS*

The influence of $TiO_2$ particle size and morphology has been addressed with the aim of increasing the photocatalytic efficiency of powdered materials. Titania nanoflowers were synthesized by oxidizing pure titanium with hydrogen peroxide solutions containing hexamethylenetetramine and nitric acid at a low temperature of 353K (Wu et al., 2006). These nanomaterials are widely utilized as photocatalyst to treat various waste waters (Balasubramaniam et al., 2004), degrade organic pollutants, and also for cleaning environment (Bhatkhande et al., 2001).

$TiO_2$ coated surfaces are increasingly studied for their ability to inactivate microorganisms. The activity of glass coated with thin films of $TiO_2$, CuO, and hybrid $TiO_2$/CuO prepared by atmospheric chemical vapor decomposition (Ap–CVD) was studied by Ditta et al. (2008). Visible light induced photocatalysis by titania particles for the reduction of environmental toxins on a global scale has been widely studied. A comparative study of $TiO_2$ supported on alumina and glass beads for the catalytic decomposition of leather was made by Sakthivel et al. (2002).

The photodegradibility of polymer like polyvinyl chloride (PVC) (Kim et al., 2006), polythene (PE), and polystyrene (PS) (Zan et al., 2004) was carried out by using $TiO_2$ nanoparticles, which is an eco-friendly method and also the need of the day, whereas Carlos et al. (2000) reported the toxicity of lignin, and Kraft effluents toward *E. coli* was completely removed by using Ag-ZnO catalyst.

Jain and Ameta (2008) reported the photocatalytic oxidation of arabinose and glucose over cadmium sulphide. Formaldehyde formed was further oxidized in the presence of hydroxyl radicals into acid and water.

$$CdS + h\nu \rightarrow e^-_{cb} + h^+_{vb}$$

$$H_2O + h^+ \rightarrow {}^{\bullet}OH + H^+$$

$$8\ {}^{\bullet}OH + C_5H_{10}O_4\,(\text{Arabinose}) \rightarrow HCHO + 4\ HCOOH + 4\ H_2O$$

$$10\ {}^{\bullet}OH + C_6H_{12}O_6\ (\text{Glucose}) \rightarrow HCHO + 5\ HCOOH + 5\ H_2O$$

Cellulose was photocatalytically degraded on supported $TiO_2$ and ZnO by Yeber et al. (2000). Cellulose bleaching effluent was completely decolorized and the total phenol content was reduced by 85% after 120min treatment with both the catalysts. Villasenor et al. (2002) carried out the photocatalytic degradation of organic matter dissolved in Kraft Black liquor, which is an important effluent from pulp and paper industries.

In recent years, production of hydrogen from water splitting using different photocatalysts has gained attention of researchers from all over the world, as the hydrogen is considered as a rich fuel source in future. Sun et al. (2013) studied photocatalytic generation of hydrogen from water using a cobalt pentapyridine complex in combination with molecular and semiconductor nanowire photosensitizers.

The treatment of waste water from different industries is left as such without any proper treatment. This pollutes the water resources nearby. There is an urgent need to find out effective ways for the treatment of polluted water. Although a number of treatment methods are available, but these are associated with some or other demerits. Photocatalysis has emerged as a promising technology for the treatment of waste water and time is not far off, when this technology may surpass other alternative techniques.

## KEYWORDS


- **Dyes**
- **Pesticides**
- **Photooxidation**
- **Photosensitizers**
- **Surfactant**


## REFERENCES


1. Alberici, R. M., Canela, M. C., Eberlin, M. N., & Jardim, W. F. (2001). *Applied Catalysis B: Environmental, 30,* 389–397.
2. Amat, A. M., Arques, A., Miranda, M. A., & Sequi, S. (2004). *Solar Energy, 77,* 559–566.
3. Ameta, S. C, Punjabi, P. B., Ameta, R., & Bhati I. (2010). Chapter 11. In A. K. Haghi (Ed.), *Recent progress in chemistry and chemical engineering research,* (pp. 173–204). New York: Nova Science Publishers.
4. Ameta, A., Bhati, I., Ameta, R., & Ameta, S. C. (2010). *Indonesian Journal of Chemistry, 10,* 20–25.
5. Ameta, R., Jain, S., Bhatt, C. V., & Ameta, S. C.(2000). *Revue Roumaine de Chimie, 45,* 49–56.
6. Ameta, R., Punjabi, P. B., & Ameta, S. C. (2011). *Journal of Serbian Chemical Society, 76,* 1049–1055.
7. Ameta, R., Vardia, J., Punjabi, P. B., & Ameta, S. C. (2006). *Indian Journal of Chemical Technology, 13,* 114–118.
8. Ameta, S. C., Ameta, R., Punjabi, P. B., Sharma, B. K., & Lodha, A. (1992). *Asian Journal of Chemical Reviews, 3,* 1–11.

9. Ameta, S. C., Choudhary, R., Ameta, R. & Vardia, J. (2003). *Journal of Indian Chemical Society, 80,* 257–265.
10. Ameta, S. C., Sharma, A., Sharma, R., & Rathore, S. S. (2002). *Journal of Indian Chemical Society,* 79, 929–931.
11. Arabatzis, M., Stergiopoulos, T., Andreeva, D., Kitova, S., Neophytides, S. G., & Falaras, P. (2003). *Journal of Catalysis, 220,* 127–135.
12. Assabbane, A., Ichou, Y. A., Tahiri, H., Guillard, C., & Herrmann, J. N. (2000). *Applied Catalysis, 24,* 71–87.
13. Augugliaro, V., Palmisano, L., Schiavello, M., & Sclafani, A. (1991). *Applied Catalysis, 69,* 323–340.
14. Balasubramaniam, G., Dionysiou, D. D., Suidan, M. T., Baudin, I., & Laine, J. M. (2004). *Applied Catalysis, 47,* 73–84.
15. Bardos, E. S., Markovics, O., Horvath, O., Toro, N., & Kiss, G. (2011). *Water Resources, 45,* 1617–1628.
16. Behnajady, M. A., Modirshala, N., & Hamzavi, R. (2006). *Journal of Hazardous Materials, 133,* 226–232.
17. Bhatkhande, D. S., Sawant, S. B., Schouten, J. C., & Pangarkar, V. G. (2004). *Journal of Chemical Technology and Biotechnology, 79,* 354–360.
18. Bhatkhande, S., Pangarkar, V. G., & Beenackers, A. A. C. M. (2001). *Journal of Chemical Technology and Biotechnology, 77,* 102–116.
19. Bilgi, S., & Demir, C. (2005). *Dyes and Pigments, 66,* 69–76.
20. Binitha, N. N., Yaakob, Z., Reshmi, M. R., Sugunan, S., Ambili, V. K., & Zetty, A. A. (2009). *Catalysis Today, 147,* 76–80.
21. Carlos, A., Gouvea, K., Wypych, F., Moraes, S. G., Duran, N., & Zamora, R. (2000). *Chemosphere, 40,* 427–432.
22. Cesareo, D., Di, D. A., Marchini, S., Passerini, L., & Tosato, M. L. (1986). *Homo-Hetero Photocatalysis, 174,* 593–627.
23. Chakrabarti, S., & Dutta, B. K. (2004). *Journal of Hazardous Materials, 112,* 269–278.
24. Campion, L. L., Giannotti, C., & Ouazzani. (1999). *Journal of Chemosphere, 38,* 1561–1570.
25. Chen, J., Eberlein, L., Cooper, C. H., & Langford, H. (2002). *Journal of Photochemistry & Photobiology, 148,* 183–189.
26. Choi, W., Hong, S. J., Chang, Y. S., & Cho, Y. (2000). *Environmental Science and Technology, 34,* 4810–4815.
27. Choi, W., & Hoffmann, M. R. (1997). *Environmental Science and Technology, 31,* 89–95.
28. Ciping, C., Daohui, L., & Guangzhi, X. (1993). *Journal of Environmental Science, 5,* 464–469.
29. Devahasdin, S., Fan, C., Li, K., & Chen, D. H. (2003). *Journal of Photochemistry and Photobiology, 156,* 161–170.
30. Dillert, R., Brandt, M., Fornefett, I., Siebers, U., & Bahnemann, D. (1995). *Chemosphere, 30,* 2333–2341.
31. Dillert, R., Fornefett, I., Siebers, U., & Bahnemann, D. (1996). *Journal of Photochemistry & Photobiology, 94,* 231–236.

32. Ditta, I. B., Steele, A., Liptrot, C., Tobin, J., Tyler, H., Yates, H. M., Sheel, D. W., & Foster, H. A. (2008). *Applied Microbiology and Biotechnology, 79,* 127–133.
33. Doong, R., & Chang, W. J. (1997). *Photochemistry and Photobiology, 107,* 239–244.
34. Fadaei, A. M., & Dehghani, M. H. (2012). *Research Journal of Chemistry and Environment, 16,* 104–109.
35. Faust, S. D., & Aly, O. M. (1999). *Chemistry of Water Treatment* (2nd edn., pp. 1–581). Boca Raton: Lewis Publishers.
36. Fuerte, A., Hernandez–Alonso, M. D., Maria, J., Martinez–Arias, A., Fernandez–Garcia, M. F., Conesa, J. C., Soria, J., & Munuera, G. (2002). *Journal of Catalysis, 212,* 1–9.
37. Fujihara, M., Satoh, Y., & Sataka,T.(1984). *Journal of Physical Chemistry, 88,* 4048.
38. Garcia, T., Solsona, B., Murphy, D. M., Antcliff, K. L., & Taylor, S. H. (2005). *Journal of Catalysis, 229,* 1–11.
39. Giahi, M., Habibi,S., Toutounchi,S., & Khavei, M. (2012). *Russian Journal of Physical Chemistry A, 86,* 689–693.
40. Gratzel, M. (1983). *Energy resources through photochemistry and catalysis*. New York.
41. Han, C., Li, Z., & Shen, J.(2009). *Journal of Hazardous Material, 168,* 215–219.
42. Hassan, M. U., Anwar. J., & Saif, M. J. (2008). *Journal of Scientific Research, 38,* 29–34.
43. Hatipoglu, A., San, N., & Cinar, Z. (2004). *Journal of Photochemistry and Photobiology, 165,* 119–129.
44. Heintz, O., Robert, D., & Weber, J. V. (2000). *Journal of Photochemistry and Photobiology, 135,* 77–80.
45. Hermann, J. M., Guillard, C., Arguello, M., Aguera, A., Tejedor, A., Piedra, L., & Fernandez-Alba, A. (1999). *Catalysis Today, 54,* 353–367.
46. Hidaka, H., Yamada, S., Suenaga, S., Zhao, J., Serpone, N., & Pelizzetti, E. (1990). *Journal of Molecular Catalysis, 59,* 279–290.
47. Houas, A., Lachheb, H., Ksibi, N., Elaloui, E., Guillard, C., & Hermann, J. M. (2001). *Applied Catalysis B: Environmental, 31,* 145–157.
48. IUPAC Glossary of Terms Used in Photochemistry. (1988). *Pure and Applied Chemistry, 60,* 1055–1106.
49. Jain, S., & Ameta, S. C. (2008). *Research Journal of Chemistry and Environment, 12,* 61–64.
50. Jing, J., Liu, M., Colvin, V. L., Li, W., & Yu, W. W. (2011). *Journal of Molecular Catalysis A: Chemical, 351,* 17–28.
51. Jirkovsky, J., Faure, V., Boule, P. (1997). *Pesticide Science, 50,* 42–52.
52. Kamat, P. V. (1993). *Chemical Reviews, 93,* 267–300.
53. Kang, M. G., Jung, H. S., & Kim, K. J. (2000). *Journal of Photochemistry and Photobiology, 136,* 117–123.
54. Kaur, P., Khant, A., & Khandelwal, R. C. (2011). *International Journal of Chemical Sciences, 9,* 980–988.
55. Kawai, T., & Sakata, T. (1980). *Journal of Chemical Society, Chemical Communication, 15,* 694–695.

56. Kerzhentsev, M., Guillard, C., Herrmann, J. M., & Pichat, P. (1996). *Catalysis Today, 27,* 215–220.
57. Kim, S. H., Kwak, S. Y., & Suzuki, T. (2006). *Polymer, 47,* 3005–3016.
58. Kimura, T., Yoshikawa, N., Matsumura, N., & Kawase, Y. (2004). *Journal of Environmental Science and Health: A Toxic Hazardous Substances and Environmental Engineering, 39,* 2867–2881.
59. Kirchnerova, J., Cohen, M. L., Guy, C., & Klvana, D. (2005). *Applied Catalysis, 282,* 321–332.
60. Klosek, S., & Raftery, D. (2001). *Journal of Physical Chemistry B, 105,* 2815–2819.
61. Konstantinou, I. K., & Albanis, T. A. (2003). *Applied Catalysis B: Environmental, 42,* 319–335.
62. Kothari, S., Ameta, P., & Ameta, R. (2007). *Indian Journal of Chemistry, 46,* 432–435.
63. Kothari, S., Jain, N., & Ameta, R. (2004). *Indian Journal of. Chemical Technology, 11,* 423–426.
64. Krysova, H., Krysa, J., Macounova, K. & Jirkovsky, J. (1998). *Journal of Chemical Technology and Biotechnology, 72,* 169–175.
65. Lair, A., Ferronato, C., Chovelon, J. M., & Herrmann, J. M. (2007). *Journal of Photochemistry and Photobiology, 193,* 193–203.
66. Li, J. H., Lu, A. H., Liu, F., & Fan, L. Z. (2008a). *Solid State Ionics, 179,* 1387–1390.
67. Li, F., Sun, S., Jiang, Y., Xia, M., Sun, M., & Xue, B. (2008b). *Journal of Hazardous Material, 152,* 1037–1044.
68. Liu, G. H., Zhu, Y. F., Zhang, X. R., & Xu, B. Q. (2002). *Analytical Chemistry, 24,* 6279–6284.
69. Lizama, C., Bravo, C., Caneo, C., & Ollino, M. (2005). *Environmental Technology, 26,* 909–914.
70. Low, G. K. C., McEvoy, R., & Mathews, R. W. (1991). *Environmetal Science and Technology, 25,* 460–467.
71. Ma, Y., & Yao, J. N. (1998). *Journal of Photochemistry and Photobiology, 116A,* 167–170.
72. Machuca-Martínez, F., & Colina-Márquez, J. Á. (2011). *Ingeniería & Desarrollo Universidad del Norte, 29,* 84–100.
73. Madhusudhana. N., Yogendra. K., Mahadevan. K. M., & Naik, S. (2011). *International Journal of Chemical Engineering Application, 2,* 294–298.
74. Magalhaes, F., & Lago, R. M. (2009). *Solar Energy, 83,* 1521–1526.
75. Mahadwad, O. K., Parikh, P. A., Jasra, R. V., & Patil, C. (2011). *Bulletin of Material Science, 34,* 551–556.
76. Makarova, O. V., Rajh, T., & Thurnauer, M. C. (2000). *Environmental Science and Technology, 34,* 4797–4803.
77. Mansoori, R. A., Kothari, S., & Ameta, R. (2004). *Journal of Indian Chemical Society, 81,* 335–337.
78. Marci, G., Addano, M., Augugliaro, V., Coluccia, S., Garcia–Lopez, E., Laddo, V., & Mastra, G. *Journal of Photochemistry and Photobiology, 160,* 105–114.
79. McManamon, C., Holmes, J. D., & Morris, M. A. (2011). *Journal of Hazardous Material, 193,* 120–127.

80. Miguel, N., Ormad, M. P., Mosteo, R., & Ovelleiro, J. L. (2012). *International Journal of Photoenergy.* doi:10.1155/2012/371714
81. Mohamed, O. S., Gaber, A. M., & Abdel-Wahab, A. A. (2002). *Journal of Photochemistry and Photobiology, 148,* 205–210.
82. Montgomery, J. M. (1985).*Consulting engineers water treatment: Design and design* (pp. 1–696). New York: John Wiley & Sons.
83. Muszkat, L., Feigelson, L., Bir, L., & Muszkat, K. A. (2002). *Pest Management Science, 58,* 1143–1148.
84. Nahen, M., Bahnemann, D., Dillert, R., & Fels, G. (1997) *Journal of Photochemistry and Photobiology, 110,* 191–199.
85. Okamoto, K., Yamanoto, Y., Tanaka, H., & Itaya, A. (1985). *Bulletin of Chemical Society of Japan, 58,* 2023–2028.
86. Ollis, D. F., Pelizzetti, E., & Serpone, N. (1991) *Environmental Science and Technology, 25,* 1522–1529.
87. Ollis, D. F. (1982). *Environmental Science and Technology, 19,* 480–484.
88. Pandiyan, T., Rivas, O. M., Martinez, J. O., Amezcua, G. B., & Martinez-Carrillo, M. A. (2002) *Journal of Photochemistry and Photobiology, 146,* 149–155.
89. Papic, S., Koprivanac, N., & Bozic, A. L. (2006). *Water Environment Research, 78,* 572–579.
90. Pare, B., Singh, P., & Jannalgadda, S. B. (2009). *Indian Journal of Chemistry, 48,* 1364–1369.
91. Park, H., & Choi, W. (2005). *Catalysis Today, 101,* 291–297.
92. Parvin, T., Keerthiraj, N., Ibrahim, I. A., Phanichphant, S., & Byrappa, K. (2012). *International Journal of Photoenergy, 44,* 1056–1059.
93. Pelizzetti, E., & Serpone, N. (1986). *Homogeneous and heterogeneous photocatalysis.* Dordrecht: Reidel Publication Company.
94. Pereira, L., Pereira, R., Oliveira, C. S., Apostol, L., Pons, M. G. M.-N., Zahraa, O., & Alves, M. M. (2013). *Journal of Photochemistry and Photobiology, 89,* 33–39.
95. Piccinini, P., Minero, C., Vincenti, M., & Pelizzetti, E. (1997). *Catalysis Today, 39,* 187–195.
96. Poulis, I., Micropoulou, E., Panou, R., & Kostopolou, E. (2003). *Applied Catalysis, 41,* 345–355.
97. Punjabi, P. B., Ameta, R., Vyas, R., & Kothari, S. (2005). *Indian Journal of Chemistry, 44,* 2266–2269.
98. Puretedal, H. R., Norozi, A., Keshararz, M. H., & Semnani, A. (2009). *Journal of Hazardous Material, 162,* 674–681.
99. Rao, N., & Dubey, S. (1996). *Journal of Molecular Catalysis A: Chemical, 104,* 197–199.
100. Sacco, O., Stoller, M., Vaiano, V., Ciambelli, P., Chianese, A., & Sannino, D. (2012). *International Journal of Photoenergy.* doi: 10.1155/2012/626759
101. Sakthivel, S., & Kisch, H. (2003). *Angewandte Chemie International Edition, 42,* 4908–4911.
102. Sakthivel, S., Shanker, M. V., Palanichamy, M., Arbindoo, B., & Murugesan, V.(2002). *Journal of Photochemistry and Photobiology, 148,* 153–159.
103. San, N., Hathipoglu, V., Kocturk, G., & Cinar, Z. (2001). *Journal of Photochemistry and Photobiology, 139,* 225–232.

104. Santhanalakshmi, J., & Komalavalli, R. (2012).*Chemical Science Transactions, 1,* 522–529.
105. Saravanan, R., Shankar, H., Rajasudha, G., Stephen, A., & Narayanan, V. (2011). *International Journal of Nanoscience, 10,* 1131–1135.
106. Schmelling, D. C., & Gray, K. A. (1995). *Water Resources Research, 29,* 2651–2662.
107. Senthilnathan, J., & Philip, L. (2010a). *Chemical Engineering Journal, 161,* 83–92.
108. Senthilnathan, J., & Philip, L. (2010b). *Water Air and Soil Pollution, 210,* 143–154.
109. Shaban, Y. A., El-Sayed, M. A., El-Maradny, A. A., Al-Farawati, R. K., & Al-Zobidi, M. I. *Chemosphere.* (PMID232611216).
110. Shanthi M., & Kuzhalosai V. (2012). *Indian Journal of Chemistry A, 51,* 428–434.
111. Singhal, B., Porwal, A., Sharma, A., Ameta, R., & Ameta, S. C. (1997). *Journal of Photochemistry and Photobiology, 108,* 85–88.
112. Soltani, N., Saion, E., Hussein, M. Z., Erfani, M., Abedini, A., Bahmanrokh, G., Navasery, M., & Vaziri, P. (2012). *International Journal of Molecular Science, 13,* 12242–12258.
113. Su, C., Hong, B. Y., & Tseng, C. M. (2004). *Catalysis Today, 96,* 119–126.
114. Sukhaser, J., Wold, A., Gao, Y. M., & Dwight, K. (1995). *Journal of Solid State Chemistry, 119,* 339–343.
115. Sun, Y., Sun, J., Jeffrey, R. L., Yang, P., & Chang, C. (2013). *Journal of Chemical Sciences, 4,* 118–124.
116. Takeda, K., & Fujiwara, K. (1996). *Water Research, 30,* 323–330.
117. Tamini, M., Qourzal, S., Assabane, A., Chovelon, J. M., Ferronato, C., & Ait-Ichou, Y. (2006). *Photochemical and Photobiological Sciences, 5,* 477–482.
118. Tanaka, K., Padermpole, K., & Hisanaga, T. (2000). *Water Research, 34,* 327–333.
119. Tang, A., Xiao, Y., Ouyang, J., & Nie, S. (2008). *Journal of Alloys Compounds, 457,* 447–451.
120. Tang, T., Zhang, J., Tian, B., Chen, F., He, D., & Anpo, M. (2007). *Journal of Colloid and Interface Science, 315,* 382–388.
121. Tatsuma, T., Tachibana, S., Miwa, T., Tryk, D. A., & Fujishima, A. (1999). *Journal of Physical Chemistry B, 103,* 8033–8035.
122. Tayade, R. J., & Key, D. L. (2010). *Material Science Forum: Current Application of Polymer and Nano Materials, 657,* 62–74.
123. Tian, B., Zhang, J., Tong, T., & Chen, F. (2008). *Applied Catalysis B: Environmental, 79,* 394–401.
124. Tong, T., Zhang, J., Tian, B., Chen, F., & He, D. (2008). *Journal of Hazardous Material, 155,* 572–579.
125. Torres-Martínez, L. M., Ruiz-Gómez, M. A., Figueroa-Torres, M. Z., Juárez--Ramírez, I., & Moctezuma, E. (2012). *International Journal of Photoenergy.* doi:10.1155/2012/939608.
126. US Environmental Protection Agency (USEPA). (1989). *Technologies for upgrading existing or designing new drinking water treatment facilities.* (EPA/625/4-89/023).
127. Vaya, D., Benjamin, S., Sharma, V. K., & Ameta, S. C. (2008). *Bulletin of Catalysis Society of India, 7,* 56–59.

128. Verma, A., Sheoran, M., Toor, A. P. (2013). *Indian Journal of Chemical Technology, 20,* 46–51.
129. Villasenor, J., Duran, N., & Mansilla, H. D. (2002). *Journal of Environmental Technology, 23,* 955–959.
130. Vohra, M. S., & Tanaka, K. (2002). *Water Research, 36,* 59–64.
131. Vyas, R., Swarnkar, H., & Ameta, S. C. (2005). *Chemical and Environmental Research, 14,* 71–78.
132. Wang, D., Zhang, J., Luo, Q., Li, X., Duan, Y., & An, J. (2009). *Journal of Hazardous Material, 169,* 546–550.
133. Wang, H., Wu, Z., Zhao, W., & Guan, B. (2007). *Chemosphere, 66,* 185–190.
134. Wang, K. H., Hsieh, Y.-H., Chou, M.-Y., & Chang, C.-Y. (1999). *Applied Catalysis B: Environmental, 21,* 1–8.
135. Wei, H., Tang, X. H., Liang, J. R., & Tan, S. Y. (2007). *Journal of Environmental Sciences, 19,* 90–96.
136. Wu, M., Huang, B., Wang, M., & Osaka, A. (2006). *Journal of American Ceramic Society, 89,* 2660–2663.
137. Xu, J. C., Shi, Y. L., Huang, J. E., Wang, B. & Li, H. L.(2004). *Journal of Molecular Catalysis A, 219,* 351–355.
138. Yamazaki, S., Yamabe, N., Nagano, S. & Fukuda, A. (2007). *Journal of Photochemistry and Photobiology, 185,* 150–155.
139. Yanagida, S., Ishimarru, Y. Mujake, Y. Shiragami, T. Hashimoto K. & Sakata, T. (1989). *Journal of Physical Chemistry, 93,* 2576–2682.
140. Yang, H., Zhang, K., Shi, R., Li, X., Dong, X., & Yu, Y. (2006). *Journal of Alloys and Compounds, 413,* 302–306.
141. Yang, Y., & Luan. (2012). *Journal of Molecules, 17,* 2752–2772.
142. Yeber, M. C., Rodriguez, J., Freer, J., Duran, N., & Mansilla, H. D. (2000). *Chemosphere, 41,* 1193–1197.
143. Yu, B., Zeng, J., Gong, L., Zhang, M., Zhang, L., & Chen, X. (2007). *Talanta, 72,* 1667–1674.
144. Yuan, Z. H., Jia, J. H., & Zhang, L. D. (2002). *Material Chemistry and Physics, 73,* 323–326.
145. Zaleska, A., Hupka, J., Wiergowski, M., & Biziuk, M. (2000). *Journal of Photochemistry and Photobiology, 135,* 213–220.
146. Zan, L., Tian, L. Liu, Z., & Peng, Z. (2004). *Applied Catalysis, 264,* 237–242.
147. Zhang, Y., Wong, J. W., Liu, P., & Yuan, M. (2011). *Journal of Hazardous Material, 191,* 136–143.
148. Zhang, P., Liang, F., Yu, G., Chen, Q., & Zhu, W. (2003). *Journal of Photochemistry and Photobiology, 156,* 189–194.
149. Zhang, R., Wang, J., Choi, J., Hu, L., & Mu, K. (2010). *Journal of Rare Earths, 28,* 353–356.

# CHAPTER 9

# PHOTO-FENTON REACTIONS: A GREEN CHEMICAL ROUTE

SURBHI BENJAMIN, NOOPUR AMETA, P. B. PUNJABI, and SURESH C. AMETA

## CONTENTS

## 9.1 INTRODUCTION

There is practically no human activity that does not produce waste products and in addition, there is a direct relationship between the standard of living in a society or country and the amount of waste products produced. These days, the criteria of determining the status of a country is well defined as developed, developing and under developed based on the production of waste products by that particular country. Until recently, the discharging waste in the environment was one of the major ways of eliminating them, until the auto purifying capacity of the environment becomes insufficient. The main problem stems from the waste coming from industry and agriculture, despite the fact that the population also plays an important role in environmental contamination, through anthropological activities.

Numerous physical, chemical and biological methods have been investigated to control organic contaminants (Gupta et al., 2009). Conventional process used to treat wastewater from textile industry includes chemical precipitation with alum or ferrous sulfate which suffers from drawbacks such as generation of a large volume of sludge leading to the disposal problem, the contamination of chemical substances in the treated wastewater, and so on. Moreover, these processes are inefficient in completely oxidizing organic compounds of complex structure. To overcome these problems, advanced oxidation processes (AOPs) have been developed to generate hydroxyl free radicals by different techniques.

Now a days, scientists are very much concerned with advanced oxidation processes (AOP's) (Legrini et al., 1993). These are based on the intermediacy of hydroxyl and other radicals to oxidize recalcitrant, toxic and non-biodegradable compounds to various by products and eventually to inert less harmful or almost harmful end products. Hydrogen peroxide is increasingly favored as an environmentally acceptable bleaching agent in both; domestic and industrial situations (Costa et al., 2004; Gould et al., 2002). Common AOPs involve photocatalytic, Fenton processes, photo-Fenton process, ozonation, photochemical and electrochemical oxidation.

Among all the advanced oxidation processes, the photo-Fenton oxidation has emerged as a very promising technology because of its high efficiency and cost-effectiveness. Photo-Fenton reaction is one of the very common advanced oxidation process used frequently for the degradation of organic pollutants. It is low cost, environment friendly, less time consuming and easy to handle.

Phenols, pesticides, fertilizers, detergents, dyes, chlororganics, and many other chemical products are disposed of directly into the environment, without being properly treated, (discharging controlled or uncontrolled) into nearby water resources. Aromatic nitro compounds are commonly used in industrial processes (manufacture of pesticides, dyes and explosives) and as a consequence, they appear as contaminants in every kind of water (especially in surface waters) and industrial waste waters. These substances produce a high toxicity, provoking serious health problems like blood infections, eye and skin irritation, affecting the central nervous system and sometimes, these are carcinogenic also. Several studies have shown the presence of these substances in surface waters (Howard, 1989) and ground waters (Duguet et al., 1989). Textile mills are major consumers of water and consequently, they are one of the largest groups of industries causing intense water pollution. The extensive use of chemicals and water results in generation of large quantities of highly polluted waste water. The dye containing colored water is almost of no use, but if this colored solution is bleached to give colorless water, it may be used for washing, cooling, irrigation and cleaning purposes. Thus, photochemical bleaching may provide a low cost method to solve the problem of water pollution.

Glaze and Chapin (1987) defined AOPs as water treatment processes at ambient temperature and pressure, which involve the generation of highly reactive radicals (especially hydroxyl radicals) in sufficient quantity to decontaminate polluted water. These treatment processes are considered as very promising methods for the remediation of contaminated ground, surface and waste waters containing non-biodegradable organic pollutants. Hydroxyl radicals are extraordinarily reactive species that attack most of the organic molecules. The kinetics of reaction is generally first order with respect to the concentration of hydroxyl radicals and to the concentration of the species to be oxidized. Hydroxyl radicals are also characterized by a little selectivity of attack, an attractive feature of an oxidant to be used in waste water treatment. Several organic compounds are susceptible to degradation by means of hydroxyl radicals. The attack by hydroxyl radical, in the presence of oxygen, initiates a complex cascade of oxidative reactions leading to mineralization.

Fenton's reagent is one of the most powerful inorganic oxidizing agents. Fenton (1894) discovered that the addition of ferrous salt and hydrogen peroxide initiated the rapid decomposition of α- hydroxy acids, such as tartaric acid and of α-glycols, but this reagent did not receive wide attention for a long period. Forty years later, the mechanism was postulated, which revealed that the effective oxidative agent in the Fenton reaction was the hydroxyl radical (Haber and Weirs, 1934). The Fenton reactions can be outlined as:

$$M^{n+} + H_2O_2 \rightarrow M^{(n+1)} + OH^- + {}^{\bullet}OH$$

Where M is a transition metal such as Fe or Cu.

## 9.2 FENTON AND PHOTO-FENTON DEGRADATION

### 9.2.1 *THE FENTON'S REAGENT*

The well known reactions of hydrogen peroxide with $Fe^{2+}$, which generates ${}^{\bullet}OH$ radical. The ferrous ion reacts with hydrogen peroxide (Fenton's reagent) stoichiometrically to give ${}^{\bullet}OH$.

$$Fe^{2+} + H_2O_2 \rightarrow Fe^{3+} + OH^- + {}^{\bullet}OHC$$

Reactions that generate hydroxyl radical in solution at low temperature have attracted interest for destruction of toxic organic compounds in waste waters.

De Laat and Gallard (1999; 2000) reported that in the absence of light and complexing ligands other than water, the most accepted mechanism of $H_2O_2$ decomposition in acidic homogeneous aqueous solution, involves the formation of hydroxyperoxyl ($HO_2^{\bullet}/O^{2-}$) and hydroxyl (${}^{\bullet}OH$) radicals. The OH radical in solution attacks almost every organic contaminant and the regeneration of metal ion can follow different paths.

It was observed that the Fenton reaction rates were increased by irradiation with UV/Visible light (Ruppert et al., 1993; Pignatello and Sun, 1993). During the Fenton reaction, the $Fe^{3+}$ ions are accumulated in the system and after all the $Fe^{2+}$ ions are consumed, the reaction practically stops. Faust and Hoigne (1990) proposed the photochemical regeneration of $Fe^{2+}$ ions by photoreduction of $Fe^{3+}$ ions. The newly generated ferrous ions react with $H_2O_2$ generating a second $^{\bullet}OH$ radical and ferric ion and this cycle continue.

$$Fe^{3+} + H_2O + h\nu \rightarrow Fe^{2+} + {}^{\bullet}OH + H^+$$

Fenton and photo-Fenton reactions depend not only on $H_2O_2$ concentration and iron added, but also on the operating pH value. Fenton process was used for the degradation of phenol, 4-chlorophenol, 2, 4-dichlorophenol and nitrobenzene by Chamarro and Esplugas (2001). The stoichiometric coefficient for the Fenton reaction was approximately 0.5mol of organic compound/mol of $H_2O_2$. The process was found to eliminate the toxic substances and to increase the biodegradability of the treated water. Some work about the treatment of textile water by means of Fenton and photo-Fenton process has been carried out and most of them showed their effectiveness for color removal and COD reduction (Balanosky et al., 1999; Kang et al., 2000).

Zepp et al. (1992) studied the reaction between Fe (II) and $H_2O_2$ in the wider pH range (3–8) for hydroxyl radical formation. Recently, two new electrochemical procedures for the detoxification of acidic waste waters, (the so called electro-Fenton and photoelectron-Fenton processes, where $H_2O_2$ is electro generated) have been developed and shown their good efficiencies for the mineralization of aniline and 4-chlorophenol (Brillas et al., 1998a, b).

### 9.2.2 *PHOTO-FENTON REACTION*

About two decades ago, it was found that the irradiation of Fenton reaction systems with UV/visible light strongly accelerated the rate of degradation of a variety of pollutants (Huston and Pignatello, 1999). This rate enhancement of Fenton oxidation has been named as photo-Fenton oxidation. Now a days, the photo-Fenton oxidation has emerged as a very promising technology because of its high efficiency and cost-effectiveness compared with other AOPs.

This behavior upon irradiation is principally due to the photochemical reduction of Fe (III) to Fe (II) in aqueous medium. Secondly, the Fe (II) so formed is then oxidized to Fe (III) on reaction with $H_2O_2$ as in dark Fenton process. Thus, process becomes cyclic generating 2 $^{\bullet}OH$ radicals per Fe-atom utilized. In addition, in the presence of light, some other reactions that produce hydroxyl radical or increase the production rate of hydroxyl radical can also occur.

The rate of reaction in the photo-Fenton process is faster than the conventional thermal Fenton process. On the basis of these reactions, it can be explained that rate of Fenton reaction is increased by irradiation because the ferrous ion required in Fenton reactions are produced during the process, which reacts with hydrogen peroxide in excess to produce more and more hydroxyl radicals.

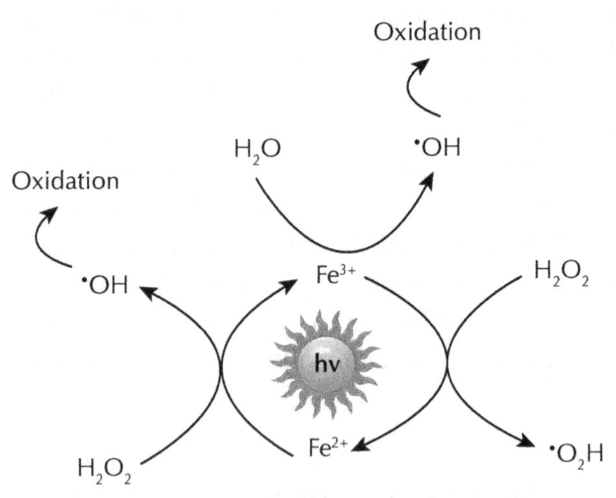


**FIGURE 9.1** Schematic diagram of photo-Fenton process.

Thus, photo-Fenton processes are better than dark Fenton processes, because photo-Fenton process consumes less $H_2O_2$ and requires only catalytic amounts of Fe (II). Any residual hydrogen peroxide that is not consumed in the process will spontaneously decompose into water and molecular oxygen and is thus a "clean" reagent in itself. These features make photo-Fenton based AOPs, a leading candidate for cost efficient, environmental friendly treatment of industrial effluents on a small to moderate scale (Pignatello et al., 2006).

An early example of an industrial scale application of the photo-Fenton process was the decontamination of 500L batches of an industrial effluent containing 2,4-dimethylaniline in a photochemical reactor fitted with a 10kW medium pressure mercury lamp (Oliveros et al., 1997).

Photo-Fenton reactions are cyclic in nature and therefore, addition of $H_2O_2$ only will keep this process continuous to generate $^{\bullet}OH$ radicals while in Fenton process, reaction stops after all the $Fe^{2+}$ ions are consumed.

For the reaction of hydroxyl radical with organic species, there are general reaction pathways. In aqueous solution, $Fe^{2+}$ and $Fe^{3+}$ states are commonly present in the form of octahedral complexes. In photo-Fenton process, the ferric iron is most critical iron species as its hydroxide precipitates at lower pH than those of ferrous iron. The main compounds, absorbing light in the Fenton system are ferric ion complexes, for example, $[Fe\ (OH)]^{2+}$ and $[Fe\ (RCO_2)]^{2+}$, which are highly photo reactive at pH below 3. These complexes give ferrous ions by undergoing ligand to metal charge transfer excitation as:

$$Fe^{3+}(L^-)_n \longrightarrow Fe^{2+}(L)_{n-1} + Lox^{\bullet}$$

Moreover, the oxidation of ligand may lead to further degradation of target compounds. Ferric complexes also undergo photoreduction. These complexes have different absorption properties and thus, photoreduction may takes place at different wavelengths and with different quantum yields. These complexes of ferric ions are commonly present with acids because acids are general by products generated during the mineralization process.

Depending upon the phase, the photo-Fenton reaction may be carried out under homogeneous or heterogeneous conditions. Under homogeneous conditions, the catalyst $Fe^{2+}/H_2O_2$ remains soluble in aqueous acidic medium, whereas under heterogeneous conditions, the metal ions like

$Fe^{2+}$, $Fe^{3+}$, $Cu^{2+}$ and so on gets anchored on some low cost carriers like bentonite, zeolite, $Fe_2O_3$, silica and so on.

### *9.2.3 PHOTODEGRADATION OF SOME ORGANIC COMPOUNDS*

#### *HYDROCARBONS*

The kinetics and mechanism of degradation of benzene derivatives with Fenton's reagent in aqueous medium was studied by Augusti et al. (1998). The active role of some transient intermediate species during the Fenton's mediated degradation of quinoline in oxidative media was suggested by Nedoloujko and Kiwi (1997). The composition of the products is considerably dependent on the presence of dioxygen, and on the pH (Stein and Loef, 1963). In the presence of dioxygen, the main product formed in a neutral medium is 2, 4-hexadiendial, while in an acidic medium, it does not arise at all. According to them, in the absence of dioxygen, the predominant product should be phenolic compounds. Dillert et al. (1996) reported the influence of $H_2O_2$ on photocatalytic degradation of trinitrotoluene and trinitrobenzene. Kuznetsova et al. (1996) studied the catalytic properties of heteropoly complexes containing Fe (III) ion in oxidation of benzene by $H_2O_2$.

The mechanism of reaction between benzene and $H_2O_2$ in the presence of dioxygen with added $Fe^{3+}$ or $Cu^{2+}$ ions was investigated by Jacob et al. (1977). Three reaction products; phenol, 2-hydroxy-2,4-hexadiendial and 3-hydroxyl-2,4–hexadiendial have been obtained. Their ratio depends on the intensity of photolytic radiations.

#### *ALCOHOLS*

The cyano complexes raise the quantum yields by as much as three folds of magnitude. Lie et al. (1998) and Braun et al. (1998) investigated the oxidative degradation of polyvinyl alcohols by photo-Fenton process. The oxidative degradation of polyvinyl alcohol by the photo-Fenton process was studied by Guardani et al. (2006).

Ovellerio et al. (2006) studied the treatment of winery waste waters by the photo-Fenton reaction in homogeneous phase, which are difficult to treat by conventional biological process. The oxidative degradation of poly(ethylene glycol) by Fenton and photo-Fenton reactions was investigated by Prousek and Duriskova (1998). Degradation of formaldehyde in the presence of methanol by photo-Fenton process was carried out by Kajitvichyankul et al. (2008).

## *CARBOXYLIC ACIDS*

The hydroxyl radicals are produced in photo-Fenton reaction and this radical attacks the aromatic ring of carboxylic acids resulting in the formation of hydroxyl derivatives. The subsequent process involves further oxidation accompanied by opening of the aromatic ring and the formation of aliphatic carboxylic acids. The formation of phenol and benzene is probably is a consequence of splitting of hydrogen from the carboxylic group, which is followed by decarboxylation.

The degradation of gallic acid (3, 4, 5-trihydroxy benzoic acid) in aqueous solution by UV/$H_2O_2$ treatment via Fenton's reagent and the photo-Fenton system was reported by Benitez et al. (2005). Nogueira et al. (2002) carried out the solar photo degradation of dichloroacetic acid and 2,4-dichlorophenol using an enhanced photo-Fenton process. The photo-assisted Fenton degradation of salicylic acid, using strongly acidic ion exchange resin (SAIER) exchanged with Fe ions as catalyst, in the presence of UV light and $H_2O_2$ was studied by Feng et al. (2004). Quici et al. (2005) studied the destruction of oxalic acid by combined heterogeneous photocatalysis and photo-Fenton reaction using UV/Fe/$H_2O_2$ and the UV/$TiO_2$/Fe/$H_2O_2$.

The inhibiting and chemiluminescent properties of benzoic acid and acetyl salicylic acid in the presence of Fenton reagent was observed by Zakhrov and Kumpan (1996). The molecular and structural characteristic of humic acid during photo-Fenton processes were studied by Fukushima et al. (2001). Varghese et al. (2007) studied the degradation of cyanuric acid with a combination of gamma radiolysis and Fenton reaction. Solozhenko et al. (1990) observed the oxidation of pyridine carboxylic acid in the presence of Fenton reagent.

Chemiluminescence in the peroxidation of linoleic acid initiated by the reaction of Fenton reagent was investigated by the Shen et al. (1991). Acero et al. (2001) reported studies on the degradation of p-hydroxyphenylacetic acid by photo-assisted Fenton reaction. The pronounced photocatalytic effects of $FeCl_3$ and $Na_2$ [Fe $(CN)_5$ NO] on the hydroxylation of some carboxylic acid initiated by $H_2O_2$/UV radiation was reported by Sedlak et al. (1989). It was concluded that the reaction proceeds via several parallel mechanisms.

## *PHENOLS AND ITS DERIVATIVES*

Molina et al. (2007) reported the incorporation of iron species over different silica support for the heterogeneous photo-Fenton oxidation of phenol whereas Sykora et al. (1997) have studied the influence of metal ions including Fenton and photo-Fenton reactions on homogeneous photoxidation of phenol. Pena et al. (2001) carried out phenol degradation by photo-Fenton reaction in highly concentrated waste water. Fenton and photo-Fenton reactions were used for phenol removal from high salinity effluent by Maciel et al. (2004). Melero et al. (2005) studied the heterogeneous photo-Fenton degradation of phenolic aqueous solution over iron-containing SBA – 15 catalysts.

Selanec et al. (2006) investigated the application of AOPs, dark Fenton and photo-assisted Fenton type processes; $Fe^{2+}$/ $H_2O_2$, $Fe^{3+}/H_2O_2$, UV/ $Fe^{2+}/H_2O_2$, UV/$Fe^{3+}/H_2O_2$ and UV/ $Fe^0/H_2O_2$, for the degradation of phenol present in waste water. The kinetics and mechanisms of these were studied by Lie and He (2004). A mechanism for the formation of phenoxy radicals during photooxidation of phenol in the presence of $Fe^{3+}$ was suggested by Nadtochenko and Kiwi (1998). Sengul et al. (2003) studied the role of ferrous ion in Fenton and photo-Fenton processes for the degradation of phenol.

The investigation of the mechanism of phenol decomposition of Fe–C–$TiO_2$ and Fe–$TiO_2$ photocatalysts via photo-Fenton process was carried out by Tryba et al. (2006) while Toyoda et al. (2006) studied the carbon coating of Fe–C–$TiO_2$ photo catalyst on phenol decomposition under UV irradiation via photo-Fenton process. Li et al. (2007) and Katsumata et al. (2004) investigated the photo-Fenton photodegradation of bisphenol A in aqueous solution with iron oxides. Yingxun et al. (2006) investigated the role of intermediate in the degradation of 4-chlorophenol, 4-nitrophenol and phenol by Fenton process and degradation of ethylene glycol and crysols in system has also been reported (Kavitha and Palaninelu 2005; McGinnis et al., 2000).

The composition of the coordination sphere of the catalyst was found to affect the selectivity of the reaction considerably and the effect was similar for both, UV and visible irradiation in the presence of $FeCl_3$ and Fe (III) (EDTA), where catechol is the main product while with ferricyanide or ferrocynide, benzoquinone predominates as the product. The effect of pH on the reaction rate of degradation of phenol was investigated by Castranats and Gibilisco (1990).

The kinetic modeling and reaction pathway of 2, 4-dichlorophenol transformation by photo-Fenton oxidation has been reported by Chu et al. (2005) while Kavitha and Palanivelu (2003) made a comparative study of the degradation of 2-chlorophenol by Fenton and photo-Fenton processes.

Feng and Cheng (2004) reported the degradation, kinetics and mechanisms of phenol in photo-Fenton process while Arana et al. (2001) carried out treatment of the highly concentrated phenolic wastewater by the photo-Fenton reaction and reported the mechanism of reaction. The photocatalytic degradation of resorcinol over titanium dioxide using photo-Fenton's related reagents was investigated by Ameta et al. (2006). Ghaly et al. (2001) reported studies on the photochemical oxidation of p-chlorophenol by UV/$H_2O_2$ and photo-Fenton process.

Photocatalytic-Fenton degradation of phenol over Fe-Ti pillared bentonite has been reported by Gao et al. (2010). Phenol degradation in water through a heterogeneous photo-Fenton process catalyzed by Fe-treated laponite was studied by Iurascu et al. (2009). Martinez et al. (2005) studied heterogeneous photo-Fenton degradation of phenolic aqueous solutions over iron-containing SBA-15 catalyst. Parida and Pradhan (2010) reported that Fe/meso-$Al_2O_3$ is an efficient photo-Fenton catalyst for the adsorptive degradation of phenol.

## *HALO COMPOUNDS*

Kwon et al. (1999) investigated the role of oxygen in the degradation pathway of 4-chlorophenol by Fenton system. Katsumata et al. (2006) carried out the photo degradation of alachor (which is one of the acetanilide herbicides) in the presence of the Fenton reagent and citrate. An attempt was made by Saritha et al. (2007) to degrade 4-chloro-2-nitrophenol (4C-2-NP), widely available in bulk drug and pesticides wastes. Sabhi and Kiwi (2001) reported the degradation of 2,4-dichlorophenol on Nafion-Fe (1.78%) under visible light irradiation in the presence of $H_2O_2$. Du et al. (2007) have carried out the oxidation of p-chlorophenol involving oxygen molecule.

The proposed mechanism is as follows:



The degradation of a phototypical halogenated aromatic pollutant, 4-chloroaniline, which was photo induced by Fe (III) species in acidic aqueous solutions (pH 2–4) of $Fe(ClO_4)_3$ was reported by Mailhot et al. (2004) A comparative investigation about the oxidation of the herbicide, 2, 4-dichlorophenoxyacetic acid (2,4-D) by $Fe^{2+}/H_2O_2/UV$ and ferrous oxalate/ $H_2O_2/UV$ processes was carried out by Kwan and Chu (2004). Yeh et al. (2002) investigated the soil organic matter during the oxidation of chlorophenols with $Fe^{2+}$- catalyzed $H_2O_2$ (Fenton oxidation) system. Homogeneous degradation of 1,2,9,10-tetrachlorodecane in aqueous solution using hydrogen peroxide, iron and UV light was observed by El-Morsi et al. (2002).

### *NITRO COMPOUNDS*

Kiwi et al. (1997) reported the combined Fenton and biological flow reactor degradation of p-nitrotoluene-ortho-sulphonic acid (p-NTS). Fenton reagent, UV/$H_2O_2$ and UV/ Fenton's regents were used to mineralize dinitrotoluene and trinitrotoluene of spent acid in toluene nitration process (Chen, 2005). Sun et al. (2007) reported the kinetic study of the degradation of p-nitroaniline by Fenton oxidation process. Fenton reagent has also been used in the oxidation of nitroaromatic explosives namely, 2,4,6-trinitrophenol ammonium picrate, 2,4-dinitrotoluenes, 2,4,6- trinitrotoluene (TNT) and hexahydro-1,3-5-trinitro -1,3,5-triazine (RDX). Liou et al. (2004) performed a series of photo-Fenton reactions for the degradation of 2, 4, 6-trinitrotoluene (TNT). Optimization of the solar photo-Fenton process in the treatment of contaminated water has been reported by Rodriguez et al. (2005). Ahmadimoghaddam et al. (2010) observed degradation of 2,4-dinitrophenol by photo-Fenton process. A comparative study of hydrogen peroxide photolysis, Fenton reagent and photo-Fenton for the degradation of nitrophenols has been carried out by Goi and Trapido (2002).

### *PESTICIDES*

The photo-Fenton process can potentially be integrated into waste water treatment process to enhance the organic compound removal. It can operate at low concentrations of contaminant and can mineralize the compound or convert it into a less toxic form. The efficiency of the process is found to be maximum at pH 2.8; however, it has been found that with addition of suitable complexing agent for $Fe^{3+}$, the process can be operated close to neutral pH. In this study, citric acid was used as a complexing agent and 2, 4-dichlorophenol (DCP) as model contaminant. pH (5–8.89

was observed to be the feasible pH range. Concludingly, it can be said that the photo-Fenton process may be used practically as treatment option for waters contaminated with pesticides and other organic compounds that are poorly biodegradable. The influence of pH on the degradation of the heribicide tebuthiuron (TBH) using *in situ* generated Fe (III)-citrate complexes by photo-Fenton process under solar irradiation was investigated by Silva et al. (2007).

The photocatalytic removal of fenitrothion in pure and natural waters by photo-Fenton reaction was reported by Sakugawa et al. (2004). The photo-Fenton degradation of the heribicide tebuthiuron, diuron and 2,4-D of heribicides in aqueous solution using ferrioxalate complex ($FeO_x$) as source of $Fe^{2+}$ under irradiation was reported by Paterlini and Nogueira (2005). Maldonado et al. (2007) investigated the photocatalytic degradation of pesticides using $TiO_2$ and Fenton as well as photo-Fenton processes. Gromboni et al. (2007) observed the microwave-assisted photo-Fenton decomposition of chlorfenvinphos and cypermethrin in residual water, while waste water containing five common pesticides methomyl, dimethoate, oxamyl, cymoxanil and pyrimethanil have been mineralized by solar AOPs–biological coupled system (Oller et al., 2007).

The photo-Fenton process was successfully applied to a mixture of ten commercial available pesticides that served as a model for a proposed recycling plant for pesticide bottles. A solution of mixed pesticides (alachor, atrazine, chlorfenvinphos, diuron and isoproturon) was considered for degradation using photo-Fenton as a preliminary step before biotreatment (Lapertot et al., 2006, 2007; Malato et al., 2002). Badawy et al. (2006) used the combination of the Fenton reaction, $UV/H_2O_2$ and the photo-Fenton process in the degradation of organophosphorous containing substrates such as fenitrothion, diazinon and profenofos. The photocatalytic degradation of two selected insecticides (dimethoate and methyl parathion) using photo-Fenton reaction was also studied by Evgenidou et al. (2007).

Fallmann et al. (1999) showed the applicability of the photo-Fenton method for treating water containing pesticides. Coupling of photo-Fenton and biological treatment for the removal of diuron and linuron from water was carried out by Farre et al. (2006). Identification of the intermediates generated during the degradation of diuron and linuron herbicides by the photo-Fenton reaction was also done by Farre et al. (2007). Flox et al.

(2007) used photoelectron-Fenton with UV-4 and solar light for mineralization of herbicide mecoprop.

Katsumata et al. (2005) observed the degradation of linuron in aqueous solution by the photo-Fenton reaction. Kinetics and products of photo-Fenton degradation of trizophos was studied by Lin et al. (2004).

## *DYES*

Carneiro et al. (2007) investigated the homogeneous photo degradation of reactive blue-4 using a photo-Fenton process under artificial and solar irradiation. Use of photo-Fenton's reagent for the photochemical bleaching of metanil yellow has been reported by Kumar et al. (2008a). He et al. (2002) investigated the photo-Fenton degradation of an azo dye at neutral pH. Muruganandham and Swaminathan (2004) reported the decolorisation of reactive orange-4 by Fenton and photo-Fenton oxidation technology. Decolorisation of azo dye reactive black-5 by Fenton and Photo-Fenton oxidation has been done by Lucas and Peres (2006). Degradation and sludge production of textile dyes by Fenton and photo-Fenton processes has been studied by Liu et al. (2007)

Decolorisation and mineralization of acid yellow 23 by Fenton and photo-Fenton processes has been reported by Modirshala et al. (2007) while the degradation of naphthol green B by photo-Fenton reagent was observed by Kumar et al. (2008b). They proposed the mechanism for the photodegradation of naphthol green B with photo-Fenton reagent as:

$$Fe^{3+} + H_2O + h\nu \longrightarrow Fe^{2+} + {}^{\bullet}OH + H^+$$

$$Fe^{3+} + H_2O_2 + h\nu \longrightarrow Fe^{2+} + HO_2^{\bullet} + H^+$$

$$Fe^{2+} + H_2O_2 \longrightarrow Fe^{3+} + {}^{\bullet}OH + OH^-$$

$${}^{\bullet}OH + H_2O_2 \longrightarrow HO_2^{\bullet} + H_2O$$

$$Fe^{2+} + {}^{\bullet}OH \longrightarrow Fe^{3+} + OH^-$$

$$Fe^{3+} + HO_2^{\bullet} \longrightarrow Fe^{2+} + O_2 + H^+$$

$${}^{\bullet}OH + {}^{\bullet}OH \longrightarrow H_2O_2$$

$$\text{Naphthol green B} + {}^{\bullet}OH \longrightarrow \text{Products}$$

A study of kinetic parameters related to the decolorisation and mineralization of reactive dyes from textile dyeing industries using Fenton and photo-Fenton processes has been made by Nunez et al. (2007). Solar photocatalytic degradation of azo dyes by photo-Fenton process was carried out by Chacon et al. (2006). Wu et al. (1998) reported the photodegradation of malachite green in the presence of $Fe^{2+}/H_2O_2$ under visible irradiation. A solar photocatalytic degradation of the azo dye, acid orange -24 was carried out by means of a photo-Fenton reaction by Bandala et al. (1997). Swaminathan et al. (2006) studied the combined homogeneous and heterogeneous photocatalytic decolorisation and degradation of a chlorotriazine reactive dye, reactive orange 4 using ferrous sulphate/ferrioxalate with $H_2O_2$ and $TiO_2$-P25 particles.

Nitampegliotis et al. (2006) studied the decolorisation kinetics of procion dye from textile dyeing industry using photo-Fenton reaction. The influence of alizarin violet-3B dye on the Fenton reaction of organic compounds under visible irradiation was examined by Ma et al. (2005). The degradation of acid orange-7 dye by three different photochemical processes photoperoxidation, Fenton and photo-Fenton was observed by Scheeren et al. (2002).

Ruppert et al. (1993) presented the photo-Fenton reaction as an effective photochemical wastewater treatment process. Mineralization of C.I. acid red 14 azo dye by UV/Fe-ZSM $5/H_2O_2$ process has been investigated by Kasiri et al. (2010). Decolourisation of reactive dyes by modified photo-Fenton Process under irradiation with sunlight has been studied by Chaudhuri and Wei (2009). Bacardit et al. (2007) observed effect of salinity on the photo-Fenton process. Degradation of organic pollutants by the photo-Fenton process has been investigated by Kim and Vogelpohl (1998). The photo-Fenton reaction and the $TiO_2$/UV process for waste water treatment has been suggested by Bauer (1999).

Factors affecting the kinetic parameters related to the degradation of direct yellow 50 by Fenton and photo-Fenton processes has been observed by Mahmoud and Ismail (2011). Performance of the photo-Fenton process in the degradation of a model azo dye mixture has been studied by Macias-Sanchez et al. (2011). Heterogeneous catalytic treatment of synthetic dyes in aqueous media using Fenton and photo-assisted Fenton process has been studied by Soon and Hameed (2011). Heterogeneous photo-Fenton oxidation of reactive azo dye solutions using iron exchanged zeolite as a catalyst was observed by Tekbas et al. (2011).

Heterogeneous photo-Fenton photo degradation of reactive brilliant orange X-GN over iron-pillared montmorillonite under visible irradiation was observed by Chen et al. (2009). Yang et al. (2009) observed degradation of methylene blue by heterogeneous Fenton reaction using titanomagnetite at neutral pH values. A study of catalytic behaviour of aromatic additives on the photo-Fenton degradation of phenol red was examined by Jain et al. (2009). Degradation of bismark brown-R using copper loaded neutral alumina as heterogeneous photo-Fenton reagent has been studied by Sharma et al. (2010). Fenton- and photo-Fenton like degradation of a textile dye by heterogeneous processes with Fe/ZSM-5 zeolite has been reported by Duarte and Madeira (2010).

Rasoulifard et al. (2011) suggested photoassisted hetero-Fenton decolorization of azo dye from contaminated water by Fe–Si mixed oxide nanocomposite. Discoloration and mineralization of reactive red HE-3B by heterogeneous photo-Fenton reaction has been observed by Feng et al. (2003). Kumar et al. (2011) carried out comparative studies of degradation of dye intermediate (H-acid) using $TiO_2$/UV/$H_2O_2$ and photo-Fenton process. Zhang et al. (2011) suggested application of heterogenous catalyst of tris (1, 10)-phenanthroline iron (II) loaded on zeolite for the photo-Fenton degradation of methylene blue. Rapid decolorization of rhodamine-B by UV/Fe (III)-penicillamine process under neutral pH has been studied by Xue et al. (2011). Saatci (2010) reported decolorization and mineralization of remazol red F3B by Fenton and photo-Fenton processes.

Fe (III) supported on ceria as an effective catalyst for the heterogeneous photo-oxidation of basic orange 2 in aqueous solution under sunlight has been investigated by Martinez et al. (2011). Photoassisted degradation of azo dyes over $FeO_xH_{2x-3}/Fe^0$ in the presence of $H_2O_2$ at neutral pH values has been investigated by Nie et al. (2007). Fernandez et al. (1999) observed the photo-assisted Fenton degradation of nonbiodegradable azo dye (orange II) in iron free solution, mediated by cation transfer membranes.

The photochemical degradation of azo dyes, namely red MX-5B, reactive black-5 and orange-G using low iron concentrations in Fenton and Fenton like systems was studied by Hsueh et al. (2005). Garcia-Martano et al. (2006a, b) assessed the environmental effect of different photo-Fenton approaches for commercial reactive dye removal. The decolorization of solution containing a common textile and leather dye, acid red 14, at pH 3 using Fenton, UV/ $H_2O_2/O_2$, $UV/H_2O_2/Fe^{2+}$, $UV/H_2O_2/Fe^{3+}$ and $UV/H_2O_2/Fe^{3+}$/oxalate processes was carried out by Daneshvar and Khataee (2006).

Different workers have investigated the use of photo-Fenton process for color removal from textile waste waters of textile industries (Kang et al., 2000, 2002; Kwan and Chu, 2003; Lloyd et al., 1997). Laponite and bentonite clay-based Fe nanocomposite have been developed as suspended photo-Fenton catalysts for the degradation of organic dyes (Serp et al., 2003).

## *MISCELLANEOUS*

Sarria et al. (2002) observed the effectiveness of photo-Fenton treatment of a biorecalcitrant waste water generated in textile activities and the biodegradability of the photo-treated solution. Vogelpohl et al. (1997) reported the landfill leachate treatment by a photo-assisted Fenton reaction while Gulyas (1997) carried out the removal of recalcitrant organics from industrial waste waters using AOPs. The intermediates in the Fenton type reagents were investigated by Walling (1998). Macfaul et al. (1998) provided a radical account of oxygenated Fenton chemistry while Gallard et al. (1998) observed the effect of pH on the oxidation rate of organic compounds by $Fe^{2+}/H_2O_2$. Eberlin et al. (1999) investigated the photolytic degradation of two common water pollutants, phenol and trichloroethylene (TCE) using Fenton's reagent/UV/ferrioxalate/$H_2O_2$/UV, and $TiO_2$/UV. Pulgarin et al. (2003) studied the mineralization of 5-amino-6-methyl-2-benzimidazolone (AMBI) using a new integrated iron (III) photo-assisted biological treatment.

The superior biodegradability of waste waters mediated by immobilized Fe-fabrics compared to Fenton homogeneous reactions was reported by Bozzi et al. (2003). A comparison of different AOPs for phenol degradation was made by Esplugas et al. (2002). Parra et al. (2004) reported the synthesis, testing and characterization of a novel Nafion membrane with superior performance in photo-assisted immobilized Fenton catalysis. The role of UV of natural sunlight in photodegradation of parathion in aqueous $TiO_2$ and Fe (0) solution in presence of $H_2O_2$ was observed by Doong and Chang (1998) whereas Luong and Lin (2000) observed the control of Fenton reactions for soil remediation. Gurses and Arslon-Alaton (2004) investigated the degradation of procaine penicillin-G (PPG) formulation effluent by Fenton-like ($Fe^{3+}$/$H_2O_2$) and UV-A Light assisted Fenton-like ($Fe^{3+}$/$H_2O_2$) UV-A processes.

The photo-Fenton process in heterogeneous phase as an alternative methodology for the treatment of winery waste waters was presented by Mosteo et al. (2006). Experimental design of Fenton and photo-Fenton reactions for the treatment of ampicillin solutions has been observed by Rozas et al. (2010).

Photo-Fenton assisted ozonation of p-coumaric acid present in olive mill wastewater was investigated by Monteagudo et al. (2005). Oxidation of some explosives by Fenton and photo-Fenton processes has been reported by Liou et al. (2003). Galvao et al. (2006) reported the application of the photo-Fenton process for the treatment of wastewater contaminated with diesel. Removal of organic contaminants by Fenton and photo-Fenton processes from paper pulp effluents was reported by Perez et al. (2002).

Tokumura et al. (2007) studied photo-Fenton process for excess sludge disintegration. Experimental design of Fenton and photo-Fenton reactions for the treatment of cellulose bleaching effluents has been reported by Torrades et al. (2003). White et al. (2003) studied the role of the photo-Fenton reaction in the production of hydroxyl radicals and photo-bleaching of colored dissolved organic matter in a coastal river of the Southeastern united

States. Multivariable approach to the photo-Fenton process as applied to the degradation winery wastewater was suggested by Ormad et al. (2006).

Park et al. (2006) made a comparison of Fenton and photo-Fenton processes for livestock wastewater treatment. Degradation of polyethylene glycol in aqueous solution by photo-Fenton and $H_2O_2$/UV processes has been reported by Giroto et al. (2010). Fe (III)-dopped zeolite Y catalyst has been used as a suspended photo-Fenton catalyst for the degradation of polyvinyl alcohol by Feng et al. (2005).

Mohajeri et al. (2010) studied the influence of Fenton reagent oxidation on mineralization and decolorisation of municipal landfill leachate while Herney-Ramirez et al. (2010) studied heterogeneous photo-Fenton oxidation with pillared clay-based catalysts for wastewater treatment. The heterogeneous photo-Fenton reaction using goethite as catalyst has been investigated by Plata et al. (2010).

$Fe_2O_3$-pillared rectorite as an efficient and stable Fenton-like heterogeneous catalyst for photodegradation of organic contaminants has been investigated by Zhang et al. (2010). Wastewater treatment by catalytic wet oxidation using $H_2O_2$ and pillared clays containing iron as heterogeneous catalyst has been widely investigated (Guelou et al., 2003; Guo and Al-Dahhan, 2003). Highly active S-modified $ZnFe_2O_4$ heterogeneous catalyst and its photo-Fenton behavior under UV-visible irradiation was studied by Liu et al. (2011). Yip et al. (2005) carried out novel heterogeneous acid-activated clay supported copper catalyzed photo bleaching and degradation of textile organic pollutant using photo-Fenton like reaction.

Elmolla and Chaudhari (2010) studied effect of photo-Fenton operating conditions on the performance of photo-Fenton-SBR process for recalcitrant wastewater treatment. Fenton and photo-Fenton processes coupled to upflow anaerobic sludge blanket (UASB) reactor to treat coffee pulping wastewater has been reported by Kondo et al. (2010). Zelmanov and Semiat (2008) studied iron (III) oxide based nanoparticles as catalysts in advanced organic aqueous oxidation of pollutants. Microstructure and photo-Fenton performance of trinuclear iron cluster intercalated montmorillonite catalyst has been investigated by Zhang et al. (2011). Ju et al. (2011) investigated sol-gel synthesis and photo-Fenton like catalytic activity of $EuFeO_3$ nanoparticles. Landfill leachate treatment by Fenton, photo-

Fenton processes and their modifications has been suggested by Agata and Jeremi (2012).

Synthesis, characterization and visible light photo-Fenton catalytic activity of hydroxy Fe/Al intercalated montmorillonite has been investigated by Li et al. (2011). Vermilyea and Voelker (2009) observed photo-Fenton reaction at near neutral pH in which oxidation of photo produced ferrous iron by hydrogen peroxide, produces reactive oxidants that may be important for degradation of biologically and chemically recalcitrant organic compounds in surface waters.

Hydroxyl radical production via the photo-Fenton reaction in the presence of fulvic acid has been studied by Southworth and Voelker (2003). Vilar et al. (2012) suggested application of Fenton and solar photo-Fenton processes for the treatment of a sanitary landfill leachate in a pilot plant with CPCs.

Evaluation of performance of a photo-Fenton process for pollutant removal from textile effluents has been made in a batch system by Modenes et al. (2012). Safarzadeh-Amiri et al. (1997) reported an improvement of photo-assisted Fenton process in the UV-vis/ferrioxalate/$H_2O_2$ system, which has been recently demonstrated to be more efficient than photo-Fenton reaction for the abatement of organic pollutants. The oxidative degradation of polyvinyl alcohols by the photo chemically enhanced Fenton reaction has been investigated by Lei et al. (1998).

Fenton reaction is well known for oxidation. Photo-Fenton reagent is a recent development in this field, which takes care of environment because there is no sludge formation because photo-Fenton reaction is reversible and does not require any addition of ferrous ions. Side wise, it will generate two hydroxyl radicals without a reasonable change in pH of the medium as compared to Fenton reaction, where only one hydroxyl ion is generated along with an increase with pH.

## KEYWORDS

- **Dyes**
- **Fenton reagent**
- **Nitro Compounds**
- **Phenols**
- **Photo-Fenton reagent**

## REFERENCES

1. Liu, L., Zhang, G., Wang, L., Huang, T., & Qin, L. (**2011**). *Industrial & Engineering Chemistry Research, 50,* 7219.
2. Acero, L. L., Real, F. J., & Leal, A. I. (2001). *Water Science and Technology, 44,* 31–38.
3. Agata, K., & Jeremi, N. (2012). *Journal of Advanced Oxidation Technology, 15,* 53–63.
4. Ahmadimoghaddam, M., Mesdaghinia, A., Naddafi, K., Nasseri, S., Mahvi, A. H., Vaezi, F., & Nabizadeh, R. (2010). *Asian Journal of Chemistry, 22,* 1009–1016.
5. Ameta, S. C., Punjabi, P. B., Kumari, C., & Yasmin. (2006). *Journal of the Indian Chemical Society, 83,* 42–48.
6. Arana, J., Rendon, E. T., Rodriguez, J. M. D., Melian, J. A. H., Diaz, O. G., & Pena, J. P. (2001). *Chemosphere, 44,* 1017–1023.
7. Augusti, R., Diar, A. O., Rocha, L. L., & Lago, R. M. (1998). *Journal of the Physical Chemistry, 102,* 10723–10727.
8. Bacardit, J., Stotzner, J., Chamarro, E., & Esplugas, S. (2007). *Industrial Engineering Chemistry Research, 46,* 7615–7619.
9. Badawy, M. I., Ghaly, M. Y., & Gad-Allah, T. A. (2006). *Desalination, 194,* 166–175.
10. Balanosky, E., Fernandez, J., Kiwi, J., & Lopez, A. (1999). *Water Science and Technology, 40,* 417–424.
11. Bandala, E. R. M., Chacon, J. M., Leal, M. T., & Sanchez, M. (1997). *Water Research, 31,* 787–798.
12. Bauer, R. (1999). *Catalysis Today, 53,* 131–144.
13. Bauer, R., Fallmann, H., Krutzler, T., Malato S., & Blanco, J. (1999). *Catalysis Today, 54,* 309–319.
14. Benitez, F. J., Real, F., Acero, J. L., Leal, A. L., & Garcia, C. (2005). *Journal of Hazardous Material, 125,* 31–39.
15. Bozzi, A., Yuranova, T., Meilczarski, E., Buffat, P. A., Lias, P., & Kiwi, J. (2003). *Appllied Catalysis, 42,* 289–303.

16. Braun, A. M., Lie, L., Hu, X., Yue, P. L., Bossmann, S. H., & Gob, S. (1998). *Journal of Photochemistry and Photobiology, 116A,* 159–166.
17. Brillas, E., Mur, E., Sauleda, R., Sanchez, L., Peral, J., Domenech, X., & Casado, J. (1998a). *Appllied Catalysis, 16,* 31–42.
18. Brillas, E., Sauleda, R., & Casado, J. (1998b). *Journal of Electrochemical Society, 145,* 759–765.
19. Carneiro, A. P., Nogueira, R. F. P., & Zanoni, M. V. B. (2007). *Dyes and Pigments, 74,* 127–132.
20. Castrantas, H. M., & Gibilisco, R. D.(1990). *ACS Symposium Series, 422,* 77–85.
21. Chacon, J. M., Leal, M. T., Sanchez, M., & Bandala, E. R. (2006). *Dyes and Pigments, 69,* 144–150.
22. Chamarro, E. A. M., & Esplugas, S. (2001). *Water Research, 35,* 1047–1051.
23. Chaudhuri, M., & Wei, T. Y. (2009). *Nature Environment Pollution Technology, 8,* 359–363.
**24.** Chen, Q., Wu, P., Li, Y., Zhu, N., & Dang, Z. (2009). *Journal of Hazardous Material,* 168, 901–908.
25. Chen, W. S., Juan, C. N., & Wei, K. M. (2005). *Chemosphere, 60,* 1072–1079.
26. Chu, W., Kwan, C. V., Chan, K. H., & Kam, S. K. (2005). *Journal of Hazardous Material, 121,* 119–126.
27. Costa, F. A. P., Reis, E. M., Azeved, J. C. R., & Nozaki, J. (2004). *Solar Energy, 77,* 29–35.
28. Daneshvar, W., & Khataee, A. R. (2006). *Journal of Environmental Science and Health A, 41,* 315–328.
29. De Laat, J., & Gallard, H. (1999). *Environmental Science and Technology, 33,* 2726–2732.
30. De Laat, J., & Gallard, H. (2000). *Water Research, 34,* 3107–3116.
31. Dillert, R., Fornefett, I., Siebus, V., & Bahnemann, D. (1996). *Journal of Photochemistry and Photobiology, 94A,* 221–229.
32. Doong, R., & Chang, W. H. (1998). *Journal of Photochemistry and Photobiology, 116,* 221–228.
33. Du, Y., Zhou, M., & Lei, L. (2007). *Journal of Hazardous Material, 139,* 108–115.
34. Duarte, F., & Madeira, L. M. (2010). *Separate the Science and. Technology, 45,* 1512–1520.
35. Duguet, J., Anselme, C., Mazounie, P., & Mallevialle, J. (1989). *Ozone Science and Engineering, 12,* 281–294.
36. Eberlin, M. N., Nogueira, R. F. P., Alberici, R. M., Mendes, M. A., & Jardim, W. F. (1999). *Industrial and Engineering Chemistry Research, 38,* 1754–1758.
37. Elmolla, E. S., & Chaudhuri, M. (2010). *Journal of Applied Sciences, 10,* 3236–3242.
38. El-Morsi, T. M., Emara, M. M., Hassan, M. H., Abd El-Bary, A., Abd-El-Aziz, S., & Friesen, K. J. (2002). *Chemosphere, 47,* 343–348.
39. Esplugas, S., Gimenez, J., Contreras, S., Pascual, E., & Rodriquez, M. (2002). *Water Research, 36,* 1034–1042.
40. Evgenidou, E., Konstantinou, I., Fytianos, K., & Poulios, I. (2007). *Water Research, 41,* 2015–2027.

41. Fallmann, H., Krutzler, T., Bauer, R., Malato, S., & BlancO, J. (1999). *Catalysis Today, 54,* 309–319.
42. Farre, M. J., Brosillon, S., Domenech, X., & Peral, J. (2007). *Journal of Photochemistry and Photobiology, 189A,* 364–373.
43. Farre, M. J., Domenech, X., & Peral, J. (2006). *Water Research, 40,* 2533–2540.
44. Faust, B., & Hoigne, J. (1990). *Atmospheric Environment, 24A,* 79–89.
45. Feng, H., & Le-Cheng, L. (2004). *Journal of Zhejiang University Science, 5,* 198–205.
46. Feng, J., Hu, X., & Yue, P. L. (2004). *Chemical Engineering Journal, 100,* 159–165.
47. Feng, J., Hu, X., & Yue, P. L. (2005). *Water Research, 39,* 89–96.
48. Feng, J., Hu, X., Yue, P. L., Zhu, H. Y., & Lu, G. Q. (2003). *Water Research, 37,* 3776–3784.
49. Fenton, H. (1894). *Journal of the Chemical Society Transactons, 65,* 899–910.
50. Fernandez, J., Bandara, J., Lopez, A., & Kiwi, J. (1999). *Langmuir: The ACS Journal of Surfaces and Colloids, 15,* 185–192.
51. Flox, C., Garrido, J. A., Rodriguez, R. M., Cabot, P. L., Centallas, F., Arias, C., & Brillas, E. (2007). *Catalysis Today, 129,* 29–36.
52. Fukushima, M., Tatsumi, K., & Nagao, S. (2001). *Environmental Science and Technology, 35,* 3683–3690.
53. Gallard, H., & De Laat, J. (2000). *Water Research, 34,* 3107–3116.
54. Gallard, H., Delaat, J., & Legube, B. (1998). *New Journal of Chemistry, 22,* 263–268.
55. Galvao, S. A. O., Mota, A. L. N., Silva, D. N., Moraes, J. E. F., Nascimento, C. A. O., & Chiavone-Filho, O. (2006). *Science of the Total Environment, 367,* 42–49.
56. Gao, J., & Wu, L. (2010). *Chinese Journal of Catalysis, 31,* 317–321.
57. Garcia-Martano, J., Munaz, I., Domenech, X., Garcia-Hortal, J. A., Torrades, F., & Peral, J. (2006a). *Journal of Hazardous Material, 138,* 218–225.
58. Garcia-Martano, J., Domenech, X., Garcia-Hortal, J. A., Torrades, F., & Peral, J. (2006b). *Journal of Hazardous Materials, 134,* 220–229.
59. Ghaly, M. Y., Hartel, G., Mayer, R., & Hasender, R. (2001). *Waste Management, 21,* 41–47.
60. Giroto, J. A., Teixeira, A. C. S. C., Nascimento, C. A. O., & Guardani, R. (2010). *Industrial Engineering and Chemical Research, 49,* 3200–3206.
61. Glaze, W., & Chapin, D. (1987). *Ozone Science and Engineering, 9,* 335–352.
62. Goi, R. F.,& Trapido, M. (2002). *Chemosphere, 46,* 913–922.
63. Gould, D. M., Griffith, W. P., & Spiro, M. (2002). *Journal of Molecular Catalysis A, 175,* 289–291.
64. Gromboni, C. F., Kamogawa, M. Y., Ferreira, A. G., Nobrega, J. A., & Nogueira, A. R. A. (2007). *Journal of Photochemistry Photobiology, 185A,* 32–37.
65. Guardani, R., Giroto, J. A., Teixeira, A. C. S. C., & Nascimento, C. A. O. (2006). *Chemical Engineering and Processing, 45,* 523–532.
66. Guelou, E., Barrault, J., Fournier, J., & Tatibouet, J. M. (2003). *Applied Catalysis B: Environmental, 44,* 1–8.
67. Gulyas, H. (1997). *Water Science and Technology, 36,* 9–16.
68. Guo, J., & Al- Dahhan, M. (2003). *Industrial and Engineering Chemistry Research, 42,* 2450–2460.

69. Gupta, V. K., Carrot, P. J. M., Carrot, M. M. L. R., & Suhas (2009). *Environmental Science and Technology, 39,* 783–842.
70. Gurses, F., & Arslan-Alaton, I. (2004). *Journal of Photochemistry and Photobiology, 165A,* 165–175.
71. Haber, F., & Weiss, J. (1934). *Journal of Proceedings of the Royal: Society London A, 147,* 332–351.
72. He, J., Tao, X., Ma, W., & Zhao, J. (2002). *Chemistry Letters, 31,* 86–87.
73. Herney-R., J., Vicente, M. A., & Madeira, L. M. (2010). *Applied Catalysis Environmental, 98,* 10–26.
74. Howard, P. (1989). *Chelsea.* Michigan: Lewis Publishers.
75. Hsueh, C. L., Huang, Y. H., Wang, C. C., & Chen, C. Y. (2005). *Chemosphere, 58,* 1409–1414.
76. Huston, L., & Pignatello, J. J. (1999). *Water Research, 33,* 1238–1246.
77. Iurascu, B., Siminiceanu, I., Vione, D., Vicente, M. A., & Gil, A. (2009). *Water Research, 43,* 1313–1322.
78. Jacob, N., Balakrishnan, J., & Reddy, M. P. (1977). *Journal of Physical Chemistry, 81,* 17–22.
79. Jain, A., Lodha, S., Punjabi, P.B., Sharma, V.K., & Ameta, S C. (2009). *Indian Academy of Sciences, 121,* 1027–1034.
80. Ju, L., Chen, Z., Fang, L., Dong, W., Zheng, F., & Shen, M. (2011). *Journal of American Ceramic Society, 94,* 3418–3424.
81. Kang, S. F., Liao, C. H., & Chen, M. C. (2002). *Chemosphere, 46,* 923–928.
82. Kang, S., Liao, C., & Po, S. (2000). *Chemosphere, 41,* 1287–1294.
83. Kasiri, M. B., Aleboyeh, H., & Aleboyeh, A. (2010). *Environmental Technology, 31,* 165–173.
84. Katsumata, H., Kaneco, S., Suzuki, T., Ohta, K., & Yobibio, Y. (2006). *Journal of Photochemistry and Photobiology, 180A,* 38–45.
85. Katsumata, H., Kaneco, S., Suzuki,T., Ohta, K., & Yobiko, Y. (2005). *Chemical Engineering Journal, 108,* 269–276.
86. Katsumata, H., Kawabe, S., Kaneco, S., Suzuki, T., & Ohta, K. (2004). *Jornal of Photochemistry and Photobiology, 162A,* 297–305.
87. Kavitha, V., & Palanivelu, K. (2004). *Chemosphere, 55,* 1235–1243.
88. Kavitha, V., & Palanivelu, K. (2003). *Journal of Environmental Science and Health A, Toxic And Hazardous Substance and Environmental Engineering, 38,* 1215–1231.
89. Kanitha, V., & Palaninelu, K. (2005). *Water Research, 39,* 3062–3072.
90. Kim, S. M., & Vogelpohl, A. (1998). *Chemical Engineering and Technology, 21,* 187–191.
91. Kiwi, J., Bandara, J., Pulgarin, C., & Peringer, P. (1997). *Journal of Photochemistry and Photobiology, 111A,* 253–263.
92. Kondo, M. M., Leite, K. U. C. G., Silva, M. R. A., & Reis, A. D. P. (2010). *Separate Science and Technology, 45,* 1506–1511.
93. Kumar, A., Paliwal, M., Ameta, R., & Ameta, S. C. (2008a). *Proceedings of National Academy of Sciences, 78,* 123–128.
94. Kumar, A., Paliwal, M., Ameta R., & Ameta, S. C. (2008b). *Collection of Czechoslovak Chemical Communication, 73,* 679–689.

95. Kumar, B. N., Anjaneyulu, Y., & Himabindu, V. (2011). *Journal of Chemical and Pharmaceutical Research, 3,* 718–731.
96. Kuznetsova, L. I., Detushiva, L. G., Fedotov, M. A., & Likholobov, V. A. (1996). *Journal of Molecular Catalisis A: Chemical, 111,* 81–90.
97. Kwan, C. Y., & Chu, W. (2003). *Water Research, 37,* 4405–4412.
98. Kwan, C. Y., & Chu, W. (2004). *Water Research, 38,* 4213–4221.
99. Kwon, B. G., Lee, D. S., Kang, W., & Yoon, Y. (1999). *Water Research, 33,* 2110–2118.
100. Lapertot, M., Ebrahimi, S., Dazio, S., Rubinelli, A., & Pulgarin, C. (2007). *Journal of Photochemistry Photobiology, 186A,* 34–40.
101. Lapertot, M., Pulgarin, C., Fernandez- Ibanez, P., Maldonado, M. I., Perez- Estrada, L., Oller, I., Gernjak, W., & Malato, S. (2006). *Water Research, 40,* 1086–1094.
102. Legrini, O., Oliveros, E., & Braun, A.M. (1993). *Chemical Reviews, 93,* 671–698.
103. Lei, L. C., Hu, X. J., & Vue, P. L. (1998). *Journal of Photochemistry and Photobiology, 116A,* 159–166.
104. Li, F. B., Li, X. Z., Li, X. M., Liu, T. X., & Dong, J. (2007). *Journal of Colloid and Interface Science, 311,* 481–490.
105. Li, H., Wu, P., Dang, Z., Zhu, N., Li, P., & Wu, J. (2011). *Clays and Clay Minerals,* 59, 466–477.
106. Lie, L. C., & He, F. (2004). *Journal of Zhejiang University Science, 5,* 198–205.
107. Lie, L., Hu, X., & Yue, P. L. (1998). *Journal of Photochemistry and Photobiology, 116A,* 159–166.
108. Lin, K., Yuan, D., Chen, M., & Deng, Y. (2004). *Journal of Agriculture and Food Chemistry, 52,* 7614–7620.
109. Liou, M. J., Lu, M. C.,& Chen, J. N. (2004). *Chemosphere, 57,* 1107–1114.
110. Liou, M. J., Lu, M. C., & Chen, J. N. (2003). *Water Research, 37,* 3172–3179.
111. Liu, L., Zhang, G., Wang, L., Huang, T., & Qin, L. (2011). *Industrial and Engineering Chemistry Research, 50*, 7219–7227.
112. Liu, R., Chiu, H. M., Shiau, C.-S., Yeh, R. Y.-L., & Hung, Y. T. (2007). *Dyes and Pigments, 73,* 1–6.
113. Lloyd, R. V., Hanna, P. M., & Mason, R. P. (1997). *Free Radical Biology and Medicine, 22,* 885–888.
114. Kajituichyanukul, P., Lu, M.-C., & Jamroensan, A. (2008). *Journal of Environmental Management, 86,* 545–553.
115. Lucas, M. S., & Peres, J. A. (2006). *Dyes and Pigments, 71,* 236–244.
116. Luong, H. V., & Lin, H. K. (2000). *Analytical Letters, 33,* 3051–3065.
117. Ma, J., Song, W., Chen, C., Ma, W., Zhao, J., & Tang, Y. (2005). *Environmental Science & Technology, 39,* 5810–5815.
118. Macfaul, P. A., Wayner, D. D. M., & Ingold, K. U. (1998). *Accounts of Chemical Research, 31,* 159–162.
119. Macias-Sanchez, J., Hinojosa-Reyes, L., Guzman-Mar, J. L., Peralta-Hernandez, J. M., & Hernandez-Ramirez, A. (2011). *Photochemical & Photo biological Sciences, 10,* 332–337.
120. Maciel, R., Sant'Anna, G. L., & Dezotti, M. J. (2004). *Chemosphere, 57,* 711–719.

121. Mahmoud, G. E. A., & Ismail, L. F. M. (2011). *Journal of Basic and Applied Chemistry, 1,* 70–79.
122. Mailhot, G., Hykrdova, L., Jirkovsky, J., Lemr, K., Grabner, G., & Bolte, M. (2004). Applied Catalysis, *50,* 25–35.
123. Malato, S., Blanco, J., Caceres, J., Fernandez-Alba, A., Aguera, R. A. & Rodriguez, A. (2002). *Catalysis Today, 76,* 209–220.
124. Maldonado, M. I., Passarinho, P. C., Oller, I., Gernjak, W., Fernandez, P., Blanco, J., & Malato, S. (2007). *Journal of Photochemistry and Photobiology, 185A,* 354–363.
125. Martinez, F., Calleja, G., Melero, J. A., & Molina, R. (2005). *Applied Catalysis B: Environmental, 60,* 181–190.
126. Martinez, S. S., Sanchez, J. V., Moreno Estrada, J. R., & Velasquez, R. F. (2010). *Solar Energy Materials & Solar Cells, 95,* 986–991.
127. McGinnis, B. D., Middlebrooks, E. T., & Adams, V. D. (2000). *Water Research, 34,* 2346–2354.
128. Melero, J. A., Martinez, F., Calleja, G., & Molina, R. (2005). *Applied Catalysis, 60,* 181–190.
129. Modenes, A. N., Espinoza-Quinones, F. R., Manenti, D. R., Borba, F. H., & Palacio, S. M. (2012). *Journal of Environmental Management, 104,* 1–8.
130. Modirshala, N., Behnajady, M. A., & Ghanbary, F. (2007). *Dyes and Pigments, 73,* 305–310.
131. Mohajeri, S., Aziz, H. A., & Isa, M. H. (2010). *Journal of Environmental Science Health A Tox. Hazard. Substance Environment Engineering, 45,* 692–698.
132. Molina, R., Martinez, F., Calleja, G., & Melero, J. A. (2007). *Applied Catalysis. B: Environtal, 70,* 452–460.
133. Monteagudo, J. M., Carmona, M., & Duran, A. (2005). *Chemosphere, 60,* 1103–1110.
134. Mosteo, R., Ormad, P., Mozas, E., Sarasa, J., & Ovelleiro, J. L. (2006). *Water Research, 40,* 1561–1568.
135. Muruganandham, M., & Swaminathan, M. (2004). *Dyes and Pigments, 63,* 15–321.
136. Nadtochenko, V. N., & Kiwi, J. (1998). *Journal of Chemical Society, 6,* 1303–1306.
137. Nedoloujko, A., & Kiwi, J. (1997). *Journal of Photochemistry and Photobiology, 110A,* 141–148.
138. Nie, Y., Hu, C. Qu, J. Zhou, L., & Hu, X. (2007). *Environtal Science and. Technology, 41,* 4715–4719.
139. Nogueira, R. F. P., Trovo, A. G. & Mode, D. F. (2002). *Chemosphere, 48,* 385–391.
140. Ntampegliotis, K., Ringa, A., Karayannis, V., Bonozoglou, V., & Papapolymerou, G. (2006). *Journal of Hazardous Material, 136,* 75–84.
141. Nunez, L., Garcia-Hortal, J. A., & Torrades, F. (2007). *Dyes and Pigments, 75,* 647–652.
142. Oliveros, E., Legrini, O., Hohl, M., Müller, T., & Braun, A. M. (1997). *Chemical Engineering and Process, 36,* 397–405.
143. Oller, I., Malato, S., Sánchez-Pérez, J. A., Maldonado, M. I., & Gassó, R. (2007). *Catalysis Today, 129,* 69–78.
144. Ormad, M. P., Mosteo, R., Ibarz, C., & Ovelleiro, J. L. (2006). *Applied Catalysis B: Environmental, 66,* 58–63.

145. Ovelleiro, J. L., Ormad, M. P., Mosteo, R., & Ibarz, C. (2006). *Applied Catalysis, 66,* 58–63.
146. Parida, K. M., & Pradhan, A. C. (2010). *Industrial and Engineering Chemistry Research, 49,* 8310–8318.
147. Park, J. H., Cho, H., & Chang, S. W. (2006). *Journal of Environtal Science and Health B, 41,* 109–120.
148. Parra, S., Henao, L., Mielczarski, E., Mielczarski, J., Albers, P., Suvorova, E., Guindet, J., & Kiwi, J. (2004). *Langmuir, 20,* 5621–5629.
149. Paterlini, W. C., & Nogueira, R. F. P. (2005). *Chemosphere, 58,* 1107–1116.
150. Pena, J. P., Arana, J., Rendon, E. T., Rodriquez, J. M. D., Melian, J. A. H., & Diaz, O. G. (2001). *Chemosphere, 44,* 1017–1023.
151. Perez, M., Torrades, F., Domenech, & Peral, J. (2002). *Water Research, 36,* 2703–2710.
152. Pignatello, J., Oliveros, E., & Mackay, A. (2006). *Critical Reviews In Environmental Science & Technology, 36,* 1–84.
153. Pignatello, J., & Sun, Y. (1993). *ACS Symposium Series, 518,* 77–84.
154. Plata, G. B. O., De La, Alfano, O. M., & Cassano, A. E. (2010). *Water Science Technology, 61,* 3109–3116.
155. Prousek, J., & Duriskova, I. (1998). *Chemicke' Listy, 92,* 218–220.
156. Pulgarin, C., Sarria, V. Deront, M., & Peringer, P. (2003). *Applied Catalysis, 40,* 231–246.
157. Quici, N., Morgada, M. E., Piperata, G., Babay, P., Gettar, R. T., & Litter, M. I. (2005). *Catalysis Today, 101,* 253–260.
158. Rasoulifard, M. H., Monfared H. H., & Masoudian, S. (2011). *Environmental Technology, 32,* 1627–1635.
159. Rodriguez, M., Malato, S., Pulgarin, C., Contreras, S., Curco, D., Gimenez, J., & Esplugas, S. (2005). *Solar Energy, 79,* 360–368.
160. Rozas, O., Contreras, D., Mondaca, M. A., Perez-Moya, M., & Mansilla, H. D. (2010). *Journal of Hazardous Material, 177,* 1025–1030.
161. Ruppert, G., Bauer, R., & Heisler, G. (1993). *Journal of Photochemistry and Photobiology, 73A,* 75–78.
162. Saatci, Y. (2010). *Journal of Environmental Engineering, 136,* 1000–1005.
163. Sabhi, S., & Kiwi, J. (2001). *Water Research, 35,* 1994–2002.
164. Safarzadeh-Amiri, A., Bolton, J. R., & Cater, S. R. (1997). *Water Research, 31,* 787–798.
165. Sakugawa, H., Nakatani, N., & Derbalah, A. S. (2004). *Chemosphere, 57,* 635–644.
166. Saritha, P., Aparna, C., Himabindu, V., & Anjaneyulu, Y. (2007). *Journal of Hazardous Material, 149,* 609–614.
167. Sarria, V., Rodriguez, M., Esplugas, S., & Pulgarin, C. (2002). *Journal of Photochemistry and Photobiolog, 151A,* 129–135.
168. Scheeren, C. W., Paniz, J. N., & Martins, A. F. (2002). *Journal of Environmental Science and Health: A, 37,* 1253–1261.
169. Schreen, C. W., Paniz, J. N., & Martins, A. F. (2002). *Journal of Environmental Scence and Health A, 37,* 1253–1261.
170. Sedlak, P., Lunak, S., Brodilov, J., & Lederer, P. (1989). *Reaction Kinetics and Catalysis Letters, 39,* 249–253.

171. Selanec, I., Kusic, H., Koprivanac, N., & Bozic, A. L. (2006). *Journal of Hazardous Materials, 136*, 632–644.
172. Sengul, F., Bali, U., & Catalkaya, E. C. (2003). *Journal of Environmental Science and Health, 38*, 2259–2275.
172. Serp, P., Corrias, M., & Kalck, P. (2003). *Applied Catalysis A, 253*, 337–358.
173. Sharma, J., Ameta, R., Sharma, V.K., & Punjabi, P. B. (2010). *Bulletin Of Catalysis Society of India, 9*, 99–106.
174. Shen, X., Tian, J. D., Thu, Z. N., & Li, X. Y. (1991). *Biopsy Chemical, 40*, 161–167.
175. Silva, M. R. A., Trova, A. G., & Nogueira, R. F. P. (2007). *Journal of Photochemistry and Photobiology, 191A*, 187–192.
176. Solozhenko, E. G., Staines, H., Soboleva, N. M., & Goncharuk, V. V. (1990). *Ukrainskii Khimichiskii ZHurnal: USSSR., 56*, 439.
177. Soon, A. N., & Hameed, B.H. (2011). *Desalination, 269*, 1–16.
178. Southworth, A., & Voelker, B. M. (2003). *Environmental Science and Technology, 37*, 1130–1136.
179. Stein, G., & Loef, J. (1963). *Journal of Chemical Society*, 2623–2633.
180. Sun, J. H., Sun, S. P., Fan, M. H., Guo, H. Q., Qiao, L. P., & Sun, R. X. (2007). *Journal of Hazardous Materials, 148*, 172–177.
181. Swaminathan, M., Selvam, K., & Muruganandham, M. (2006). *Solar Energy Materials and Solar Cells, 89*, 61–74.
182. Sykora, J., Pado, M., Tatarko, M., & Izakovic, M. (1997). *Journal of Photochemistry and Photobiology, 110A*, 167–175.
183. Tekbaş, M., Yatmaz, H. C., & Bektaş, N. (2011). *Clays and Clay Minerals, 59*, 466–477.
184. Tokumura, M., Sekine, M., Yoshinari, M., Znad, H.-T., & Kawase, Y. (2007). *Process Biochemistry, 42*, 627–633.
185. Torrades, F., Perez, M., Mansilla, H. D., & Peral, J. (2003). *Chemosphere, 53*, 1211–1220.
186. Toyoda, M., Tryba, B., Morawski, A. W., & Inagaki, M. (2006). *Chemosphere, 64*, 1225–1232.
187. Tryba, B., Morawski, A. W., Inagaki, M., & Toyoda, M. (2006). *Journal of Photochemistry Photobiology, 179A*, 224–228.
188. Varghese, R., Arvind, U. K., & Arvindakumar, C. T. (2007). *Journal of Hazardous Materials, 142*, 555–558.
189. Vermilyea, A.W., & Voelker, B. M. (2009). *Environmental Science and Technology, 43*, 6927–6933.
190. Vilar, V. J. P., Moreira, J. M. S., Fonseca, A., Saraiva, I., & Boaventura, R. A. R. (2012). *Journal of Advanced Oxidation Technologies, 15*, 107–116.
191. Vogelpohl, A., Kim, S. M., & Geissen, S. U. (1997). *Water Science and Technology, 35*, 239–248.
192. Walling, C. (1998). *Accounts of Chemical Research, 31*, 155–157.
193. White, E. M., Vaughan, P. P., & Zepp, R. G. (2003). *Aquatic Sciences, 65*, 402–414.
194. Wu, K. G., Zhang, T. Y., Zhao, J. C., & Hidaka, H. (1998). *Chemistry Letters, 27*, 857–858.
195. Xue, X. F., Liu, Y. X., Shao,Y. Q., & Deng, N. S. (2011). *Advanced Material Research, 130*, 183–185.

196. Yang, S., He, H., Wu, D., Chen, D., Ma, Y., Li, X., Zhu, J., & Yuan, P. (2009). *Industrial Engineering Chemistry Research*, *48*, 9915–9921.
197. Yeh, C. K. J., Kao, Y. A., & Cheng, C. P. (2002). *Chemosphere*, *46*, 67–73.
198. Yingxun, D., Zhou, M., & Lie, L. (2006). *Journal of Hazardous Materials*, *136*, 859–865.
200. Yip, A. C. K., Lam, F. L. Y., & Hu, X. (2005). *Chemical Communications*, *25*, 3218–3220.
201. Zakharov, I. V., & Kumpan, V. V. (1996). *Kinetics and Catalysis*, *37*, 174–178.
203. Zelmanov, G., & Semiat, R. (2008). *Water Research*, *42*, 492–498.
204. Zepp, R. G., Faust, B. C., & Hoigne, J. (1992). *Environmental Science and Technology*, *26*, 313–319.
205. Zhang, G., Gao, Y., Zhang, Y., & Guo, Y. (2010). *Environmental Science and Technology*, *44*, 6384–6389.
206. Zhang, J., Hu, F. T., Liu, Q. Q., Zhao X., & Liu, S. Q. (2011). *Reaction Kinetics Mechanism and Catalysis*, *103*, 299–310.
207. Zhang, S., Liang, S., Wang, X., Long, J., Li, Z., & Wu, L. (2011). *Catalysis Today*, *175*, 362–369.

CHAPTER 10

# SONOCHEMISTRY: A POLLUTION FREE PATHWAY

GARIMA AMETA, SURBHI BENJAMIN, VIKAS SHARMA, and SHIPRA BHARDWAJ

## CONTENTS

## 10.1 INTRODUCTION

Different forms of energy, like heat, light, and so on can drive some chemical reactions. But in past few decades, ultrasound has emerged as a potential source to enhance the chemical reactivity, besides being used in non chemical situations, such as the medical diagnosis, navigation bats, cleaning and drilling of teeth, SONAR, and material testing. It has been found that when ultrasound in low frequency range is passed through a chemical system, it influences its chemical reactivity. The study of effects of ultrasound on chemical reactivity is termed as sonochemistry.

The chemical effect of ultrasound was first reported by Richards and Loomis (1927). The basis for the present day generation of ultrasounds was established as far back as 1880 with the discovery of piezoelectric effect by the Curie brothers (1880, 1881). Crystalline materials showing this effect are known as piezoelectric materials. Ultrasonic devices consist transducers (energy converters), which are composed of these piezoelectric materials. An inverse piezoelectric effect is used in transducers that is. a rapidly alternating potential is placed across the faces of piezoelectric crystal, which generates dimensional change and thus, converts electrical energy into sound energy. The first ultrasonic transducer was a whistle developed by Galton, (1883) who was then investigating the frequency of human hearing.

## 10.2 SONOCHEMICAL REACTIONS

### *10.2.1 ULTRASOUND*

#### *CLASSIFICATION*

Scientifically, sound is the transmission of energy through the generation of acoustic pressure waves in the medium. These are mechanical in nature that is, require a medium and the particles in the medium vibrate to transfer the waves. The frequency of the wave determines its regime and these are classified in different kinds:

- ***Infrasound***: Sound waves having frequency less than 20Hz.
- ***Audible sound***: Sound waves having frequency between 20 and 20KHz.
- ***Ultrasound***: Sound waves with frequency more than 20KHz.
- ***Hypersound***: Sound waves with frequency higher than 10GHz.

## *PRINCIPLES OF SONOCHEMISTRY*

Sonochemical effects are due to the phenomenon of acoustic cavitation that is, the creation, growth and implosive collapse of gas filled bubbles in a liquid in response to an applied ultrasonic field. Cavitation was first identified and reported by Thornycraft and Barnaby (1895). When a liquid is irradiated by ultrasound, micro bubbles will appear, grow, and oscillate extreme quickly and even collapse near a solid surface generating micro jets and shock waves. Moreover, in the liquid phase surrounding the particles, high micro mixing will increase the heat and mass transfer and even the diffusion of species inside the pores of the solid.

## *PHENOMENON OF CAVITATION*

Sound waves are basically pressure waves. It consists of alternating compression and expansion cycles. When an acoustic field is applied to a liquid, the sonic vibrations create an acoustic pressure ($P_a$) at any time (t), which is given by the equation

$$P_a = P_A \sin 2\pi ft$$

where, $P_A$ is the maximum pressure amplitude of the wave and f is the frequency of the sound wave.

When ultrasound is passed through a liquid, the total pressure (P) in the liquid is given by

$$P = P_a - P_h$$

where, $P_h$ is the hydrostatic pressure

During the negative cycle of the wave, the distance between the molecules of the liquid will vary (oscillate) about a mean position. It the distance between the molecules exceeds the critical molecular distance, R (for example for water, the value of R is $10^{-8}$cm), then the liquid will breakdown and voids will be created; that is, formation of cavitation bubbles. During the positive cycle of the wave, the bubbles grew in size due to the positive acoustic pressure and then finally collapse, leading to the formation of new nuclei for the next cavitation (Lickies and McGrowth, 1996, Figure 10.1).

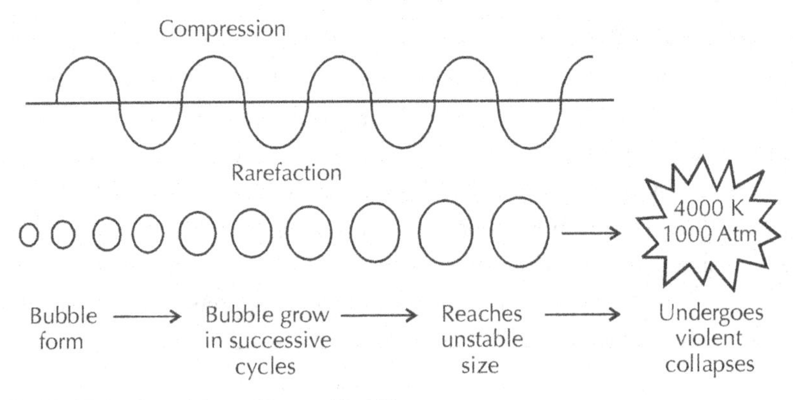


**FIGURE 10.1** Impulsive collapse of bubbles.

This is the mechanism for a cavity formation. If attempts are made to remove contaminants, particulates, precipitates, and so on from a solution, there are generally small dust moles or crystalline materials present in it. These irregular surfaces allow gas to be trapped. Acoustic waves consist of alternating compression and rarefaction waves. Upon rarefaction, the gas, which is trapped in the dust molecule is pulled out to become a free-standing bubble.

The occurrence of these collapses near a solid surface will generate microjets and shock wave (Suslick and Casadonte, 1987). Moreover, in the liquid phase surrounding the particles, high micromixing will increase the heat and mass transfer and even the diffusion of species inside the pores of the solid.

## FACTORS AFFECTING CAVITATION

(i) ***Ultrasound frequency*:** Mostly sonochemical work involves frequencies between 20 and 50KHz. It has been observed that sonochemical effect is limited at higher frequencies. It is due to the reason that the bubble has less time to grow and collapse at higher frequencies.

(ii) ***Presence of gas*:** Dissolved gases act as nucleation sites for cavitation. As gases are removed from the reaction mixture because of the implosion of the cavitation bubbles, initiation of new cavitational events becomes increasingly difficult. Bubbling gases through the mixture facilitate the production of cavitation bubbles.

(iii) ***Effect of external pressure*:** On increasing external pressure, $P_h$, it requires the application of greater acoustic pressure, $P_a$ and hence, the system requires higher ultrasonic intensities, and so on to generate cavitation bubbles.

(iv) ***Temperature*:** Sonication proceeds more efficiently at lower temperatures. It is due to the effect on the solvent properties like densities, surface tensions, viscosities, and so on.

## SOURCES OF ULTRASOUND IN SONOCHEMISTRY

- *Ultrasonic cleaning bath:* The simple ultrasonic bath is economic and easily available source of ultrasonic irradiation in the chemical laboratories. It has several drawbacks like ultrasonic intensity is limited for every ultrasonic cleaning bath, temperature control is also difficult and reproducibility of results is relatively poor because inter-laboratory comparisons are difficult. Different baths have different frequencies and power outputs.
- *Ultrasonic probe system*: In this type of system, a metal 'probe' is used, which is attached to the transducer introduced directly into a reaction system itself for increasing the amount of ultrasonic power.

  The horn is part of an assembly which amplifies small vibrations of the piezoelecric crystal to larger amplitude. This probe system has several advantages over a cleaning bath.

The ultrasonic probe system is of two types:

1. Cup-horn system
2. Flow cell system

- *Submersible transducer:* In this type of system, transducer is directly immersed into the reaction. It is an alternative to the cleaning bath and also used for batch sonication.
- *Whistle reactor:* In this system, ultrasonic intensities are very low. Generally, this equipment is used for emulsification, polymerization, and phase transfer reactions. Liquids are pumped at a high rate through a narrow gap onto a thin metal blade. This sets the blade into vibration with a sufficiently high frequency to cause cavitation, which causes the reaction to proceed.
- *Tube reactor:* Tube is surrounded by a radical transducer. Reaction mixture flows through this tube. The ultrasonic energy is focused toward the middle of the tube with much lower powers at the inner surface. In this way, erosion problems are reduced.

### 10.2.2 ORGANIC SYNTHESIS

The first report about the effect of ultrasound to chemical reactions is by Richards and Loomis (1927) involving rate studies on the hydrolysis of dimethyl sulphate and the iodine 'clock' reaction (the reduction of potassium iodate by sulfurous acid). Later Porter and Young (1938) reported that ultrasound increases the rate of the Curtius rearrangement.

The driving force for developments in the field of use of ultrasound in organic synthesis has many facets. The increasing requirement for environmental technology is that, which minimizes productions of waste at source (Cains et al., 1998). Ultrasound may offer cleaner reactions by improving product yields and selectivities, enhancing product recovery and quality through application to crystallization, product recovery and purification processes. Ultrasound enhances the rate particularly of those reactions, which involve free radical intermediates (Singh et al., 2001). Sonication allows the use of non activated and crude reagents as well as an aqueous solvent system and therefore, it is eco-friendly and non toxic. Ultrasound is widely used for improving the conditions, reducing long reaction times, avoiding high temperatures, unsatisfactory yield, and incompatibility with other functional groups.

Han and Boudjouk (1982) found significant increases yields and rates of Reformatsky reactions by ultrasound.

$$CH_3(CH_2)_6CHO + BrCH_2COOEt \xrightarrow[\bullet)))]{\text{Zn, Dioxane}} CH_3(CH_2)_6CHOH\ CH_2COOEt$$

It was observed that ultrasound promoted synthesis and gave 95% yield within 5min, whereas the conventional method gives only 61% yield with 12hr at 80°C on stirring.

Ceric ammonium nitrate (CAN) effectively catalyzes the three component condensation of an aldehyde, $\beta$-ketoester and urea in methanol to afford the corresponding dihydropyrimidinones in excellent yields under sonication. The reaction of benzaldehyde, ethyl acetoacetate, and urea offer 3.5 hr sonication results in the formation of 3,4-dihydropyrimidine-2(1H)-one in 92% yield, and it is believed to proceed through a single electron transfer with initial formation of a $\beta$-ketoester radical that adds to the imine intermediate (Yadav et al., 2001).



Li et al. (1999) improved Bucherer–Bergs method for synthesis of 5,5-disubstituted hydantoins using ultrasound and Claisen–Schmidt condensation for cycloalkanenes and chalcones.



| | Traditional | US |
|---|---|---|
| Time (h) | 4 | 3.5 |
| T (°C) | 150 (Autoclave) | 45 |
| Yield (%) | 70 | 89 |

Sonication methods were used to synthesize selenium heterocycles and anhydrides by Hu et al. (1996) and Wang and Zhao (1997). Ultrasound increased 5–10% yield of cis- and trans-2,6-diphenyl-1,4-diselenofulvenes. Traditionally, the anhydrides were produced under phase transfer condition at 10°C using aqueous NaOH as non organic phase. Using ultrasound, the anhydrides can be prepared in a single organic phase even at 45°C. The synthesis of indoles was also improved under sonication (Koulocheri and Harautounain, 2001). Ultrasound increased the yield by about 10–15% and reduced the reaction time to one-fifth under a polyphosphoric acid (PPA) condition.

Ultrasound can seriously affect photocatalytic ring opening of $\alpha$-epoxyketones by 1-benzyl-2,4,6-triphenylpyridinium tetrafluoroborate (NBTPT) as photocatalyst in methanol because of the efficient mass transfer of the reactions and the excited state of NBTPT (Memarian et al., 2007). The higher yields and shorter reaction times are the main advantages of this method.

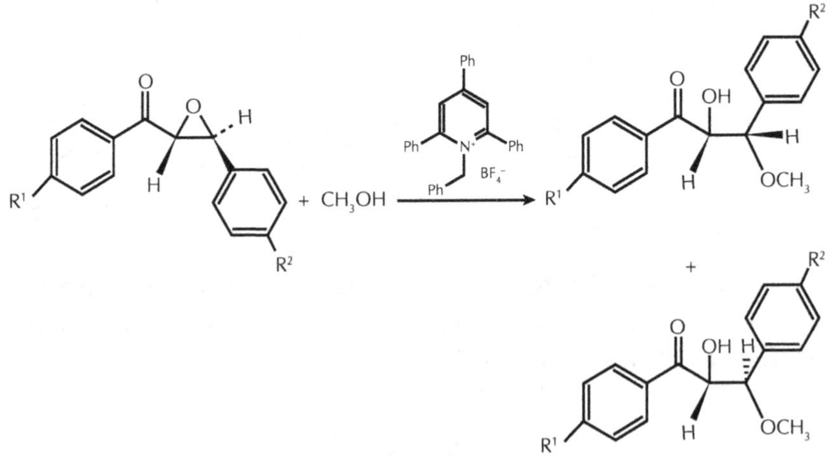

A multi-component synthesis of spiro-oxindoles was carried out in the presence of a catalytic amount of p-TSA as an inexpensive and available catalyst in EtOH under ultrasound irradiation (Dabiri et al., 2011). This method is quite simple. It starts from readily accessible commercial starting materials, and provides biologically interesting products in good yields and shorter reaction times.

When ultrasound is applied to an Ullmann reaction that normally requires a 10-fold excess of copper and 48hr of reaction time, it can be reduced to only 4-fold excess of copper and a reaction time of hardly 10hr. The particle size of the copper shrinks from 87 to 25μm, but the increase in the surface area cannot fully explain the increase in reactivity. It was suggested that sonication also assists the breakdown of intermediates and desorption of the products form the surface (Mason, 1997).

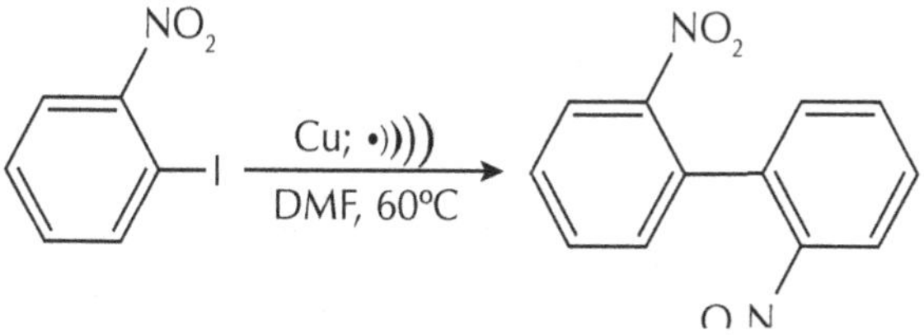

$NO_2$ $NO_2$
Cu; •)))) DMF, 60°C
$O_2N$

Typically, ionic reactions are accelerated by physical effects like better mass transport, which is also called "False sonochemistry". If the extreme conditions within the bubbles lead to totally new reaction pathways, for example via radicals generated in the vapor phase that would only have a transient existence in the bulk liquid, one speaks about "sonochemical switching". Such a switching has been observed in the following Kornbulum–Russel reaction, where sonication favors a single electron transfer SET pathway (Mason, 1997).

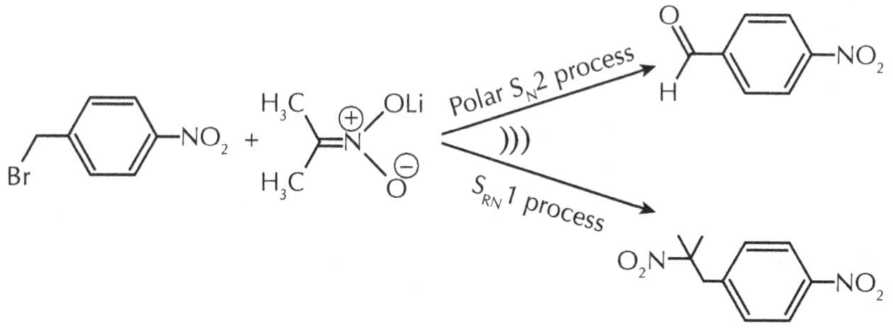

Br $NO_2$ + $H_3C$ $H_3C$ OLi N O
Polar $S_N2$ process
)))
$S_{RN}1$ process
O H $NO_2$
$O_2N$ $NO_2$

*β*-Amino-*α*,*β*-unsaturated esters are produced by a sonochemical reaction. Blaise reaction of nitriles, zinc powder, zinc oxide and ethyl bromoacetate in THF was reported in a commercial ultrasonic cleaning bath (Lee and Chang, 1997).

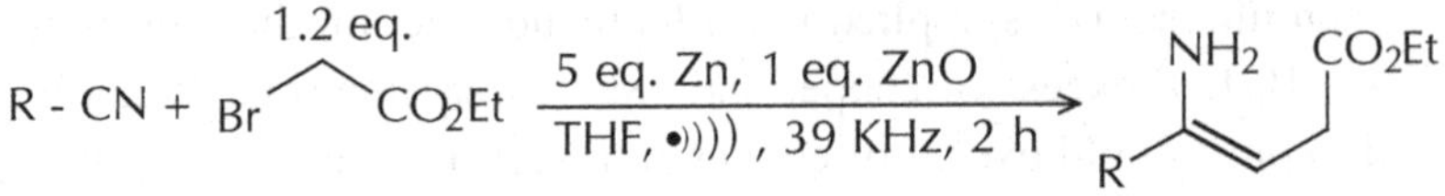

R - CN +
1.2 eq.
Br
CO2Et
5 eq. Zn, 1 eq. ZnO
THF, •))) , 39 KHz, 2 h
NH2
CO2Et
R

A palladium catalyzed and ultrasonic promoted Sonogashira coupling (1,3-dipolar cycloaddition of acid chlorides, terminal acetylenes and sodium azide) in one pot enables an efficient synthesis of 4,5-disubstituted-1,2,3-(NH)-triozoles in excellent yield (Li et al., 2009).

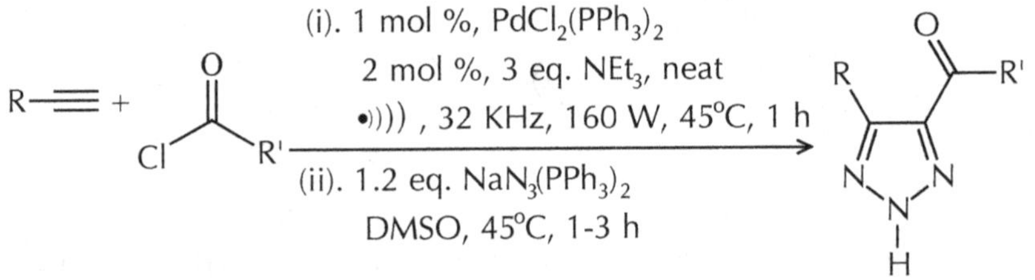

R
O
Cl
R'
(i). 1 mol %, PdCl2(PPh3)2
2 mol %, 3 eq. NEt3, neat
•))) , 32 KHz, 160 W, 45°C, 1 h
(ii). 1.2 eq. NaN3(PPh3)2
DMSO, 45°C, 1-3 h
O
R
R'
N
N
N
H

Structurally and functionally diverse N-carbamoylamino acids were obtained through the alkylation of monosubstituted parabanic acid followed by hydrolysis of the intermediate products in very good yields and excellent purity (Bogolubsky et al., 2008).

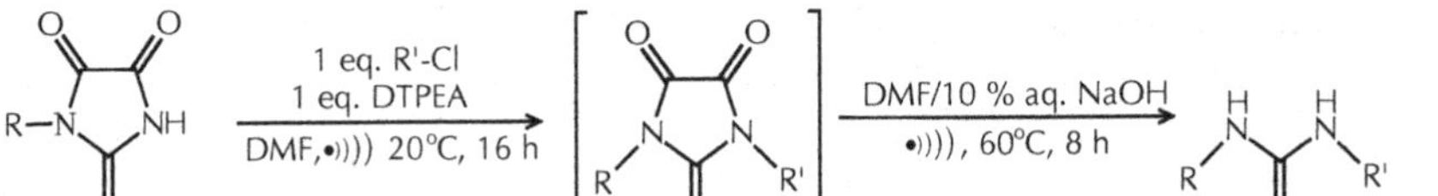

O
O
R—N
NH
O
1 eq. R'-Cl
1 eq. DTPEA
DMF,•))) 20°C, 16 h
O
O
N
N
R
R'
O
DMF/10 % aq. NaOH
•))), 60°C, 8 h
H
N
H
N
R
R'
O

Parabanic acid

N-Carbamoylamino acid

Hantzsch 1,4-dihyropyridine and polyhydroquinoline derivatives were synthesized in excellent yields in aqueous micelles. The reaction is catalyzed by PTSA and strongly accelerated by ultrasonic irradiation (Kumar and Mayura, 2008).

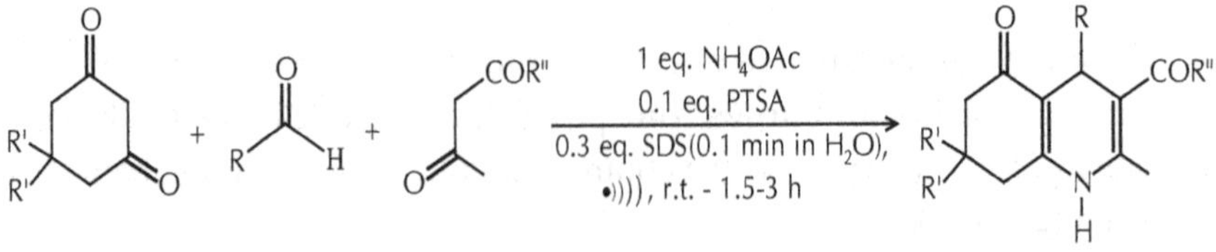

O
R'
R'
O
O
R
H
COR"
O
1 eq. NH4OAc
0.1 eq. PTSA
0.3 eq. SDS(0.1 min in H2O),
•))), r.t. - 1.5-3 h
O
R
COR"
R'
R'
N
H

Application of ultrasound accelerates the conversion of hydroxamic acid from carboxylic acids in the presence of 1-proponephosphonic acid cyclic anhydride (Vasantha et al., 2010). Further, T3P has also been employed to activate the hydroxamates, leading to isocyanates via Lossen rearrangement. Trapping with suitable nucleophile affords the corresponding ureas and carbonates (Vasantha et al., 2010).

$$\mathrm{RCOOH} \xrightarrow[\substack{2.3\ \text{eq. NMM} \\ \text{MeCN, ·))))\ , r.t.} \sim 60\ \text{h}}]{\substack{1.2\ \text{eq. } NH_2OH.HCl \\ 1.1\ \text{eq. T3P (50\% in EtOAc)}}} \mathrm{RCONHOH}$$

Hydroxamic acid

### *10.2.3 BIOLOGICAL REACTIONS*

Enzymes are being used for enhancing the reactivity of biological materials or organic synthesis because of the regio- and stereospecific nature of their reactions. Yeast cells can have the cuticle removed and enzyme released in active form by the use of ultrasound (Bujons et al., 1988). Attaching enzymes to hydrophobic surfaces of polymers or macromolecules presents a problem because reactants in aqueous solution will encounter the enzyme. Stirring can solve this problem but not very efficiently.

One of the most well known uses of ultrasound in biotechnology is for microbial cell disruption (Suslick, 1988). When a cellular material is placed in an ultrasonic field, the shock waves produced by surrounding cavitationsal events are capable of causing mechanical damage to the surrounding cellular materials. Other important use of ultrasound is to synthesize N-acetylamino acids from the amino acids and acetic anhydride without racemization (Reddy and Ravindranath, 1992). This reaction was later advantageously incorporated as a synthesis step in the production of $\alpha$, $\beta$- and cyclic spaglumic acids (Reddy and Ravindranath, 1992).

Lee et al. (1989, 1990a, 1990b) used the ultrasound promoted cycloadditions to synthesize series of natural compounds. The cycloaddition of diene **(1)** with enone **(2)** gave a 76% yield of cycloadducts with the desired regiomers in the ratio of 5:1 under sonication for 2hr at 45°C as compared with 15% yield of cycloadducts in a ratio of 1:1 when refluxed in benzene for 8hr. Deprotection of **(3)** yields natural compound tanshindiol B.

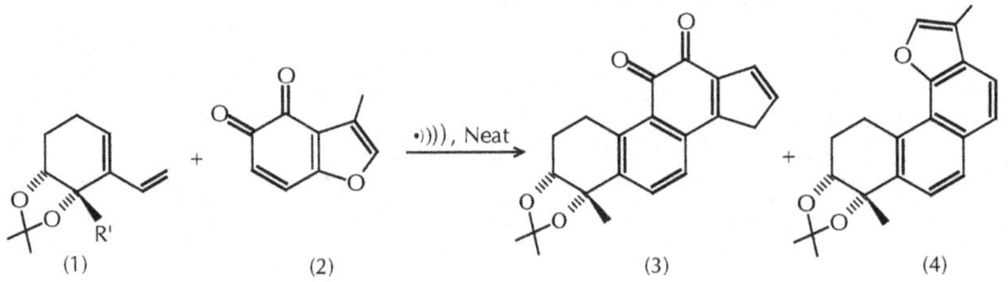

•))), Neat

(1) (2) (3) (4)

•)))), 2hr, 45°C, 76%, (3)/(4) = 5:1, Reflux in $C_6H_6$, 8hr, 15%, (3)/(4) = 1:1

Curcumin is a natural yellow colorant of turmeric (a spice used primarily as a food colorant and also to flavor several foods). It is a nutraceutical compound used worldwide for medicinal as well as food purposes. It has attracted special attention due to its potential pharmacological activities such as to protect cells from $\beta$-amyloid in Alzheimer's disease and cancer preventive properties. Curcumin is insoluble in water at acidic and neutral pH. In order to make it water soluble, it is envisaged that attachment of a polar group and molecule would enhance the hydrophilicity of the molecule. This can be achieved by making its suitable sugar derivatives.

Ultrasound brought about acceleration and increases the yields of the curcumin glucosides in the Koenigs–Knorr type reaction of 2,3,4,6-tetra-O-acetyl-$\alpha$-1-bromoglucose with the potassium salt of curcumin [bis-1,7-(3'-methoxy-4'-hydroxy) phenyl-5-hydroxy-1,4,6-heptatrien-3-one] under the biphasic reaction conditions in the presence of benzyltributyl ammonium chloride as a phase transfer catalyst. The reaction was stereoselective leading to the preponderant formation of either mono or di-$\beta$-glucoside tetraacetates of curcumin under controlled conditions in mono- and biphasic reactions, respectively.

α-Tetra-O-acetyl-α-D-glucopyranosyl bromide + $PhO^{\ominus}K^{+}$ (Potassium salt of curcumin) → ($PTC, CHCl_3/H_2O, 25°C$, •)))) → Curcumin-di-β-glucoside tetraacetate

2-Methoxy-6-alkyl-1,4-benzoquinones occur widely in nature; particularly in plants and most of them show potent biological activities, such as anticancer activity, radiosensitization activity, and 5-1 ipoxygenase inhibitory activity. Ultrasound assisted Witting reaction of alkyltriphenyl phosphonium bromides with O-vanillin in basic aqueous conditions followed by reduction with Na/n-BuOH gave 2-methoxy-6-alkylphenols. Oxidation of 2-methoxy-6-alkyl phenols with Fremy's salt produced 2-methoxy-6-alkyl-1,4-benzoquinenes (Wu et al., 2009).

O-Vanillin → ($RCH_2PPh_3Br$, $K_2CO_3/H_2O$, DMSO, •)))), 90-100°C) → ($90-100°C$, 1 h, Na/n-BuOH) → 2-Methoxy-6-alkyl phenol → ($(KSO_3)_2NO$ (Fremy's salt)) → 2-Methoxy-6-alkyl-1,4-benzoquinane

N-Arylhyroxylamines are an important class of compound frequently used as key intermediates in the synthesis of fine chemical, natural products and some promising biologically active compounds. They also display a wide range of physiological and pharmacological activities. Zn/$HCOONH_4/CH_3CN$ system is used for preparing N-arylhydroxylamines by the reduction of the corresponding nitroarenes under ultrasound (Xun et al., 2009). This method is quite efficient, environmentally benign, highly chemoselective, and especially simple and most practical.

$NO_2$ —— •)))), 25°C; Zn, $CH_3CN/HCOONH_4$ ——→ NHOH

R R

Nitroarene N-Arylhydroxylanine

Carbazole and especially heterocycle-containing carbazole derivatives are embodied in many naturally occurring products and possess a broad spectrum of useful biological activities such as antitumor, antimitotic, and antioxidative activities. They are also widely used as building blocks for new organic materials and play a very important role in electroactive and photoactive devices. On the other hand, the benzofuran derivatives also show some important biological properties, such as antimicrobial, anticonvulsant, anti-inflammatory, antitumor, and antifungal activities.

### *10.2.4 POLYMER SYNTHESIS*

Ultrasound has been successfully applied to a wide range of polymer synthesis. High frequency ultrasound in the range of 1–10MHz has been applied to the determination of structure and conformation of polymers as reviewed by Pethrick (1991). The chemical effects of ultrasound arise from cavitation that is, the collapse of microscopic bubbles in a liquid. Upon implosion of a cavity, extreme conditions in the bubble occur (5,000K and 200bar) and high strain rates are generated outside the bubble ($10^{-1}$ $s^{-1}$). Monomer molecules are dissociated by the high temperature inside the hot spot, whereas polymer chains are fractured by the high strain rates outside the cavitation bubble (Madress et al., 2002). These two reactions lead to the formation of radicals, which can initiate a free radical polymerization. The majority of the radicals are generated by scission of polymer chains (Kuijpers et al., 2001).

An important parameter in ultrasound induced bulk polymerizations is the viscosity. The polymerization reactions yielded high molecular weight polymers. In a pen reaction, the long chains formed cause a drastic increase in the viscosity. A high viscosity hinders cavitation and consequently reduces the production rate of radicals. Emulsion and precipitation polymerization provide a potential solution to this problem. Ultrasound induced bulk polymerizations are usually performed at room temperature. This low temperature is chosen because radical formation induced by ultrasound is more efficient at lower temperatures. In emulsion polymerizations, a heterogeneous reaction system is involved. The polymers are insoluble in the continuous aqueous phase and therefore, the viscosity of the water phase does not increase upon reaction. Ultrasound induced emulsion polymerization is a well studied system, where indeed high conversions can be obtained (Cooper et al., 1996; Zhang et al., 2002). During precipitation polymerizations, the polymer precipitates from the reaction mixture, resulting in a constant viscosity and thus, a constant radical formation rate by ultrasound. In this perspective, high pressure carbon dioxide ($CO_2$) is an interesting medium as most monomers have a high solubility in $CO_2$, whereas it exhibits an antisolvent effect for most polymers.

## *(I) RADICAL POLYMERIZATION*

Generally, free radical polymerization consists of four elementary steps; initiation, propagation, chain transfer, and termination (Rudin, 1990). When ultrasound is used to initiate polymerization, radicals can be formed from monomer and from polymer molecules both (Feldman, 1995).

Fracture of the polymer chain occurs at a random site in degradation of polymers. An alternative method is ultrasound induced polymer scission, which involves a much better controlled, non random process (Niezette and Linkens, 1978). Ultrasound induced polymer breakage is a direct consequence of cavitation, because under conditions that suppress cavitation, no degradation was observed. In this non random scission process, the polymer is fractured at the centre of the chain (Glyn et al., 1972; Vander Hoff and Gall, 1977).

One of the applications of ultrasound induced polymer scission is the production of block copolymers. The synthesis of block copolymers by ultrasound starts with the dissolution of a homopolymer in a different monomer. Subsequently, ultrasonic scission of the polymer chains generates polymeric radicals, which initiate the polymerization reaction with the monomer present. In this way, ultrasound provides the controlled formation of block copolymers. Solution of two different polymers in a non reactive solvent can also lead to the formation of block copolymers. In this case, the generated polymeric radicals have to undergo termination by cross combination.

During the synthesis of alkyl silicon network polymers, the poly(alkylsilynes), and $(SiR)_n$ were greatly facilitated by the sonochemically generated NaK emulsions in hydrocarbon solvent (Bianceni et al., 1989). By preventing the passivation of the reductant (NaK) with salt and growing polymer, sonication initiates the reductive condensations of alkyltrichlorosilanes in inert saturated hydrocarbon solvents; thereby, preventing the complications and side reactions often associated with ethereal solvent and electron transfer reagents.

The preparation of polyurethanes from a number of diisocyanides and diols under sonication was also reported (Price et al., 2002). The sonication made the reactions faster at the early stages and led to higher molecular weights in all cases. In the process of searching for clean and low emission polymerization techniques, supercritical fluid technology, sonochemistry, and microemulsion techniques have attracted more and more interest because of their unique advantages over conventional techniques.

Ultrasound enhanced atomization of viscoelastic liquids such as gel forming xanthan gum (xan) solutions and applications to nanoparticles synthesis was reported by Liu et al. (2000). Zhang et al. (2009) reported sonochemical preparation of polymer nanocomposites. Oxygen permeability of the samples was studied and it was found that the oxygen flow rate was reduced by the combined effect of clay loading and ultrasound. The flame retardant property of the nanocomposites due to clay dispersion was also investigated by measurement of limiting oxygen index (LOI) (Patra et al., 2012).

### 10.2.5 HETEROGENEOUS CATALYSIS

(i) *Liquid–liquid systems:* Ultrasound forms very fine emulsions in systems with two immiscible liquids, which is very beneficial when working with phase transfer catalyzed or biphasic systems. When very fine emulsions are formed, the surface area available for reaction between the two phases is significantly increased; thus, increasing the rate of the reaction. This aspect of ultrasound has also been used for coal, oil, and water mixtures to increase the efficiency of combustion, as well as to decrease the amount of pollutants produced during the combustion process (Dooher et al., 1980).

(ii) *Liquid–solid systems:* The most pertinent effects of ultrasound on liquid–solid systems are mechanical and are attributed to symmetric and asymmetric cavitations. When a bubble is able to collapse symmetrically, localized area of high temperatures and pressures are generated in the fluid. In addition, shock waves are produced, which have the potential of creating microscopic turbulence within interfacial films surrounding nearby solid particles, also referred to as microstreaming (Elder, 1959).

The allylation of ketones and aldehydes by allylic alcohols has been improved using ultrasonic irradiation of a palladium-tin dichloride catalyst in less polar solvents. Inverted regioselectivity was observed as compared with homogeneous carbonyl allylation in polar solvents.

$$H_3OCH = CHCH_2OH + H_5C_6CHO \xrightarrow[Pd/SnCl_2]{\bullet))))} H_3CCH = CHCH_2CH\,(OH)\,C_6H_5$$

Carbon sonogels have been produced by Tonanon's group (2005a, 2005b). Irradiation with high intensity ultrasound promotes the sol-gel polycondensation of resorcinol and formaldehyde, typical precursors for carbon supports. As with the oxides, gelation occurs more rapidly, when exposed to ultrasonic irradiation. The resulting carbon gels have increased mesoporosity as compared with those prepared without ultrasound and moderate surface areas (500–800m$^2$ g$^{-1}$). Such materials could be ideal supports for Platinum-based fuel cell catalysts.

A nanostructured, bifunctional catalyst, $Mo_2C$/ZSM-5 was prepared by irradiation of $Mo(CO)_6$ and HZSM-5 in a slurry with hexadecane (Dontsin and Suslick, 2000). Dhas et al. (2001) prepared Co and Ni promoted $MoS_2$ supported on alumina through high intensity ultrasonic irradiation of isobutene slurries containing $Mo(CO)_6$, $Co_2(CO)_8$, elemental sulfur and $Al_2O_3$ or Ni-$Al_2O_3$ under argon flow. Lee et al. (2003) have also prepared $MoS_2/Al_2O_3$. Through the use of ultrasound, higher loadings of Mo can be achieved, resulting in a more active catalyst. A CoMoS/$Al_2O_3$ catalyst has also been prepared by combining sonochemical and CVD techniques (Lee et al., 2005).

The reactivity of p-phenyl substituted $\beta$-enamino compounds using the acidic clay montmorillonite, K-10, as a solid support under sonication was investigated (Valduga et al., 1997; 1998). The results indicated the influence of the K-10 support on the regiochemistry of these reactions. For steric reasons in the first step of this reaction, the interaction of K-10 with the nitrogen of the amino group makes the carboxylic carbon more electrophilic and the addition of the methyl hydrazine occurs by initial addition of the unsubstituted nitrogen followed by cyclization to give pyrazole. However, for compound **(8)**, the regiochemistry was inverted. It was believed that this is due to a stronger interaction between K-10 and the nitro group than that between K-10 and the nitrogen or oxygen atoms of enamino ketone; thereby, moving the reaction path to the more conventional one.

$NH_2NHCH_3$, K-10/US or EtOH/reflux

(5) - (8) → (9) and/or (10)

| | R | Ratio of isomers EtOH/ reflux | | 17 and 18 K-10/US | |
|---|---|---|---|---|---|
| | | **(9)** | **(10)** | **(9)** | **(10)** |
| **(5)** | H | 0 | 100 | 100 | 0 |

**TABLE** *(Continued)*

| | | | | | | |
|---|---|---|---|---|---|---|
| **(6)** | Me | 55 | 45 | 93 | 7 | |
| **(7)** | OMe | 45 | 55 | 94 | 6 | |
| **(8)** | $NO_2$ | 0 | 100 | 20 | 80 | |
| **(5)** | | **(6)** | **(7)** | | | **(8)** |
| R=H | | Me | OMe | | | $NO_2$ |

A comparison between conventional and ultrasound assisted heterogeneous catalysis that is, hydrogenation of 3-buten-1-ol in aqueous solution was reported by DieselKamp et al. (2004). Iron containing SBA-15 material (Fe-SBA-15) is a promising catalyst for the treatment of phenolic aqueous solutions by coupling ultrasound with heterogeneous catalytic wet peroxide oxidation (Molina et al., 2006). The optimal hydrogen peroxide concentration was two times the stoichiometric amount, achieving results of TOC degradation ranging from 30 to 40%, which represents a low oxidant dosage. The catalyst loading plays more important role than the oxidant concentration in the activity of the sono-Fenton catalytic system. The remarkable stability (less than 4ppm loss for the best reaction conditions) of the Fe-SBA-15 heterogonous catalyst under ultrasonic irradiation is particularly noteworthy.

A diverse set of applications of ultrasound have been explored in the synthesis of nanostructured materials including both; direct sonochemical reactions and ultrasonic spray pyrolyses (Bang and Suslick, 2010). The usefulness of sonochemistry as a synthetic tool resides in its versatility. One can produce spherical nanoparticles with narrow particle size distribution. Final particle sizes depend on generated precursor droplet size distribution, if one assumes that every droplet undergoes same process steps and transform to a particle.

### 10.2.6 ORGANOMETALLIC PROCESS

The use of high intensity ultrasound to enhance the reactivity of metals as stoichiometric reagents has became an important synthetic technique for many heterogeneous organometallic reactions, especially those involving reactive metals, such as magnesium, lithium, and zinc. This development originated form the early work of Renaud (1950) in France, when Grignard reagent was formed under the influence of ultrasonic waves.

$$R-X+Mg \xrightarrow{\bullet)))} R-Mg-X$$

A few examples of other sonochemical reactions are

$$C_6H_5Br + Li \xrightarrow{\bullet)))} C_6H_5Li + LiBr$$

$$R-Br + Li + R_2'NCHO \xrightarrow[H_2O]{\bullet)))} RCHO + R_2'NH$$

$$2\ o\text{-}C_6H_4(NO_2)I + Cu \xrightarrow{\bullet)))} o\text{-}(O_2N)H_4C_6\text{-}C_6H_4(NO_2) + 2\ CuI$$

$$R\,R'HC-OH + KMnO_4\ (s) \xrightarrow{\bullet)))} R\,R'C=O$$

$$C_6H_5CH_2Br + KCN \xrightarrow[Al_2O_3]{\bullet)))} C_6H_5CH_2CN$$

$$MCl_5 + Na + CO \xrightarrow{\bullet)))} M(CO)_6 \quad (M = V, Nb, Ta)$$

Another application of sonochemistry involves the preparation of amorphous metals. If one can cool a molten metal alloy quickly enough, it can be frozen into a solid before it has a chance to crystallize. Such amorphous metallic alloys lack long range crystalline order and have unique electronic, magnetic, and corrosion resistant properties. The production of amorphous metals; however, is difficult because extremely rapid cooling of molten metals is necessary to prevent crystallization.

Dioxane is not a suitable solvent for common Reformatsky reaction, because it has a tendency to promote enolization. Neither in ether nor benzene, the common Reformatsky solvents produced high yields even after several hours of sonication. A insonated Reformatsky reaction for the preparation of a $\beta$-lactam from Zn, ethylbromoacetate and a diaryl Schiff base gave 95% yield with 4hr at room temperature, whereas the conventional method of refluxing in toluene gives only 60% yield (Bose et al., 1984). An ultrasound promoted synthesis of hydroesters by Reformatsky reactions using indium metals was reported by Lee et al. (2001). THF is another good solvent. The reaction of benzaldehyde with ethyl bromoacetate in the presence of indium in THF afforded ethyl 3-hydroxy-3-phenylpropanoate in 97% yield in 2hr as compared to 70% yield in 17hr with stirring.

Luche et al. investigated the use of sonication in a variety of organometallic reactions (1982). Sonication plays an essential role in the formation on stirred or low density irradiation of a mixture of n-butyl bromide, lithium and 2-cyclohexanone in presence of copper (I) iodide, yielding mostly 1-n-butyl-2-cyclohexan-1-ol. They investigated the sonicated preparation of organozinc reagents and their conjugate addition to $\alpha$-enones (Luche et al., 1982; Petrier et al., 1984). Later it was found that these organometallic reagents, prepared by sonication, gave rise to clean and selective conjugate additions to $\alpha,\beta$-unsaturated aldehydes and ketones in the presence of catalytic amounts of nickel acetylacetonate (Petrier et al., 1985). Einhorn et al. (1986a; 1986b; 1988) also investigated the Bouveault reactions. They discovered that reaction intermediate prepared under sonochemical conditions easily undergoes ortho-directed lithiation. The use of tetrahydropyran as solvent dramatically increases the rates and yields of metallation, which can be accomplished with an *in situ* generated alkyl lithium.

Br + Li HOC N US THP OLi N N

12 eq. n-BuLi THP; RT 0.25-0.5 h

LiO N Li N (i). DMF (ii). $H_3O^+$ CHO CHO

62 % Yield

Another impressive results are *in situ* generation and use of butyllithium reagents under sonochemical conditions (Einhorn and Luche, 1987;

Jayne et al., 1988). Lithium diisopropylamide (LDA) can be prepared without much effort from diisopropylamine, lithium, and butyl halides by sonicating the mixture in dry THF at 15–18°C. Much better yield was obtained from butyl chloride (92%) than bromide (40%).

The Simmons–Smith cyclopropanation of alkenes with diiodomethane and Zn under sonication has been investigated by Repic and Vogt (1982). Methyl oleate gave a 99% yield of cyclopropanated adduct after 2hr sonication as compared to 50% yield without sonication, while (-)-$\alpha$-pinene gave a 90% yield under 4hr insonication and 12% without sonication.

$CH_2I_2$ / Zn, •))) → 90%

Arylzinc compounds containing electron withdrawing groups, such as $CO_2CH_3$, $CON(CH_3)_2$, CN, Br, Cl, or $CF_3$ at ortho position prepared under sonication were applied to palladium (o)-catalyzed cross-coupling with aryl halides (Takagi, 1993). Methyl 2-iodobenzoate reacted with Zn under sonication gave 87% aryl zinc compound in 1,1,3,3-tetramethylurea, which was coupled with ethyl 2-bromobenzoate in 100% yield. However, electron donating groups diminished the reactivities of aryl iodides greatly, so that the reaction did not reach to completion under sonication even at 50°C.

$COCH_3$, I → Zn, TMU, •))), 87% → $COCH_3$, ZnI → Br-$C_6H_4$-$COCH_2CH_3$, Pd (O), 100% → $COCH_3$ / $COCH_2CH_3$

Phenylenedizinc (II) compounds were also prepared in N,N,N,N'-tetramethy-lethylenediamine and 1,1,3,3-tetramethylurea under sonication

(Takagi et al., 1994), which can be used for Pd (o) catalyzed synthesis of symmetrically 1,2-disubstituted benzenes. However, the reaction did not succeed, when coupled with non-aryl group.

The ultrasound was used for the conjugate addition to α-enones in the presence of zinc-copper couple. Ultrasound increased the yields from poor (20–30%) to good or excellent (Luche and Allavena, 1988, Luche et al., 1988). The ultrasound irradiation was effective in enhancing the reactivity of organomagnesium reagents towards ethylene ketals of *α,β*-unsaturated aldehydes and it is an efficient alternative to traditional heating or using Lewis acid catalysis (Lu et al., 1998).

$H_2C = CHCH_2MgBr$, Toluene → (11) + (12) + (13)

| | ))) | Stirred |
|---|---|---|
| Yield | 100% | 17% |
| ratio **11** : **(12)** + **(13)** | 85 : 15 | 84 : 16 |

## 10.2.7 SCALE UP CONSIDERATION

The exposure of a chemical reaction to ultrasound creates bubbles, which will explode creating very high temperature and pressure. This drives a chemical reaction and yields some product. The ultrasound can be used to synthesize some organic compounds, polymerized some monomers as well as destroy some of the organic contaminants. Sonochemistry has a lot of possibilities, but it has not been investigated in detail so far.

## KEYWORDS


- **Heterogeneous catalysis**
- **Organometallic process**
- **Polymer synthesis**
- **Sonochemistry**
- **Ultrasound**


## REFERENCES


1. Abedini, R., & Mousavi, S. M. (2010). *Petroleum and Coal, 52,* 81–98.
2. Bang, B. J. H., & Suslick, K. S. *(2010). Advance Material, 22,* 1039–1059.
3. Bianceni, P. A., Schilling, F. C., & Weidman, T. W. (1989). *Macromolecules, 22,* 1697–1704.
4. Bogolubsky, A. V., Ryabukhin, S. V., Pakhomo, G. G., Ostapchuk, E. N., Shivanyuk, A. N., & Tolmachev, A. A. (2008). *Synlett,* 2279–2282.
5. Bose, A. K., Gupta, K., & Manhas, M. S. (1984). *Journal of the Chemical Society, Chemical. Communication,* 86–87.
6. Bujons, J., Guajardo, R., & Kyler, K. B. (1988). *Journal of the American Chemical Society, 110,* 604–606.
7. Cains, P. W., Martin, P. D., & Price, C. J. (1998). *Organic Process Research and Development, 2,* 34–48 .
8. Cooper, G., Grieser, F., & Biggs, S. *(1996). Journal of the Colloid and Interface Science, 184,* 52–63.
9. Curie, J., & Curie, P. (1881). *Comptes Rendus, 93,* 1137–1140.
10. Curie, J., & Curie, P. (1880). *Comptes Rendus, 91,* 294–295.
11. Dabiri, M., Tisseh, Z. N., Bahramnejad, M., & Bazgir, A. (2011). *Ultrasonics Sonochemistry, 18,* 1153–1159.
12. Deiselkamp, R. S., Judd, K. M., Hart, T. R., Peden, C. H. F.; Posakeny, G. J., & Bend, L. J. (2004). *Journal of Catalysis, 221,* 347–353.
13. Dhas, N. A., Ekhtiarzadeh, A., & Suslick, K. S. (2001). *Journal of the American Chemical Society, 123,* 8310–8316.
14. Dontsin,V. & Suslick, K. S. (2000). *Journal of the American Chemical Society, 122,* 5214–5215.
15. Dooher,V., Gengerg,V., Moon, R., Gilmartin, S., Jakatt, B., Skura, S., & Weight, J. (1980). *Fuel, 59,* 883–891.
16. Einhorn, J., Einhorn, C., & Luche, J. L. (1988). *Tetrahedron letters, 29,* 2183–2184.
17. Einhorn, J., & Luche, J. L. (1987). *Journal of Organic Chemistry, 52,* 4124–4126.
18. Einhorn, J., Luche, J. L. (1986a). *Tetrahedron letters, 27,* 1791–1792.

19. Einhorn, J., & Luche, J. L. (1986b). *Tetrahedron Letters, 27,* 1793–1796.
20. Elder, S. A. (1959). *Journal of the Acoustical Society America, 31,* 54–64.
21. Feldman, D. (1995). *Polymer News, 20,* 138.
22. Galton, F. (1883). *Inquiries into human faculty and development.* London: McMillan.
23. Glyn, P. A. R., Vander Hoff, B. M. E., & Rully, P. M. (1972). *Journal of Macromolecular Science Chemistry, A6,* 1653–1664.
24. Han, B. H., & Boudjouk, P. J. (1982). *Journal of Organic Chemistry, 47,* 5030–5032.
25. Hu, Y., Wang, J., & Li, S. (1997). *Synthetic Communication, 27,* 243–248.
26. Jayne, C. S. B., Petrier, C., & Luche, J. L. (1988). *Journal of Organic Chemistry, 50,* 1212–1218 .
27. Koulocheri, S. D., & Harautounain, S. A. (2001). *European Journal of Organic Chemistry,* 1723–1729.
28. Kuijpers, M. W. A., Kemmere, M. F., & Keurentjes, J. T. F. (2001). *Annual AICHE Meeting, 341B.*
29. Kumar, A., & Mayura, R. A. (2008). *Synlett,* 883–885.
30. Lee, A. S. Y. & Chang, R. Y. (1997). *Tetrahedron Letters, 38,* 443–446.
31. Lee, J. J., Kim, H., Koh, J. H., Jo, A., & Meon, S. H. (2005). *Applied Catalysis B, 58,* 89–95.
32. Lee, J. J., Kim, H., & Moon, S. H. (2003). *Applied Catalysis B, 41,* 171–180.
33. Lee, J., Mei, H. S., & Snyder, J. K. (1990b). *Journal of Organic Chemistry, 55,* 5013–5016.
34. Lee, J., & Snyder, J. K. (1990a). *Journal of Organic Chemistry, 55,* 4995–5008.
35. Lee, P. H., Bang, K. Lee, K., Sung, S. Y., & Chang, S. (2001). *Synthetic Communication, 31,* 3781–3789.
36. Lee, J. & Snyder, J. K. (1989). *Journal of the American Chemical Society, 111,* 1522–1524.
37. Li, J., Chen, G., Wang, J., & Li, T. (1999). *Synthetic Communication, 29,* 965–971.
38. Li, J., Wang, D., Zhang, Y., Li, J., & Chen, B. (2009). *Organic Letters, 11,* 3024–3027.
39. Lickiss, P. D. & McGrath, V. E. (1996). *Chemical Bromine, 32,* 47–50.
40. Liu,V., Song, V. L., Tsai, C. S., & Lin, H. M. (2000). *Ultrasonics Symposium, 1,* 687–690.
41. Lu, T., Cheng, S., & Sheu, L. (1998). *Journal of Organic Chemistry, 63,* 2738–2741.
42. Luche, J. L. & Allavena, C. (1988). *Tetrahedron Letters, 29,* 5369–5372.
43. Luche, J. L., Allavena, C., Petrier, C. & Dupuy, C. (1988). *Tetrahedron Letters, 29,* 5373–5374.
44. Luche, J. L., Petrier, C., Gemal, A. L., & Zikra, N. (1982). *Journal of Organic Chemistry, 47,* 3805–3806.
45. Luche, J. L., Petrier, C., Landsard, J. P., Greene, A. E. (1983). *Journal of Organic Chemistry, 48,* 3837–3839.
46. Madress, G., Kumar, S., & Chattopadhyay, S. (2002). *Polymer Degradation Stability, 69,* 73–78.
47. Mason, T. J. (1997). *Chemical Society Reviews, 26,* 443–451.
48. Memarian, H. R., & Teluri, A. S. (2007). *Beilstein Journal of Organic Chemistry, 3.* doi: 10.1186/1860-5397-3-2
49. Molina, R., Martinez, F., Melero, J. A., Bremner, O. H., & Chakinala, A. G. (2006). *Applied Catalysis B Environment, 66,* 198–207.

50. Niezette, J., & Linkens, A. (1978). *Polymer, 19,* 939–942.
51. Patra, S. K., Prusty, G., Swain, S. K. (2012). *Bulletin of Material Science, 35,* 27–32.
52. Pethrick, R. A. (1991). *Advances in Sonochemistry, 2,* 65–133.
53. Petrier, C., Luche, J. L., & Deepuy, C. (1984). *Tetrahedron Letters, 25,* 3463–3466.
54. Petrier, C., Tayne, C. S. B., Dupuy, C., & Luche, J. L. (1985). *Journal of Organic Chemistry, 50,* 5761–5765.
55. Porter, C. W., & Young, L. (1938). *Journalof the American Chemical Society, 60,* 1497–1500.
56. Price, G. J., Lenz, V., & Ansell, C. W. G. (2002). *European Polymer Journal, 38,* 1531–1536.
57. Reddy, A. V., & Ravindranath, B. (1992). *International Journal of Peptide and Protein Research, 40,* 472–476.
58. Reddy, A. V., & Ravindranath, B. (1992). *Synthetic Communication, 22,* 257–264.
59. Renaud, P. (1950). *Bulletin de la Société Chimique de* (France), 1044–1045.
60. Repic, O., & Vogt, S. (1982). *Tetrahedron Letters, 23,* 2729–2732.
61. Richards, W. T. & Loomis, A. L. (1927). *Journal of the American Chemical Society, 49,* 3086–3100.
62. Rudin, A. (1990). *The elements of Polymer Science and Engineering.* San Diego: Academic Press.
63. Singh, J., Kaur, J., Nayyar, S., Bhandari, M., & Kaal, G. L. (2001). *Indian Journal of Chemistry, 40B,* 386–390.
64. Suslick, K. S. (1990). *Science, 247,* 1439–1445.
65. Suslick, K. S. (1988). *Ultrasound: It's chemical, physical and biological effects.* New York: VCH.
66. Suslick, K. S.; Casadonte, D. J. (1987) *Journal Of the American Chemical Society, 109,* 3459–3461.
67. Suslick, K. S., Doktyoz,V., Flint, V. (1990) *Ultrasonics, 28,* 280–290.
68. Takagi, K. (1993). *Chemistry Letters,* 469–472.
69. Takagi, K., Shimoishi, V., Sasaki, K. (1994). *Chemistry Letters,* 2055–2058.
70. Thornycraft, J., Barnaby, S. W. (1895). *Tospedo Boot Destroyers, Proceeding-Institution Civil Engineers, 122,* 51–103.
72. Tonanon, N., Siyasukh, A., Tanthapanichaken, W., Nishihara, H., Mukei,V., Tamon, H. (2005a) *Carbon, 43,* 525–531.
73. Tonanon, N., Siyasukh, A., Wareenin, Y., Charinpenitkul, T., Chaken,V., Nishihara,V. Mukai, S. R., & Tamen, H. (2005b). *Carbon, 43,* 2808–2811.
74. Valduga, C. J., Breibante, H. S., & Braibante, M. E. F. (1997). *Journal of Heterocyclic Chemistry, 34,* 1453–1457.
75. Valduga, C. J., Santis, D. B., Brai, H. S., & Braibente, M. E. F. (1998). *Journal Of Heterocyclic Chemistry, 36,* 505–508.
76. Vander Hoff, B. M. E., & Gall, C. E. (1977). *Journal of Macromolecular Science, A11,* 1739–1758.
77. Vasantha, B., Hemantha, H. P., & Sureshbabu, V. V. (2010). *Synthesis,* 2990–2996.
78. Wada, Y., Yin, H., Kitamura, T., & Yanagida, S. (2000). *Chemistry. Letters, 6,* 632–633.
79. Wang D., & Sakakibaro M. (1997). *Ultrasonic Chemistry, 4,* 255–261.

80. Wang, J., Zhao, K. (1996). *Synthetic. Communications, 26,* 1617–1622.
81. Wu, L. Q., Yang, C. G., Yang, L. M., & Yang, L. J. (2009). *Journal of the Chinese Chemical. Society, 56,* 47–50.
82. Xun, S. Q., Wen, L. R., Yu, H. X., Hai, L. L., & Zhang, F. (2009). *Chemical Research Chinese Universities, 25,* 183–188.
83. Yadav, S. J., Reddy, B. V. S., Reddy, K. B., Raj, K. R., Prasad, A. R. (2001). *Journal of the Chemical Society, Perkin Transaction, 1,* 1939–1941.
84. Yuji, W., Hengbo, Y., Shozo, Y. (2002). *Catalysis Surveys Japan, 5,* 127–138.
85. Zhang, C., Wang, Q., Xia, H., & Qiu, G. (2002). *European Polymer Journal, 38,* 1769–1776.
86. Zhang, K., Park, B. J., Fang, F. F., & Choi, H. J. (2009). *Molecules, 14,* 2095–2110.

CHAPTER 11

# MICROWAVE ASSISTED ORGANIC SYNTHESIS: A NEED OF THE DAY

CHETNA AMETA, K. L. AMETA, B. K. SHARMA, and RAJAT AMETA

## CONTENTS

## 11.1 INTRODUCTION

After almost one and half century of the first chemical revolution, a new kind of chemical revolution has come up that is, "Green chemistry". The fundamental idea of green chemistry is that the manufacturer of any chemical needs to consider, what will be the fate of human life after this particular chemical is generated and used in society.

Twelve principles of green chemistry can be used to access a particular synthetic protocol's greenness (Anastas and Warner, 2000). These principles address some basic aspects, such as use of various solvents; the amount of chemical waste produced; the use of catalyst and reagents (quantity and reusability); the amount of chemical step (energy efficiency) and atom economy; and the use of safer chemicals and reaction conditions. It is very difficult for a new synthetic protocol to satisfy all the 12 principles, which is not expected also, but if most of the principles are satisfied by a protocol, the developed process will be considered as a green route.

There are two alternative ways to categorize different approaches from green chemistry point of view.

- Synthesis via an environment friendly synthetic pathways or processes.
- To develop some new benign replacements, which are capable of achieving the desired performance without any adverse ecological impacts.

One can achieve greener protocol through a proper choice of starting materials (feed stock), atom economic methodologies with minimum chemical steps, the use of appropriate greener solvents and reagents, and efficient strategies for isolation of product and its purification. Thus, a major goal of this protocol should be to maximize simultaneously the efficient use of safer raw materials on one hand and to reduce the wastes produced in a particular process on the other.

Therefore, there is need to find solutions to problems faced in developing green and sustainable synthetic methods in the fields of healthcare and fine chemicals like posing significant challenges to the synthetic organic chemists, the pressure to produce these substances expeditiously and that too in an environmentally benign fashion. It is worthwhile to note that rapid development of "Green Organic Chemistry" is due to the recognition that ecofriendly products and processes will be economical in the long term as these do not require treatment of "end-of-the-pipe" pollutants, and byproducts so commonly generated by conventional synthetic procedures and one such technology is microwave assisted organic synthesis (MAOS) (Kingston and Haswell, 1997; Lidstrom and Tiernery, 2005).

Microwave heating under controlled conditions is an invaluable technology because it not only dramatically reduces reaction times, typically from days or hours to minutes or even seconds, that is speed up the reaction, but it also fulfills the aim of green chemistry by reducing side reactions increasing yields and improves reproducibility (Hayes, 2002). This approach has now become a central tool in this rapid paced, time sensitive field and it has also blossomed into a useful technique for a variety of applications in organic synthesis, where high yielding protocols and facility of purification are highly desirable (Bradley, 2001; Kuhmert, 2002). Furthermore, this technique is energy efficient and the possibilities for application in combinatorial, parallel and automated environmentally benign chemistry are obvious.

## 11.2 MICROWAVE ASSISTED CHEMISTRY

### *11.2.1 MICROWAVES AS ENERGY SOURCE*

Microwave (MW) radiations are electromagnetic radiations, which are widely used as a source of heating in organic synthesis. Microwaves have enough momentum to activate reaction mixture to cross energy barrier and complete the reaction in lesser time. A microwave oven consists of a magnetron, a wave guide feed and an oven cavity. A magnetron is a thermionic diode that works on the principle of dielectric heating by converting part of the electric power into electromagnetic energy and the rest of it into heat energy. Microwaves occupy a place in the electromagnetic spectrum between infrared waves and radio waves, ranging in wavelengths between 0.01 and 1m, and operate in a frequency range between 0.3 and 30GHz. The typical bands for industrial applications are 915 ± 15 and

2450 ± 50MHz. The wavelength between 1 and 25cm are extensively used for RADAR transmissions and the remaining wavelength range is used for telecommunications. The entire microwave region is therefore not available for heating applications and the equipment operating at 2.45GHz, corresponding to a wavelength of 12.2cm, is quite commonly used. The energy carried by microwave at 2.45GHz is 1 joule per mole of quanta, which is relatively very small energy.

### *11.2.2 MICROWAVES AS A TOOL FOR SYNTHETIC CHEMISTRY*

The earliest description of the magnetron (the high-power generator of microwave power), a diode with a cylindrical anode was reported by Hull (1921a, 1921b). The potential of microwave heating for organic synthesis has been explored in last two and half decades after the first reports appeared in 1986 (Gedye, 1986; Giguere et al., 1986). Initially, reactions were performed in domestic microwave ovens using appropriate solvents. After that, several groups started investigating reactions in solvent free conditions including "dry media" usually with open vessels (Stadler and Kappe, 2000; Vidal et al., 2000). The use of microwave units specially designed for synthesis, which are expensive and it becomes rather difficult at times to procure. Thus, unmodified home microwave units are suitable in some cases. However, simple modifications (for example, a reflux condenser) can enhance the safety factor. High-pressure chemistry should only be carried out in special reactors with a microwave oven specifically designed for this purpose. A further point in favor of using the more expensive apparatus is the question of reproducibility, since only these specialized machines can achieve good field homogeneity and in some cases, these can even be directed on the reaction vessel.

It has long been known that molecules undergo excitation with electromagnetic radiation. This effect is utilized in household microwave ovens to heat up food. However, chemists have been using microwaves only as a reaction methodology for a few years. Some of the first examples gave amazing results, which led to a flood of interest in microwave-accelerated synthesis (Banik et al., 1992; Banik et al., 1993; Bose et al., 1991; Bose et al., 1994).

The MW heating has not been restricted to organic chemistry only, but its application to various aspects of inorganic chemistry and polymer chemistry has also been investigated with several advantage of an ecofriendly approach. In the past few decades, this technique has found a valuable place in the synthetic chemist's tool box, which is evident from a large number of publications (Lidstrom et al., 2001; Loupy, 2006; Perreux and Loupy, 2001; Strauss, 2002; Watkins, 2002), particularly acetylation reaction (Moghaddam and Sharifi, 1995), addition reaction, elimination reaction (Mogilaiah et al., 2003), alkylation reaction (Abramovitch et al., 1995), alkynes metathesis (Miljanic et al., 2003), allylation reaction (Motorina et al., 1996), amination reaction (Mccarroll et al., 2003), aromatic nucleophillic substitution reaction (Robeiro and Khandikar, 2003), arylation reaction (Wali et al., 1995), carbonylation reaction (Yamazaki and Kondo, 2002), combinatorial reaction (Al-obeidi et al., 2003), condensation reaction (Ameta et al., 2010), coupling reaction (Burton et al., 2003), cyanation reaction (Arevela and Leadbeater 2003), cyclization reaction (Shrimali et al., 2009), cyclo-addition reaction (Lerestif et al., 1995), deacetylation reaction (Kumar et al., 2003), dehalogenation reaction (Calinescu et al., 2003), Diel's-Alder reaction (Mavoral et al., 1995), dimerization reaction (Santagada et al., 2003), transesterification reaction (Roy and Gupta, 2003), enantioselective reaction (Diaz-Ortiz et al., 2003), halogenation reaction (Inagaki et al., 2003), hydrolysis reaction (Plazl et al., 2003), Mannich reaction (Sitha et al., 2010), oxidation reaction (Kiasat et al., 2003), phosphorylation synthesis (Gospondinova et al., 2002), polymerization reaction (Vu et al., 2003), rearrangement reaction (Srikrishna and Kumar, 1995) and so on.

The initial slow development of this technology in the last 1980's and early 1990's has been attributed to lack of its controllability and reproducibility coupled with detail understanding of the basics of MW dielectric heating.

## *PRINCIPLE*

The basic principle behind heating by microwaves is the interaction of charged particle of the reaction material with electromagnetic wavelength of a particular frequency. The phenomenon of producing heat by electromagnetic irradiation involves either collision or conduction and some time both. Two basic principles are involved in heating the materials by microwaves.

## *DIPOLAR POLARIZATION*

Dipolar polarization is the phenomenon responsible for the majority of microwave heating. It depends upon polarity of solvent and compound. In polar molecules, different electronegativities of individual atoms results in a permanent electric dipole, which is sensitive to external electric fields and will attempt to align with them by rotation. This realignment is rapid for a free molecule, but in liquid, the instantaneous alignment is prohibited by the presence of other molecules. A limit is, therefore, placed on the ability of the dipole to respond to an electric field, which affects the behaviour of the molecule with different frequencies of electric field for example, under low frequency irradiation, the dipole may react by aligning itself in phase with the electric field. Molecules will polarize uniformly and thus, no random motion results.

Under high frequency irradiation, the polar molecule will attempt to follow the field, but intermolecular inertia stops any significant motion before the field has reversed, in this case, the dipole do not have sufficient time to respond the field, and it does not rotate. As no motion is induced in the molecules, no energy transfers will take place, and therefore, no heating results. In case of intermediate frequency, the field will be such that the molecule is almost (but not quite) able to keep in phase with the field polarity. The microwave frequency is low enough so that the dipoles have enough time to respond to the alternating field, and therefore to rotate, but high enough so that the rotation does not precisely follow the field. As the dipole reoriented to align itself with the field, the field is already changing, and a phase difference causes energy to be lost from the dipole in random collisions. Thus, giving rise to dielectric heating.

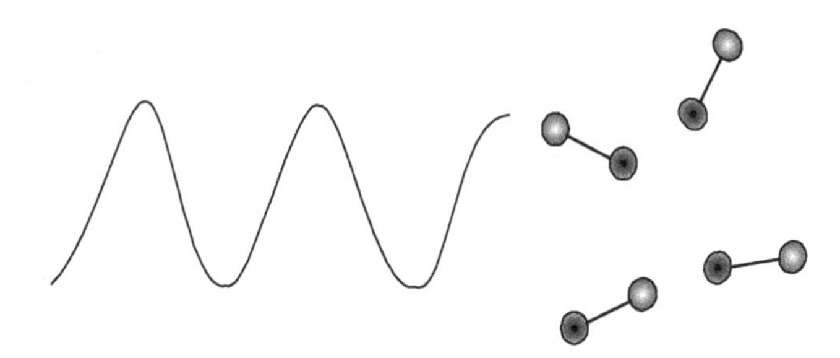

**FIGURE 11.1** Microwave heating by dipolar polarization mechanism.

### *CONDUCTION MECHANISM*

The conduction mechanism generates heat through resistance to an electric current. The oscillating electromagnetic field generates an oscillation of electrons or ions in a conductor resulting in an electric current. This current faces internal resistance, which heats the conductor.

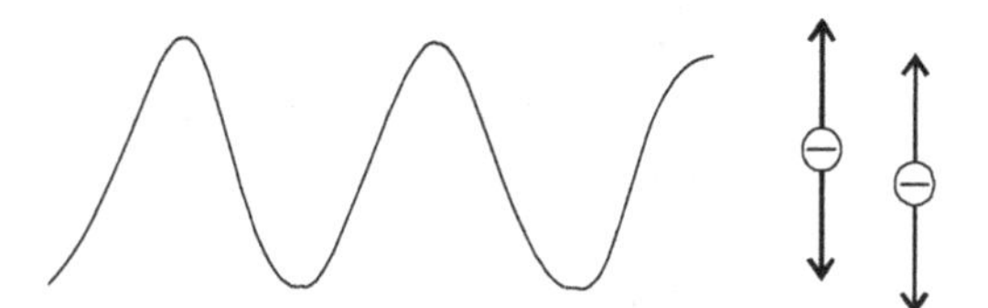

**FIGURE 11.2** Conduction mechanism.

When the irradiated sample is an electrical conductor, the charge carriers (electrons, ions etc.) are moved through the material under the influence of the electric field, resulting in a polarization. These induced currents will cause heating in the sample due to electrical resistance.

## *11.2.3 MICROWAVE CHEMISTRY APPARATUS*

Microwave synthesis has started with a kitchen microwave oven with good results. Now a days, many types of advanced microwave ovens have been introduced in the market. These consist of a microwave source (Magnetron), a microwave cavity or an applicator (multimode cavity or single mode cavity,), mode stirrer, sensor probe (thermocouples or IR sensor) and software with digital display.

Two types of reactors are used for microwave assisted organic synthesis; these are multimode and monomode reactors. The differentiating feature of a single mode apparatus is its ability to create a new standing wave pattern, which is generated by the interference of fields that have the same amplitude but different oscillating directions. This interface generates an array of nodes where microwave intensity is zero and an array of antinodes with the magnitude of microwave energy is at its highest. Therefore, sample should be placed at the antinode at appropriate distance from the magnetron.

Multimode reactors (domestic microwave ovens) are the most common instruments used in organic synthesis since they are comparatively inexpensive and readily available. Lot of satisfying organic synthesis has been done with domestic microwave. Multimode reactors provide a field pattern with areas of high and low field strength, commonly referred to as "hot and cold spots". This non-uniformity of the field leads to the heating efficiency varying drastically between different positions of the sample. This drawback is overcome by the use of mode stirrer.

The mode stirrer is a periodically moving metal vane that continuously changes the instantaneous field pattern inside the cavity, and therefore the field intensity is homogeneous everywhere throughout the cavity. Thus samples can be placed anywhere inside the cavity, because the field is homogeneous throughout the cavity. In modern microwave reactors, preinstalled digital thermometers (sensors and probes) are used for temperature. Moreover, some sophisticated ovens are equipped with computers also (Barlow and Marder, 2003; Ranger et al., 1995).

### *11.2.4 REACTION VESSELS AND REACTION MEDIUM*

The reaction vessel must be transparent to the microwaves. These are preferably being made up of teflon, polystyrene and glass (Strauss and Trainor, 1995). Metallic containers are not used as it gets heated soon due to preferential absorption and reflection of rays. For reactions in solvents, the solvent must have a dipole moment and a boiling point higher than the desired reaction temperature and a dielectric constant. Some of the solvents used commonly as microwave absorber are N,N-dimethyl formamide or DMF (b.p.154°C, $\varepsilon = 36.7$), formamide (b.p. 216°C, $\varepsilon = 111$), methanol (b.p. 65°C, $\varepsilon = 32.7$), ethanol (b.p. 78°C, $\varepsilon = 24.6$), chlorobenzene (b.p. 132°C, $\varepsilon = 5.6$), 1,2-dichlorobenzene (b.p. 180°C, $\varepsilon = 10.5$), 1,2-dichloroethane (b.p. 83°C), ethylene glycol (b.p. 196°C, $\varepsilon = 37.7$), dioxane (b.p. 101°C, $\varepsilon = 2.25$) and diglyme (b.p. 162°C, $\varepsilon = 7.23$). The presence of salts in polar solvents can frequently enhance microwave coupling. Hydrocarbon solvents such as toluene ($\varepsilon = 2.4$), hexane ($\varepsilon = 1.9$) and benzene ($\varepsilon = 2.3$), because of less dipole moment, are unsuitable as they absorb microwave radiation poorly. For solid state reactions, mineral oxides such as zeolite, alumina, silica, montmorillonite K-10 clay and so on, are used as absorbents.

### *11.2.5 MICROWAVE EFFECT*

The microwave effect applies to a range of observations in microwave chemistry. These may be classified in two categories: specific microwave and non-thermal microwave effects (Stuerga and Gaillard, 1996, 1996).

#### *SPECIFIC MICROWAVE EFFECTS*

These are those effects that cannot be easily done by conventional heating methods. Examples include (i) selective heating of specific reaction components, (ii) rapid heating rates and temperature gradients, (iii) the elimination of wall effects and (iv) superheating of solvents.

#### *NON-THERMAL MICROWAVE EFFECTS*

Excitation with microwave radiation results in the molecules aligning their dipoles within the external field. Strong agitation, provided by the reorientation of molecules, in phase with the electrical field excitation, causes an intense internal heating. The question of whether a nonthermal process is operating can be answered simply by comparing the reaction rates between the cases where the reaction is carried out under irradiation versus under conventional heating. In fact, no nonthermal effect has been found in the majority of reactions (De la Hoz et al., 2005), and the acceleration is attributed to superheating alone. It is clear that nonthermal effects do play a role in some reactions.

### *11.2.6 COMPARISON BETWEEN MICROWAVE HEATING AND CONVENTIONAL HEATING*

Microwave radiation provides rapid and homogeneous heating, which has certain advantages such as reaction rate acceleration, milder reaction conditions and higher chemical yields. In short, microwave enhanced chemical reactions are safer, faster, cleaner and more economical than conventional reactions. It helps in developing cleaner and greener synthetic routes (Chemat and Esveld, 2001; Lidstrom et al., 2001; Nuchter et al., 2003, 2004).

#### *INCREASED RATE OF REACTION*

Microwave heating enhances the rate of certain chemical reactions by 10 to 1,000 times compared to conventional heating. This is due to its ability to increase the temperature of a reaction. For instance, synthesis of fluorescein, which usually takes about 10hrs by conventional heating methods, can be completed in only 35min by means of microwave heating. The rate acceleration caused by microwaves has been attributed to superheating of solvents (liquid phase reactions) and high temperature on the surface of catalyst or other solid reactants. The water molecule is the target for microwave ovens in the home; like any other molecule with a dipole, it absorbs microwave radiation. Microwave radiation is converted into heat with high efficiency, so that "superheating" becomes possible at ambient pressure. Enormous accelerations in reaction time can be achieved, if superheating is performed in closed vessels under high pressure; a reaction that takes several hours under conventional conditions can be completed over the course of minutes.

### *EFFICIENT SOURCE OF HEATING*

Microwave assisted heating is a highly efficient process and results in a significant energy saving. This is because microwaves heat up just the sample and not the apparatus and therefore, energy consumption is less.

### *HIGHER YIELDS*

In certain chemical reactions, microwave radiation produces higher yields as compared to conventional heating methods. For example, microwave synthesis of fluorescein results in an increase in the yield of the product from 70 to 82%.

### *UNIFORM HEATING*

Microwave radiation provides uniform heating throughout a reaction mixture unlike conventional heating methods. It is because in conventional heating, the walls of the oil bath gets heated first and then the solvent. As a result of this, there is always a temperature difference between the walls and the solvent. In the case of microwave heating, only the solvent and the solute particles are excited, which results in uniform heating of the solvent.

### *SELECTIVE HEATING*

Selective heating is based on the principle that different materials absorb microwaves to different extent. Some materials are transparent where as others absorb microwaves. Therefore, microwaves can be used to heat a combination of such materials for example, the production of metal sulfide with conventional heating requires weeks because of the volatility of sulfur vapours while rapid heating of sulfur in a closed tube results in the generation of sulfur fumes, which can cause an explosion. However, in microwave heating, since sulfur is transparent to microwaves, only the metal gets heated. Therefore, reaction can be carried out at a much faster rate with rapid heating, without the threat of an explosion.

#### *ENVIRONMENTAL-FRIENDLY CHEMISTRY*

Reactions conducted through microwaves are cleaner and more environment friendly than conventional heating methods. Microwaves heat the compounds directly and therefore, use of solvents in the chemical reaction can be reduced or eliminated. An approach was developed to carry out a solvent free chemical reaction on solid support like clay, alumina, zeolite and so on. The reactants adsorbed on solid support under microwave; react at a faster rate than conventional heating. The use of microwaves has also reduced the extent of purification required for the end products of chemical reactions involving toxic reagents.

#### *GREATER REPRODUCIBILITY OF CHEMICAL REACTIONS*

Reactions under microwave irradiation show more reproducibility as compared to conventional heating because of uniform heating and better control of process parameters. The temperature of chemical reactions can also be easily monitored. This is of particular relevance in the lead optimization phase of the drug development process in pharmaceutical companies.

### 11.2.7 LIMITATIONS OF MICROWAVE CHEMISTRY

The limitations associated with microwave heating are its scalability, limited application and the hazards involved in its use.

## *LACK OF SCALABILITY*

The yield obtained by using domestic microwave apparatus is limited to a few grams. Although there have been developments in the recent past relating to the scalability of microwave equipment, still there is a gap that needs to be spelled to make this technology scalable.

## *LIMITED APPLICABILITY*

Microwaves can be used for heating only those materials, which absorb them. Microwaves cannot heat materials, which are transparent to these radiations.

## *SAFETY HAZARDS RELATING TO THE USE OF MICROWAVE HEATING APPARATUS*

Although manufacturers of microwave heating apparatus have directed their research to make microwaves a safe source of heating. Uncontrolled reaction conditions may result in undesirable results for example, chemical reactions involving volatile reactants under superheated conditions may result in explosive conditions. Moreover, improper use of microwave heating for rate enhancement of chemical reactions involving radioisotopes may result in uncontrolled radioactive decay.

## *HEALTH HAZARDS*

Health hazards related to microwaves are caused by the penetration of microwaves. The microwaves operating at a low frequency range are only able to penetrate the human skin white, higher frequency range microwaves can reach body organs. Research has proved that prolonged exposure to microwaves may result in the complete degeneration of body tissues and cells. It has also been established that constant exposure of DNA to high frequency microwaves during a biochemical reaction may result in complete degeneration of the DNA strand.

### 11.2.8 CLASSIFICATION OF MICROWAVE REACTIONS

Broadly microwave assisted organic synthesis can be classified into solvent assisted and solvent free synthesis

*SOLVENT ASSISTED SYNTHESIS*

The black assisted synthesis proceeds in the presence of solvent with good polarity, high boiling point and sufficient chemical stability. N, N-dimethylformamide (DMF) is a good solvent for microwave assisted organic synthesis as it has a high boiling point (154°C), high dielectric constant and is miscible with water, making it relatively easy to be removed from the reaction mixture.

Erdelyi and Gogoll (2001) showed that the reaction of aryl halides with trimethylsilyl acetylene in DMF as solvent under microwave radiations. They reported 85–95% yield of product in 5–25min.

$$\text{Ar-X} + \text{H}-\equiv-\text{Si(CH}_3)_3 \xrightarrow[\text{Et}_2\text{NH, DMF/MW}]{\text{Pd(PPh}_3)_2\text{Cl}_2\text{, CuI}} \text{Ar}-\equiv-\text{Si(CH}_3)_3$$

Yield 80-95%

Hu et al. (1999) carried out substitution of nitro group in presence of tetrahydrofuran in good yields (83%).

$$\text{R}_1\text{-ZnI} + \text{R}_2\text{-C}_6\text{H}_4\text{-CH=CH-NO}_2 \xrightarrow{\text{THF}} \text{R}_2\text{-C}_6\text{H}_4\text{-CH=CH-R}_1$$

A rapid and efficient methods for the preparation of isoxazole, pyrazole and pyrimidine derivatives of imidazolinone and quinazolinone under MWI has been reported by Vyas et al. (2008). Microwave assisted synthesis and characterization of some new annulated pyrimidinone derivatives has been done by Sancheti et al. (2007).

## *SOLVENT FREE SYNTHESIS*

Solvent-free method offer safe and efficient reaction pathways, which are time and money saving and often enable elimination of waste treatment. Solvent-free syntheses are divided into solid support reaction and neat reactions.

## *SOLID SUPPORT REACTIONS*

Microwave assisted solid support reaction increases the rate of reactions and decreases the reaction time. It also gives high conversion with high selectivity, avoiding solvent in most of cases, therefore workup also becomes easier. In MW promoted deprotection, condensation, rapid one-pot synthesis, synthesis of heterocyclic compounds using recyclable mineral oxides supported reagents such as $Fe(NO_3)_3$-clay (clayfen), $NH_2OH$-clay, $PhI(OAc)_2$-alumina, $NaIO_2$-silica, $CrO_3$-alumina, $MnO_2$-silica and $NaBH_2$-clay, zeolite and so on. proved themselves as a base for the growth of a newer branch of green chemistry. Kann has studied use of polymer supported organometallic catalysts in organic synthesis (2010). The KF-alumina-mediated Bargellini reaction has been reported by Rohman and Myrboh (2010).

KF/$Al_2O_3$, $CHCl_3$, Toluene, r.t.

Lalitha and Sivakamasundari (2010) studied the synthesis of few vinyl quinolones on solid support. Self-condensation of hydroxybenzene derivatives under microwave using heterogeneous has been reported by Gomez et al. (2010).

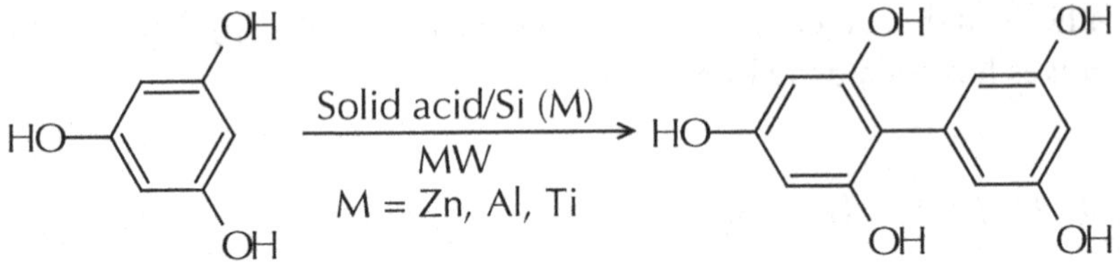
OH
HO
OH
Solid acid/Si (M)
MW
M = Zn, Al, Ti
OH
OH
HO
OH
OH

Yang et al. (2012) described microwave-assisted solid-phase synthesis, biological evaluation and molecular docking of angiotensin i-converting enzyme inhibitors.

## *NEAT REACTIONS*

Such reactions are performed between neat reagents. Here, at least one of reagent must be a polar liquid. There are liquid-liquid or liquid-solid systems; the latter implies that the solid is soluble in the liquid phase or at least the liquid counterpart is adsorbed on the solid and hence, the reaction occurs at the interface. Ameta et al. (2011) reported solvent-free synthesis of pthalimide derivatives under microwave irradiation. A series of NH-pyrazoles were efficiently synthesized from the reaction of dimethylaminovinylketones and hydrazine hydrate in solid state by Longhi et al. (2010).

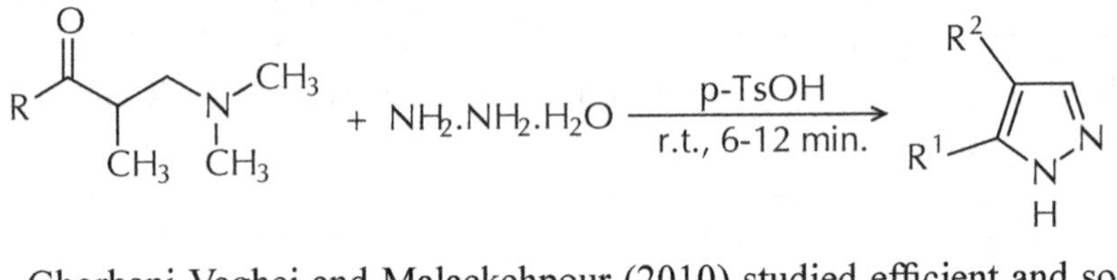
O
R
N
CH3
CH3
CH3
+ NH2.NH2.H2O
p-TsOH
r.t., 6-12 min.
R2
R1
N
N
H

Ghorbani-Vaghei and Malaekehpour (2010) studied efficient and solvent-free synthesis of 1-amidoalkyl-2-naphthols using N,N, N', N',-tetrabromobenzene-1,3-disulphonamide. LiBr has been used as a catalyst for solvent-free synthesis of fused thiazoloquinazoline derivatives (Ameta et al., 2010). Chang et al. (2010) carried out efficient solvent-free catalytic hydrogenation of solid alkenes and nitro-aromatics using Pd nanoparticles entrapped in aluminium oxy-hydroxide. $AlCl_3$ has been used as a catalyst for microwave enhanced and solvent-free green protocol for the production of dihydropyrimidine-2-(1H)-ones (Kumar et al., 2010).

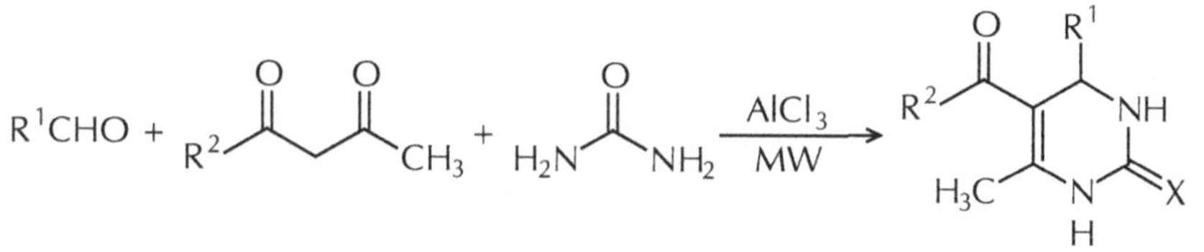

Microwave assisted solvent free synthesis of 1,3-diphenylpropenones has been reported (Kakati and Sarma, 2011). Solvent-free microwave-assisted synthesis of E-(1)-(6-benzoyl-3,5-dimethylfuro [3′,2′:4,5] benzo[b] furan-2-yl)-3-(aryl)-2-propen-1-ones and their antibacterial activity was studied by Ashok et al. (2011).

### *11.2.9 APPLICATIONS IN ORGANIC SYNTHESIS*

Green chemistry involves design of chemical synthesis to prevent pollution and thereby solve the environmental problems. The microwave chemistry is the current approach in green chemistry, as it follows number of principles of green chemistry. Here some applications of microwave assisted organic synthesis are reported.

#### *OXIDATION REACTIONS*

A remarkably fast microwave assisted selective oxidation of benzylic alcohols with calcium hypochlorite under solvent free conditions has been reported (Mohammad et al., 2001). Alcohols are readily adsorbed on moist alumina and rapidly oxidized to corresponding carbonyl compounds upon exposure to microwaves under solvent free conditions.

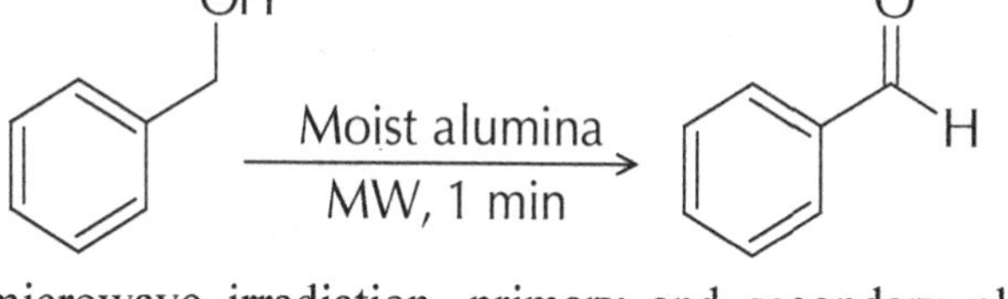

Under microwave irradiation, primary and secondary alcohols are selectively oxidized to the corresponding aldehydes and ketones within

5–25s. using commercially available and magnetically retrievable magtrieve (Mojtahedi et al., 2000).

R⌒OH —(Magtrieve™; MW, 5-25 sec.)→ RC(=O)H

The oxidations of some simple secondary alcohols by hydrogen peroxide-urea adduct using microwaves as an energy source has also been described (Bogdal et al., 2003). Tetravalent chromium dioxide as a magnetically retrievable oxidant has been shown to be a very useful oxidant for microwave assisted and conventional transformation of aromatic and alkyl aromatic molecules into the corresponding aryl ketones, quinones or lactones (Lukasiewicz et al., 2004). Microwave assisted synthesis of luminescent cyanobipyridyl zinc (II) bis (thiolate) complexes by Bagley et al. (2010). Liu et al.(2010) studied size effect of silica-supported gold clusters in the microwave assisted oxidation of benzyl alcohol with $H_2O_2$.

## *REDUCTION REACTION*

Bose et al. (1993; 1999) were first to report the use of microwaves for transfer hydrogenation in organic synthesis. A series of β-lactam derivatives were hydrogenated using formates and Pd/C or Raney Ni as catalyst. Schmoger et al. (2011) carried out microwave assisted reduction of organic compounds using catalytic and stoichiometric reactions. Ethylene glycol was used as the solvent and electron source for the microwave-assisted reduction reaction. An improved microwave-assisted one-pot tandem Staudinger/aza-Witting/reduction was reported by Chen et al.(2011) for the synthesis and biological activity of novel 5'-arylamino-nucleosides. Wada et al. (2000; 2002) carried out the reductive dehalogenation of chlorinated phenols to phenol, cyclohexanol and other chlorine-free compounds within 20min using microwave irradiation. Microwave-assisted reduction of carbonyl compounds in solid state using sodium borohydride supported on alumina has been studied by Varma and Saini (1997).

$$R\text{-}C_6H_4\text{-}CO\text{-}R^1 \xrightarrow[\text{MW}]{NaBH_4\text{-}Al_2O_3} R\text{-}C_6H_4\text{-}CH(OH)\text{-}R^1$$

## *ALKYLATION*

The MW assisted selective N-alkylation of 6-amino-2-thiouracil in DMF using different alkyl halides has been investigated by Loupy et al. (2001) whereas no reaction was observed under the same conditions in a thermo-regulated oil bath.

$$\text{6-amino-2-thiouracil} \xrightarrow[Na_2CO_3,\ n\text{-}Bu_4N]{\text{RX/DMF/MW}} \text{N-alkylated product}$$

A series of diethers has been obtained by alkylation of dianhydrohexanes under MW conditions at 140°C in 90% yield (Chatti et al., 2001). Matondo et al. (2003) carried out O-alkylation of o-bromophenol under MW conditions using TBAB as catalyst. The reaction of diethyl ethoxycarbonylmethylphosphonate with a series of alkyl halides was carried out under microwave and solventless conditions at 120°C, in the presence of $Cs_2CO_3$. In the absence of a phase transfer catalyst, it afforded the corresponding monoalkylated products in yields of >70%. The thermal variant carried out in boiling acetonitrile was slow and led to incomplete conversions. In the MW method, the phase transfer catalyst is substituted by MW irradiation and there is no need for a solvent (Grun et al., 2012).

$$\text{o-bromophenol} + R\text{-}Br \xrightarrow[\text{MW, 40°-60°C}]{\text{KOH.TBAB}} \text{product (OR, Br)}$$

## REARRANGEMENT

A general and efficient procedure for the selective Meyer-Schuster isomerization of both, terminal and internal alkynols, has been developed by using catalytic amounts of the readily accessible oxovanadium (V) complex [V(O)Cl(OEt)$_2$]. Reactions proceeded smoothly in toluene at 80°C under microwave irradiation to provide the corresponding α,β-unsaturated carbonyl compounds in excellent yields and short times without the assistance of any additive (Antinolo et al., 2012). Li et al. (2006) carried out Beckmann rearrangement in acetone under microwave irradiation using silica sulfate as an efficient and recyclable catalyst.

HO–N=C($R_1$)($R_2$) → ($SiO_2$-$OSO_3H$; Acetone, MW, 2-6 min.) → $R_1$–C(=O)–NH–$R_2$

A microwave-assisted Domino rearrangement of propargyl vinyl ethers to multifunctionalized aromatic platforms has been described by Tejedor et al. (2011). An aza-Cope rearrangement of N-allylanilines in few minutes using $BF_3 - OEt_2$ under microwave activation has been studied by Gonzalez et al. (2008). Microwave assisted Hoffmann rearrangement has been reported by Miranda et al. (2011).

R–$C_6H_4$–C(=O)–$CH_3$ → (MeOH/TBCA/KOH; MW, 5 min., 60°C) → R–$C_6H_4$–NH–C(=O)–O–$CH_3$

## CYCLOADDITION

Microwave assisted Diels-Alder cycloaddition reaction between 2-fluoro-3-methoxy-1,3 butadiene and ethylene was studied by Patrick et al. (2007).

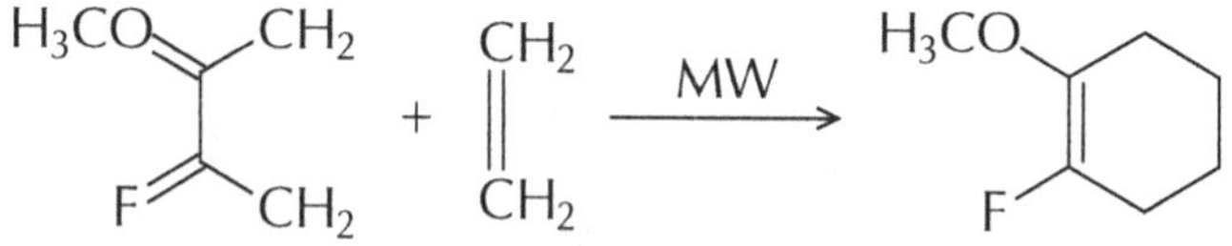

$$H_3CO,\ CH_2,\ F,\ CH_2 \quad + \quad CH_2{=}CH_2 \quad \xrightarrow{MW} \quad H_3CO,\ F$$

This reaction gives moderate yield under thermal conditions, but very good yield under microwave radiation. Similarly, intramolecular hetero Diels-Alder cycloaddition of acetylenic pyrimidines to give bicyclic pyridine was reported by Shao (2005). The microwave-assisted functionalization of carbon nanohorns (CNHs) via [2+1] nitrenes cycloaddition, providing well dispersible hybrid materials possessing aziridino-rings covalently grafted onto the graphitic network of CNHs, was also studied (Karousis et al., 2011). Microwave-assisted stereoselective 1,3-dipolar cycloaddition of C,N-diarylnitrone and bis(arylmethylidene)acetone was reported by Paul et al. (2012).

## *CONDENSATION*

The Claisen condensation and its intramolecular variant the Dieckmann condensation are classic reactions studied in organic chemistry because of their importance in organic synthesis and biochemical transformations. The growth in the use of microwave technology in both; the synthetic and teaching laboratories warrants the modification of existing methodologies to incorporate this technology. Simple microwave-assisted procedures for carrying out Claisen and Dieckmann condensation reactions are suitable for organic chemistry teaching laboratories that utilize microwave technology. Although solvents can be used, the procedure is amenable to solvent-free conditions that promote green chemistry (Horta et al., 2011). Ighilahriz et al. (2008) synthesized 4(3H)-quinazolinones by cyclocondensation of anthranilic acid, aniline and orthoester (or formic acid) in presence of HPA as a catalyst under MWI.

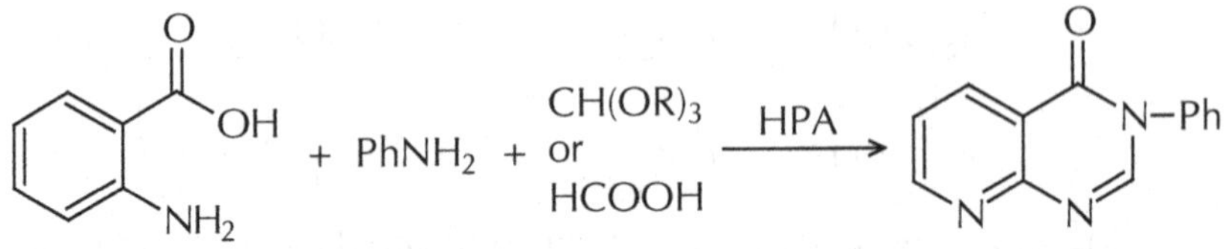

Agrawal and Joshipura (2005) studied Friedlander condensation under microwave irradiation. They also made a comparison over conventional method. In Friedlander condensation, a mixture of 2-aminobenzophenone and ethyl acetoacetate was irradiated for 5–7min. without catalyst and the products were obtained is good yields (80–91%).

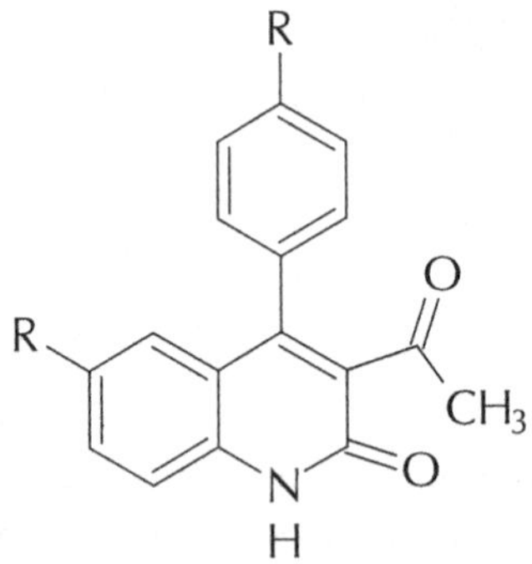

Thiourea is an inexpensive, efficient and mild catalyst for the synthesis of Knoevenagel condensation of pyrozoles derivative. In the presence of 10mol% of thiourea, pyrazole aldehyde react with active methylene compound under microwave-assisted solvent-free conditions at 300W for 2–5min to give corresponding products in good yields (Li et al., 2011).

## *PROTECTION REACTION*

Microwave in addition to its massive applications in rapid organic synthesis, may also be used for simple and fast protection reactions. Microwave assisted selective protection of glutaraldehyde and its monoacetals and thioacetals was carried out by Flink et al. (2010) They reported that reactions proceed efficiently and only traces of deprotected materials were left.

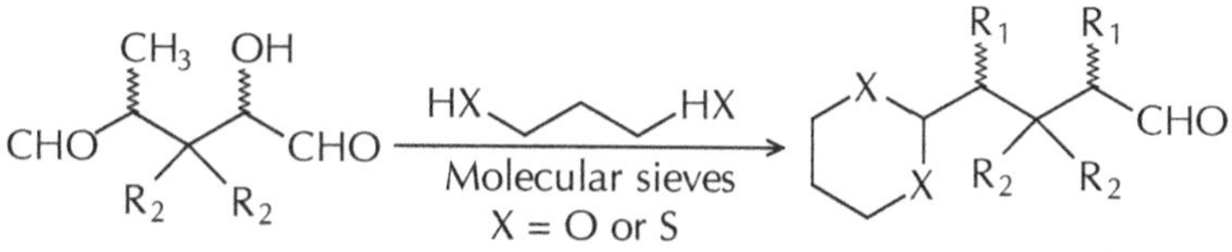

Liu et al. (2005) carried out graft polymerization of ε-caprolactone on chitosan under microwave irradiation via a protection-graft deprotection procedure with phthaloylchitosan in presence of stannous octoate as a catalyst. After deprotection, the phthaloyl group was removed and amino group was regenerated. N-acylbenzotriazole mediated microwave assisted syntheses of protected and novel unprotected ferrocenoylamidoamino acids have been prepared (Hura et al., 2011). These amino acids undergo substitution reaction with 1-(ferrocenylcarbonyl)-1H-benzotriazole in partially aqueous media under microwave irradiation.

## *TRANSITION METAL CATALYZED COUPLING REACTIONS*

Organometallic catalysis in general and palladium catalyzed reactions in particular have emerged as the success stories in the growing heterogeneous field of microwave assisted chemistry (Larhed et al., 2002; Olofsson et al., 2002). The first MW-promoted Suzuki coupling was reported in 1996 (Larhed and Hallberg). Microwave accelerated metal-catalyzed transformation of aryl and vinyl halides by carboxylations, Heck and Sonogashira reactions and cross coupling have been reported by Nilsson et al. (2006).

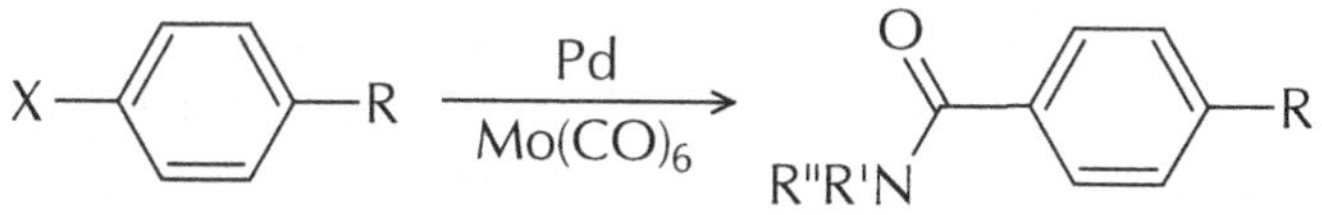

Perbenzylated pyranoid glycal couples with aryl bromides under microwave irradiation in the presence of 5% mol palladium (II) acetate in DMF to produce 2',3'-unsaturated C-aryl-d-glycopyranosides in a rapid and stereospecific manner (Lei et al., 2009).

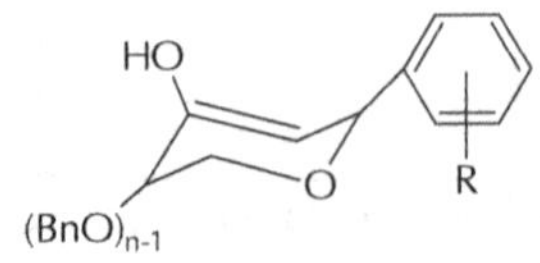

Leadbeater (2003; 2005) reported Suzuki-Miyaura coupling in aqueous medium and isolated products in good yields. Fluoride free cross-coupling reactions under ligand free conditions with low Pd loadings in water using NaOH under microwave heating has been carried out by Alacid and Najera (2008).

## *SYNTHESIS OF HETEROCYCLES*

The efficient use of MW heating approach in several organic synthesis as an emerging green technique has been proposed by several workers (Molteni and Ellis, 2005; Polshettiwar and Varma, 2008). A rapid, efficient and environmental benign methodology for the preparation of 2,5-disubstituted indole analogues was developed (Biradar and Sasidhar, 2011). The synthesis of pyrroles by reaction of hexane-2,5- dione with primary amines has been carried out by Danks (1999) under microwave irradiation.

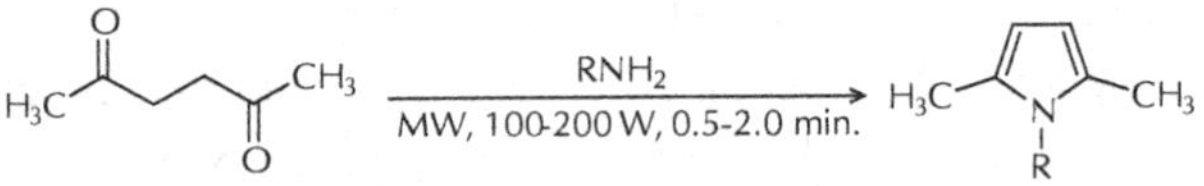

The MW assisted synthesis of benzimidazoles as potential HIV-1 integrase inhibitors by condensation of R-hydroxycinnamic acids with 1,2-phenylenediamine in water was performed by Ferro et al. (2004). A simple, high yielding synthesis of 2,4,5-trisubstituted imidazoles from 1,2-diketones and aldehydes in the presence of ammonium acetale was reported by Wolkenberg et al. (2004) under microwave irradiation.

## *SYNTHESIS OF NANOCOMPOSITES*

Wang and Lee (2005) carried out microwave-assisted synthesis of $SnO_2$-graphite nanocomposite for Li-ion battery applications. Polyacrylamidemetal nanocomposites with metal nanoparticles homogeneously dispersed in the polymer matrix have been successfully prepared using corresponding metal salt and acrylamide monomer in ethylene glycol by microwave heating (Zhu and Zhu, 2006). Mallikarjuna and Varma (2007) described shape controlled bulk synthesis of noble nanocrystals and their catalytic properties under microwave irradiation. One-step microwave-assisted method was used for the synthesis of small gold nanoclusters, Au16NCs@ BSA, which were used as fluorescence enhanced sensor for detection of silver (I) ions with high selectivity and sensitivity (Yue, et al., 2012). The UV curable organic/inorganic hybrid nanocomposites with high refractive indices, moderate hardness, good adhesive strength and excellent gas blocking performances have been rapidly synthesized by *in situ* microwave heating process (Lin et al., 2011). Chen and Wang (2010) reported microwave assisted synthesis of a $Co_3O_4$-graphene sheet-on-sheet nanocomposite as a superior anode material four Li-ion batteries. In addition, number of publication related to microwave assisted synthesis of nanocomposites have appeared in recent past years (Jia et al., 2011; Li et al., 2012; Ma et al., 2010; Motshekga et al., 2012; Zhang et al., 2010).

## *SYNTHESIS OF IONIC LIQUIDS*

A microwave assisted preparation of a series of ambient temperature ionic liquid, 1-alkyl-3-methylimidazolium halides via efficient reaction of 1-methylimidazole with alkyl halide under solvent-free conditions has been described by Varma and Namboodiri (2001). Khadilkar and Rebeiro (2002) reported the synthesis of various alkyl pyridinium and 1-alkyl-3-methylimidazolium halide under microwave exposure in a closed vessel.

X, R, $H_3C$, NH, N, + R—X, MW, or, $N^+$, $X^-$, R

Law et al. (2002) reported solvent-free route to ionic liquid precursors using a water-moderated microwave process.

$$\text{H-imidazole} \xrightarrow[\text{R—X}]{\text{MW}} \text{H-N}(+)\text{N-R } X^-$$

Several Bronsted acid ionic liquids have been synthesized and used as solvents and catalysts for three-component Mannich reactions of aldehydes and amines at 25°C (Zhao et al., 2004). A microwave-assisted ionic liquid solvothermal method was demonstrated to synthesize $CaF_2$ double-shelled hollow microspheres. This method was simple and time saving and can also be extended to prepare hollow microspheres of $MgF_2$ and $SrF_2$ (Xu and Zhu, 2012). Unique role of ionic liquid in microwave assisted synthesis of monodispersed magnetic nanoparticles have been studied by Hu et al. (2010). Rahman et al. (2011) carried out microwave-assisted bronsted acidic ionic liquid-protonated one-pot synthesis of heterobicyclic dihydropyrimidinones by a three-component coupling of cyclopentanone, aldehydes, and urea.

### 11.2.10 MISCELLANEOUS

In addition to above, microwave heating process have been employed for regioselective and chemoselective synthesis, polymer synthesis, ceramic products, intercalation products, organometallic and coordination compounds, macromolecules, radiopharmaceuticals and so on. Faghihi et al. (2004) investigated facile synthesis of novel optically active poly (amide-imide)s containing N, N'-(pyromellitoyl)-bis-I-phenylalanine diacid chloride and 5,5-disubstituted hydantoin derivatives under microwave irradiation.

Various molecules like carboranes (Armstrong and Valliant, 2007), phenol-formaldehyde resole (Deetz et al., 2001), phenothiazine (Bajia et al., 2007), 4,5-disubstituted pyrazolopyrimidines (Gaina et al., 2007), and curcumin analogs (Wu et al., 2003), have been synthesized by microwave heating method with high efficiency. The synthesis of a variety of transition metal coordination compounds under atmospheric conditions has been carried out under MW radiations (Nichols et al., 2006). A particularly useful application of microwave assisted synthesis at elevated pressure has been the preparation of radiopharmaceuticals containing isotopes with short half lives such as C–11 and F–18 (Haung et al., 1987; Thorell et al., 1992). In addition, microwave assisted synthesis of macromolecules like epoxyresins (Bogdal and Gorczyk, 2003), polyesters (Chatti et al., 2003), polyethers (Pielichowski et al., 2003) and poly (aspartic acid) (Burazyk et al., 2003) have been reported.

Microwave assisted chemical synthesis has advantages like reaction acceleration, yield improvement, enhanced physicochemical properties and the evolvement of new material phases. Shi and Hwang (2003) demonstrated the significance of these advantages in industrial applications.

Ultimately, it is concluded microwave mediated synthesis is a green chemical technology because microwave not only accelerate chemical processes but also improves yield, selectivity, reduces pollution and enable reactions to occur in solvent free conditions.

Microwave assisted organic synthesis is a green chemical technique and reactions can be completed in less time, with increased yield and more purity. Many of these reactions can be carried out without solvents or on solid support. Time is not far off, when these microwave assisted reactions will replace majority of organic reactions, even on industrial scale.

## KEYWORDS


- **Atom Economy**
- **Carboranes**
- **Feed Stock**
- **Green Route**
- **Nanocomposites**


## REFERENCES


1. Abramovitch, R. A., Shi, Q., & Bogdal, D. (1995). *Synthetic Communication, 25*, 1–8.
2. Agrawal, Y. K. & Joshipura, H. M. (2005). *Indian Journal of Chemistry, 44B*, 1649–1652.

3. Alacid, E. & Najera, C. (2008). *Journal of Organic Chemistry, 73,* 2315–2322.
4. Al-obeidi, F., Austin, R. E., Okonoya, J. F., & Bond, D. R. S. (2003) *Mini-Reviews in Medicinal Chemistry, 3,* 449–460.
5. Ameta, C., Ameta, R., Tiwari, U., Punjabi, P. B., & Ameta, S. C. (2011). *Journal Of the Indian Chemical Society, 88,* 827–833.
6. Ameta, C., Ameta, R., Tiwari, U., Punjabi, P. B., & Ameta, S. C. (2010). *Afinidad, LXVII,* 393–400.
7. Ameta, C., Sitha, D., Ameta, R., & Ameta, S. C. (2010). *Indonesian Journal of Chemistry, 10,* 376–381.
8. Anastas, P. T. & Warner, J. C. (2000). *Green chemistry: Theory and practice*. Oxford: Oxford University Press
9. Antinolo, A., Carrillo-Hermosilla, F., Cadierno, V., García-Álvarez J., & Otero, A. (2012). *ChemCatChem, 4,* 123–128.
10. Arevela, R. K. & Leadbeater, N. E. (2003). *Journal of Organic Chemistry, 68,* 9122–9125.
11. Armstrong, A. F. & Valliant, J. F. (2007). *Inorganic Chemistry, 46,* 2148–2158.
12. Ashok, D., Sudershan, K., & Khalilullah, M. (2011). *Heterocyclic Letters, 1,* 311–317.
13. Bagley, M. C., Lin Z., & Simon, J. A. (2010). *Dalton Transactions, 39,* 3163–3166.
14. Bajia, S. C., Swarnkar, P., Kumar, S., & Bajia, B. (2007). Impact Factor. *Chemistry—A European Journal, 4,* 457–460.
15. Banik, B. K., Manhas, M. S., Kaluza, Z., Barakat, K. J., & Bose, A. K. (1992). *Tetrahedron Letters, 33,* 3603–3606.
16. Banik, B. K., Manhas, M. S., Newaz, S. N., & Bose, A. K. (1993). *Bioorganic and Medicinal Chemistry Letters, 3,* 2363–2368.
17. Barlow, S. & Marder, S. R. (2003). *Advvanced Functional Materials, 1B,* 517–518.
18. Biradar, J. S. & Sasidhar, B. S. (2011). *European Journal of Medicinal Chemistry, 46,* 6112–6118.
19. Bogdal, D. & Gorczyk, S. (2003). *Polish Journal of Chemical Technology, 5,* 88–89.
20. Bogdal, D., Lukasiewicz, M., Pielichowski, J., Miciak, A., & Bednarz, S. (2003). *Tetrahedron, 59,* 649–653.
21. Bose, A. K., Banik, B. K., Barakat, K. J., & Manhas, M. S. (1993). *Synlett, 25,* 575–576.
22. Bose, A. K., Banik, B. K., Wagle, D. R., & Manhas, M. S. (1999). *Journal of Organic Chemistry, 64,* 5746–5753.
23. Bose, A. K., Manhas, M. S., Banik, B. K., & Robb, E. W. (1994). *Research on Chemical Intermediates, 20,* 1–12.
24. Bose, A. K., Manhas, M. S., Ghosh, M. Shah, M., Raju, V. S., Bari, S. S., Newaz, S. N., Banik, B. K., Choudhary, A. G., & Barakat, K. J. (1991). *Journal of Organic. Chemistry, 56,* 6968–6970.
25. Bradley, D. (2001). *Modern Drug Discovery, 4,* 32–36.
26. Burazyk, A., Bogdal, D., & Pielichowski, J. (2003). *Polish Journal of Chemical Technology, 5,* 3–4.
27. Burton, G., Cao, P., Li, G., & Rivero, R. (2003). *Organic Letters, 5,* 4373–4376.

28. Calinescu, I., Calinescu, R., Martin, D. I., & Radoiv, M. T. (2003). *Researchon Chemical Intermediates, 29,* 71–81.
29. Chang, F., Kim, H., Lee, B. Park, S., & Park, J. (2010) *Tetrahedron Letters, 51,* 4250–4252.
30. Chatti, S., Bartolussi, M., Loupy, A., Blais, J. C. Bogdal, D., & Rogr, P. (2003). *Journal of Applied Polymer Science, 90,* 1255–1266.
31. Chatti, S., Bortolussi, M., & Loupy, A. (2001). *Tetrahedron, 57,* 4365–4370.
32. Chemat, F. & Esveld, E. (2001). *Chemical Engineering & Technology, 7,* 735–744.
33. Chen, S. Q. & Wang, Y. (2010). *Journal of Material Chemistry, 20,* 9735–9739.
34. Chen, H., Zhao, J., Li, Y., Shen, F., Li, X., Yin, Q., Qin, Z., Yan, X., Wang, Y., Zhang, P., & Zhang, J. (2011). *Bioorganic Medicinal and Chemical Letters, 21,* 574–576.
35. Danks, T. N. (1999). *Tetrahedron Letters, 40,* 3957–3960.
36. De la Hoz, A., Diaz-Ortiz, A., & Moreno, A. (2005). *Chemical Society Reviews, 34,* 164–178.
37. Deetz, M. J., Malerich, J. P., Beatty, A. M., & Smith, B. D. (2001). *Tetrahedron Letters, 42,* 1851–1854.
38. Diaz-Ortiz, A., Hoz, A. D., Merrero, M. A. Prieto, P., Sanchez-Migallon, A., Cassio, F. P., Arriela, A., Vivanco, S., & Foces, C. (2003). *Molecular Diversity, 7,* 165–169.
39. Erdelyi, M. & Gogoll, A. (2001). *The Journal of Organic Chemistry, 66,* 4165–4169.
40. Faghihi, K., Zamani, K., Mirgamie, A., & Mallakpour, S. (2004). *Journal of Applied Polymer Science, 91,* 516–524.
41. Ferro, S., Rao, A., Zappala, M., Chimirri, A., Barreca, M. L., Witwouw, M., Debyser, Z., & Monforte, P. (2004). *Heterocycles, 63,* 2727–2734.
42. Flink, H., Putkonen, T., Sipos, A., & Jokela, R. (2010). *Tetrahedron, 66,* 887–890.
43. Gaina, L., Cristea, C., Moldovan, C., Porumb, D., Surducan, E., Deleanu, C., Mahamoud, A., Barbe, J., & Silberg, I. A. (2007). *International Journalof Molecular Sciences, 8,* 70–80.
44. Gedye, R., Smith, F., Westway, K., Baldisera, H. A. L., Laberage, L., & Rousell, J. (1986). *Tetrahedron Letters, 27,* 279–282.
45. Ghorbani-Vaghei, R. & Malaekehpour, S. M. (2010). *Central European Journal of Chemistry, 8,* 1086–1089.
46. Giguere, R. J., Bray, T. L., Duncan, S. M., & Majetich, G. (1986). *Tetrahedron Letters, 27,* 4945–4948.
47. Gomez, M. V., Moreno, A., Vazquez, E. De la Hoz, A., Aranda, A. I., & Diaz-Ortiz, A. (2010). *ARKIVOC,* (iii), 264–273.
48. Gonzalez, I., Bellas, I., Souto, A., Rodriguez, R., & Cruces, J. (2008). *Tetrahedron Letters, 49,* 2002–2004.
49. Gospondinova, M., Gredard, A., Jeannin, M., Chitanv, G. C., Carpov, A., Thiery V., & Besson, T. (2002). *Green Chemistry, 4,* 220–222.
50. Grun, A., Blastik, Z., Drahos, L., & Keglevich, G. (2012). *Heteroatom Chemistry, 23,* 241–246.
51. Haung, D. R., Moerlein, S. M., Lang, L., & Welch, M. J. (1987). *Journal of the Chemical Society, Chemical Communication,* 1799–1801.
52. Hayes, B. L. (2002). *Microwave synthesis: Chemistry at the speed of light.* Mathews NC: CEM Publishing.
53. Horta, J. E. (2011). *Journal of Chemical Education, 88,* 1014–1015.

54. Hu, H. L., Yu, J. H., Yang, S. Y., Wang J. X., & Yin, Y. A. (1999). *Synthetic Communications, 29,* 1157–1164.
55. Hu, H., Yang, H., Heng, P., Cui, D., Peng, Y., Zhang, J., Lu, F., Lian, J., & Shi, D. (2010). *Chemical Communications, 46,* 3866–3868.
56. Hull, A. W. (1921a). *Physical Review, 18,* 31.
57. Hull, A. W. (1921b) *Journal of American Institute of Electrical Engineers, 40,* 715–723.
58. Hur, D., Dal, S. F. E., Varol, G. A., & Hur, E. (2011). *Journal of Organometallic Chemistry, 696,* 2543–2548.
59. Ighilahriz, K., Bouteameur, B., Chami, F., Rabia, C., Hamdi, M., & Hamdi, S. M. A. (2008). *Molecules, 13,* 779–789.
60. Inagaki, T., Fukuhara, T., & Hara, S. (2003). *Synthesis, 8,* 1157–1159.
61. Jia, N., Li, S. M., Ma, M. G. Sun, R. C., & Zhu, L. (2011). *Carbohydrate Research, 346,* 2970–2974.
62. Kakati D. & Sarma, J. C. (2011). *Chemistry Central Journal, 5,* 8–12.
63. Kann, N. (2010). *Molecules, 15,* 6306–6331.
64. Karousis, N., Ichihashi, T., Yudasaka, M., Iijima, S., & Tagmatarchis, N. (2011). *Chemical Communications, 47,* 1604–1606.
65. Khadilkar, B. M. & Rebeiro, G. L. (2002). *Organic Process Research Developement, 6,* 826–828.
66. Kiasat, A. R., Kazemi, F., & Rastogi, S. (2003). *Synthetic Communications, 33,* 601.
67. Kingston, H. M. & Haswell, S. J. (Eds.) (1997). *Microwave–enhanced chemistry. Fundamentals, sample preparation and applications.* Washington, DC: American Chemical Society.
68. Kuhmert, N. (2002). *Angewandte Chemie International Edition, 41,* 1863–1866.
69. Kumar, D., Suresh, & Sandhu, J. S. (2010). *Indian Journal of Chemistry, 49B,* 360–363.
70. Kumar, G. J., Ajithabai, M. D., Santhosh, B., Veena, C. S., & Nair, M. S. (2003). *Indian Journal of Chemistry, 42B,* 429–431.
71. Lalitha, P. & Sivakamasundari, S. (2010). *Journal of Chemical and Pharmaceutical Research, 2,* 387–393.
72. Larhed, M. & Hallberg, A. (1996). *Journal of Organic Chemistry, 61,* 9582–9584.
73. Larhed, M., Moberg, C., & Hallberg, A. (2002). *Accounts of Chemical Research, 35,* 717–727.
74. Law, M. C., Wong, K. Y., & Chan, T. H. (2002). *Green Chemistry, 4,* 328–330.
75. Leadbeater, N. E. & Marco, M. (2003). *Journal of Organic Chemistry, 68,* 888–892.
76. Leadbeater, N. E. (2005). *Chemical Communication, 745,* 2881–2902.
77. Lei, M., Gao, L., & Yang, J. S. (2009). *Tetrahedron Letters, 5,* 5135–5138.
78. Lerestif, J. M., Perocheav, J., Tonnard, F., Bazareav, J. P., & Hamelin, J. (1995). *Tetrahedron Letters, 51,* 6757–6774.
79. Li, J. P., Qiu, J. K., Li, H. -J., & Zhang, G. -S. (2011). *Journal of Chinese Chemical Society, 58,* 268–271.
80. Li, L., Guo, Z., Du, A., & Liu, H. (2012). *Journal-Of Material–Chemistry, 22,* 3600–3605.
81. Li, Z., Ding, R., Lu, Z., Xiao, S., & Ma, X. J. (2006). *Journal of Molecular Catalysis A: Chemical, 250,* 100–103.

82. Lidstrom, P. & Tiernery, J. P. (Eds.) (2005). *Microwave assisted organic synthesis*. Oxford: Blackwell Publishing.
82. Lidstrom, P., Tierney, J., Wathey, B., & Westman, J. (2001). *Microwave Assisted Organic Synthesis. -A Review Tetrahedron, 57,* 9225–9283.
83. Lin, J. S., Chung, M. H., Chen, C. M., Juang, F. S., & Liu, L. E. (2011). *Journal of Physical Organic Chemistry, 24,* 193–202.
84. Liu, L., Li, Y., Fong, Y., & Chen, L. (2005). *Carbohydrate Polymers, 60,* 351–356.
85. Liu, Y., Tsunoyama, H. Akita, T., & Tsukuda, T. (2010). *Chemistry Letters, 39,* 159–161.
86. Longhi, K., Moreira, D. N., Marzari, M. R. B., Floss, V. M., Bonacorso, H. G., Zanatta, N., & Martins, M. A. P. (2010). *Tetrahedron Letters, 51,* 3193–3196.
87. Loupy, A. (Ed.) (2006). *Microwaves in organic synthesis*, (2nd Edn.). Weinheim: Wiley-VCH.
88. Loupy, A., Cabrales, N., Lam, A., Suarez, M., Perez, R., & Rodriguez, H. (2001). *Heterocycles, 55,* 291–301.
89. Mohammad, M., Mohammad, R. S., Mohammad, B., & Jafar, S. S. (2001). *Montshefte Fur Chemie, 132,* 655-658.
90. Lukasiewicz, M., Bogdal, D., & Pielichowski, J. (2004). *International Electronic Conference on Synthetic Organic Chemistry ECSOC-8.*
91. Mojtahedi, M. M., Saidi, M. R., Bolourchian, M., & Shirazi, J. S. (2000). *Monatshefte fur chemie, 132,* 655–658.
92. Ma, M. G., Zhu, J. F., Jia, N., Li, S. M., Sun, R. C., Cao, S. W., & Chen, F. (2010). *Carbohydrate Research, 345,* 1046–1050.
93. Mallikarjuna, N. N. & Varma, R. S. (2007). *Journal of Crystal Growth, 7,* 686–690.
94. Matondo, H., Baboulene, M., & Rico-Lattes, I. (2003). *Applied Organometalic Chemistry, 17,* 239–243.
95. Mavoral, J. A., Cativicla, C., Garcia, J. I., Pires, E., Rovo, A. J., & Figueras, F. (1995). *Applied Catalysis, 131,* 159–166.
96. Mccarroll, A. J., Sandham, D. A., Titumb, L. R., Lewis, A. K. D., Cloke, F. G. N., Davies, B. P., Desantand, A. P., Hiller, W., & Caddicks, S. (2003). *Molecular Diversity, 7,* 115–123.
97. Miljanic, O. S., Volhardt, K. P. C., & Whitener, G. D. (2003). *Synlett, 1,* 29–34.
98. Miranda, L. S. M., La Silva, T. R., Crespo, L. T., Esteves, P. M., De Matos, L. F., Diederichs, C. C., & Alvesdesouza, R. O. M. (2011). *Tetrahedron Letters, 52,* 1639–1640.
99. Moghaddam, F. M. & Sharifi, A. (1995). *Synthetic Communications, 25,* 2457–2461.
100. Mogilaiah, K., Kavitha, S., & Babu, H. R. (2003). *Indian Journal of Chemistry, 42B,* 1750–1752.
101. Molteni, V. & Ellis, A. D. (2005). *Current Organic Synthesis, 2,* 333–375.
102. Motorina, I. A., Parly F., & Grierson, S. (1996). *Synlett, 4,* 389–391.
103. Motshekga, S. C., Pillai, S. K., Ray, S. S., Jalama, K., & Krause, R. W. M. (2012). *Journal of Nanomaterials*, doi:10.1155/2012/691503.
104. Nichols, C. E., Youssef, D., Harris, R. G., & Jha, A. (2006). *ARKIVOC, 13,* 64–72.
105. Nilsson, P., Olofssen, K., & Larhed, M. (2006). *Chemistry of Material Science, 266,* 103–144.

106. Nuchter, M., Muller, U., Ondruschka, B., Tied, A., & Lautenschlager, W. (2003). *Chemical Engineering and Technology, 26,* 1208–1216.
107. Nuchter, M., Ondruschka, B., Bonrath, W., & Gum, A. (2004). *Green Chemistry, 6,* 128–141.
108. Olofsson, K., Hallberg, A., Larhed, M., & Loupy, A. (2002). *Microwaves in Organic Synthesis* (p. 37). Weineim: Wiley.
109. Patrick, T. B., Gorrell, K., & Rogers, J. (2007). *Journal of Fluorine Chemistry, 128,* 710–713.
110. Paul, N., Kaladevi, S., & Muthusubramanian, S. (2012). *Helvetica Chimica Acta, 95,* 173–184.
111. Perreux, L. & Loupy, A. (2001). *Tetrahedron, 57,* 9199–9223.
112. Pielichowski, J., Dziki, E., & Polaczek, J. (2003). *Polish Journal of Chemical Technology, 5,* 3–4.
113. Plazl, I., Leskovesek, S., & Koloini, T. (2003). *Chemical Engineering Journal, 59,* 253–257.
114. Polshettiwar, P. V. & Varma, R. S. (2008). *Accounts of Chemical Research, 41,* 629–639.
115. Rahman, M., Majee, A., & Hajra, A. (2011). *Chem Inform. Abstract,* 42.
116. Ranger, K. D., Strauss, C. R., Trainer, R. W., & Thorn, J. S. (1995). *Journal of Organic Chemistry, 60,* 2456–2460.
117. Robeiro, G. L. & Khandikar, B. M. (2003). *Synthetic Communications, 33,* 10405–10410.
118. Rohman, M. R. & Myrboh, B. (2010). *Tetrahedron Letters, 51,* 4772–4775.
119. Roy, I. & Gupta, M. N. (2003). *Tetrahedron Letters,* **5***9,* 5431–5436.
120. Sancheti, A., Swarnkar, N., Soni, M. D., Vardia, J., Punjabi, P. B., & Ameta, S. C. (2007) *Journal of Indian Chemical Society, 84,* 1234–1238.
121. Santagada, V., Fiorino, F., Perissuti, E., Severino, B., Terracciano, S., Cirino, G., & Caliendo, G. (2003). *Tetrahedron Letters, 5,* 2131–2134.
122. Schmoger, C., Stolle, A., Bonrath, W., & Ondruschka, B. (2011). *Current Organic Chemistry, 15,* 151–167.
123. Shao, B. (2005). *Tetrahedron Letters, 46,* 3423–3427.
124. Shi, S. & Hwang, J. Y. (2003). *Journal of Mineral Material Characterization Engineering, 2,* 101–110.
125. Shrimali, K., Sitha, D., Vardia, J., & Ameta, S. C. (2009). *Afinidad, 66,* 173–176.
126. Sitha, D., Ameta, R., Punjabi, P. B., & Ameta, S. C. (2010). *International Journal of Chemical Science, 8*(3), 1973–1982.
127. Srikrishna, A. & Kumar, P. P. (1995). *Tetrahedron Letters, 36,* 6313–6316.
128. Stadler, A. & Kappe, C. O. (2000). *Journal of Chemical Society* Perkin Transactions, *2,* 1363–1368.
129. Strauss, C. R. & Trainor, R. W. (1995). *Australian Journal of Chemistry, 48,* 1665–1692.
130. Strauss, C. R. (2002). *Angewandte Chemie International Edition, 41,* 3589–3591.
131. Stuerga, D. & Gaillard, P. (1996). *Journal of Microwave Power and Electromagnetic Energy, 31,* 101–113.
132. Stuerga, D. & Gaillard, P. (1996). *Journal of Microwave Power and Electromagnetic Energy, 31,* 87–99.

133. Tejedor, D., Mendez-Abt, G., Cotos, L., Ramirez, M. A., & García-Tellado, F. A. (2011). *Chemistry- A European Journal, 17,* 3318–3321.
134. Thorell, J. O., Stone–Elander, S., & Elander, N. (1992). *Journal Of Compounds and Radiopharmaceuticals, 31,* 207–217.
135. Varma, R. S. & Namboodiri, V. V. (2001). *Chemical Communications, 7,* 643–644.
136. Varma, R. S. & Saini, R. K. (1997). *Tetrahedron Letters, 38,* 4337–4338.
137. Vidal, T., Pefit, A., Loupy, A., & Gedye, R. N. (2000). *Tetrahedron Letters, 56,* 5473–5478.
138. Vu, Z. T., Liu, L. J., & Zhuo, R. X. (2003). *Journal of Polymer Chemistry Edition, 41,* 13–21.
139. Vyas, R., Swarnkar, N., Sancheti, A., Vardia, J., & Punjabi, P. B. (2008). *Journal Indian Chemical Society, 85,* 1217–1226.
140. Wada, Y., Yin, H. B., Kitamura, T., & Yanagida, S. (2000). *Chemistry Letters,* 632–633.
141. Wada, Y., Yin, H., & Yanagida, S. (2002). *Catalysis Surveys from Japan, 5,* 127–138.
142. Wali, A., Paillai, S. M., & Satish, S. (1995). *Indian Petrochemical Corporation Limited Communication,* 294.
143. Wang, Y. & Lee, J. Y. (2005). *Journal of Power Sources, 144,* 220–225.
144. Watkins, K. J. (2002). Fighting the Clock. *Chemical and Engineering News, 80,* 27–34.
145. Wolkenberg, S. E., Wisnoski, D. D., Leister, W. H., Wang, Y., Zhao, Z., & Lindsley, C. W. (2004). *Organic Letters, 6,* 1453–1456.
146. Wu, T. Y. H., Schultz, P. G., & Ding, S. (2003). *Organic Letters, 5,* 3827–3830.
147. Xu, J. S. & Zhu, Y. J. (2012). *Crystal Engineering Communication, 14,* 2630–2634.
148. Yamazaki, K. & Kondo, V. (2002). *Journal of Combinatorial Chemistry, 4,* 191–192.
149. Yang, S., Da-Wei, H., Xiao-Hui, L., Jian-En, H., & Zhi-Long, X. (2012). *Chemical Research Chinese Universities, 28,* 108–113.
150. Yue, Y., Liu, T. Y., Li, H.-W., Liu, Z., & Wu, Y. (2012) *Nanoscale Research Letters, 4,* 2251–2254.
151. Zhang, W., Chen, J., Swiegers, G. F., Ma, Z. F., & Wallace, G. G. (2010). *Nanoscale Research Letters, 2,* 282–286.
152. Zhao, G., Jiang, T., Gao, H., Han, B., Huang, J., & Sun, D. (2004). *Green Chemistry Articles, 6,* 75–77.
153. Zhu, J. F. & Zhu, Y. J. (2006). *Journal of Physical Chemistry, 110B,* 8593–8597.

# CHAPTER 12

# GREEN COMPOSITES

YASMIN, N. P. S. CHAUHAN, and ROHIT AMETA

## CONTENTS

## 12.1 INTRODUCTION

The demand of environment friendly materials is growing and will continue to grow by growing public concern about environmental pollution, which has led to development and design of biodegradable composite materials. Green chemistry concepts may be utilized for environment friendly materials and to minimize the toxic effects of organic compounds, which pose potential risks to human health. The serious efforts are needed for the development of green composites for minimization of environmental impact of polymer composite production.

Assessment of the environmental impact arising from activities such as construction, packaging or transport is an essential activity, when attempting to design sustainable approaches to our development needs. In this respect, life cycle assessment (LCA) has emerged as a widely accepted technique for evaluating the environmental aspects associated with a wide variety of products, processes or activities from initial synthesis to final disposal for a sustainable environment. The regulatory assessment and monitoring procedures must be updated time to time depending upon the composition, intended usage conditions in order to promote clean processing, applications, biodegradation, recycling and reprocessing.

Most of the commercial fibers and resins like plastics and polymers are derived from petroleum feedstock. The major problem associated with this is the high rate of depletion of petroleum resources. By one estimate, the current consumption rate of petroleum is about 100,000 times the rate at which the earth can generate it. Another problem is most of the composites and plastics derived from petroleum are non-biodegradable under normal environmental conditions. Composites made by thermoset resins cannot be reused or recycled and most of these composites end up in the landfills at the end of their life. They last for several decades without degrading and make that land unusable. Incineration produces large amount of the toxic gases that require costly scrubbers; hence both; land filling and incineration are expensive and environmentally undesirable.

Growing global environmental and social concern, the high rate of depletion of petroleum resources and now environmental regulations have forced the search for green composites, compatible with the environment.

Green composite combines plant fibers with natural resins to form natural composite materials. Natural fibers such as kenaf, flax, jute, hemp, sisal, banana fiber, cotton, kapok and coir are emerging as low cost, light weight and eco-friendly alternative to synthetic fibers such as carbon fibers, glass fibers and kevlar fibers. The resin and fibers used in green composites are biodegradable, on dumping, they are decomposed by the action of microorganisms. They are converted into $H_2O$ and $CO_2$, which are absorbed into the plant system. Moreover because of their moderate mechanical strength, they can be used to reinforce plastics and fabricate composites for various applications for example, packaging, product casing, housing and automotive panels, furniture and so on.

## 12.2 GREEN COMPOSITES

### *12.2.1 DESIGNING FOR COMPOSITES*

The best way to get the ideas of designing for composites is from natural processes. It has maked functional products for millennia.

A leaf is a good example of natural processes. A real leaf is passing through different stages in its life that is; changing form, changing process, response to adverse conditions and to damage and attack, aerodynamics and thermal control, chemistry and physics, deployment and retraction, disposability and recycling, even the role of environment for other life forms after it shades off from tree. Natural processes give ideas about what valuable means and what enhances the environment and life processes? Technology provides us the tools to see these attributes of nature and to design composite material properties from nano to micro and micro to macro scale in terms of fabrication and assembly techniques.

The observable visual aspects of the physical formation and patterns that are characteristics of natural material engineering are network construction for example, bird nest, leaf and web. Interfacial issues like differential properties arranged in 3D to accommodate complex characteristics, for example: tendon, spider web, foot, braiding, twisting, binding, orientation of fibers, layers of fiber orientation, combination of different fiber materials, fiber matrix relation and so on… help us conceptualize the principles of our design that is, no waste, relate to environment and local available resources. Let us discuss some examples to learn from natural material structure in which the visible shapes of the object, surface detail and texture, and structural features, display some of the shapes of essential functional attributes.

i. Leaf networks (Figure 12.1). Form fluid variation
ii. Crab shell architecture (Figure 12.2). For improving damage control
iii. Spider web (Figure 12.3). Three dimensional arrangement to accommodate complex characters.
iv. Corn husk leaf packaging (Figure 12.4). A nature food wrapper.

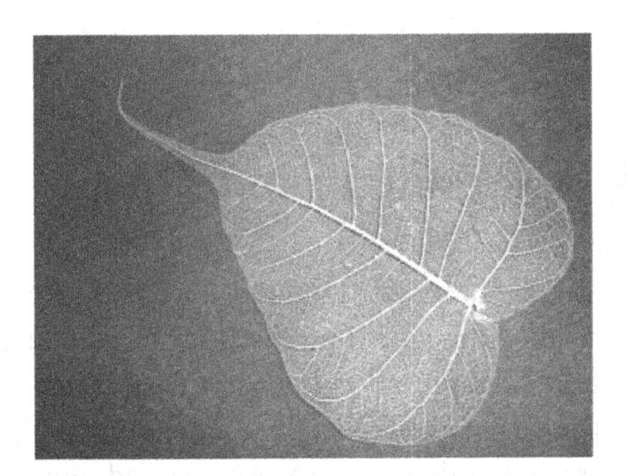

**FIGURE 12.1** Leaf networks.

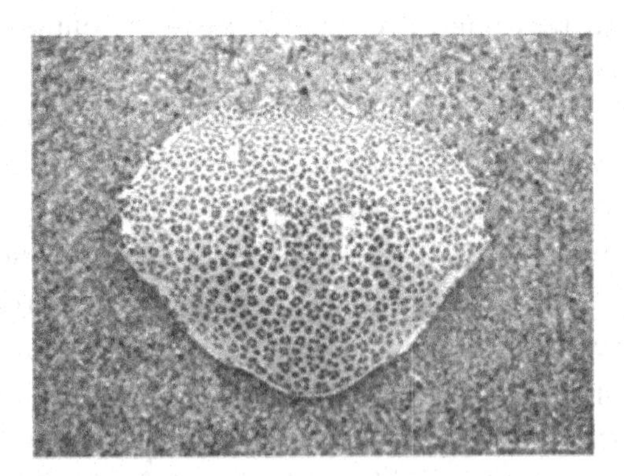

**FIGURE 12.2** Crab shell structure.

**FIGURE 12.3** Spider web.

**FIGURE 12.4** Corn husk packing.

## *12.2.2 LIFE CYCLE ASSESSMENT*

The life cycle assessment is a technique for evaluating environmental impact at all stages of a products life from initial synthesis to final disposal such as recycling, incineration and disposal for a sustainable environment. The sustainability parameter includes energy use, price, transportation, health hazards, renewable resources, waste prevention, biodegradability, recycling and so on. In recent years, life cycle assessment work is easily supported by using a number of commercial LCA software packages as well as authoritative public access datasets for example: ISO standard 14040 to 14043 provide detailed guidelines for conducting LCA but attention to the selection of appropriate data for the goal and scope of the study will always remains an important issue in LCA.

The LCA methodology consists of four independent elements (ISO 14040 series) that is:

i. The definition of goal and scope.
ii. The life cycle inventory analysis.
iii. The life cycle impact assessment and
iv. The life cycle interpretation of results.

### 12.2.3 NATURAL FIBER SOURCES

The natural composite materials are created by combing plant fibers with natural resins. Natural fibers are emerging as low cost, light weight and recyclables. Easy availability, thermal insulation, $CO_2$ neutrality, acceptable strength, stiffness and biodegradability are the added advantages that make biocomposites environmentally superior alternative to synthetic fibers.

- Fibers such as jute, cotton kenaf, banana fiber, pineapple leaf, sisal, henequen, kapek, flax, coir are being used as biofiller in thermoplastic and thermosetting polymers to develop composites through various techniques.
- Chicken feather, bone, scales of marine animals, silk and lamb wool are also used as biofiller in composites.
- Recently agricultural waste fibers like wheat straw, grass fibers, soy stalk and corn stalk like miscanthus are also promoted as reinforcing filler in biocomposites.

#### *KENAF FIBER*

In recent years, cellulose materials are also used as reinforcement fibers, not only for ecological and economical reasons, but also because of their high mechanical and thermal performance. Kenaf is well known as a cellulose source and it can be a good alternative to glass fibers because:

- It fixes $CO_2$ effectively.
- Kenaf absorbs $N_2$ and P found in the soil and in waste water. (Abe and Ozaki, 1998).
- It has low density and high specific mechanical properties and it is biodegradable.

Han et al. concluded that kenaf has been used as an alternative raw material to wood in the pulp and paper industries to avoid the destruction of forests (1999). Serizawa et al. reported that adding kenaf fiber to poly(lactic acid) PLA greatly increases its heat resistance and modulus and also enhances its crystallization and therefore, the ease of moulding this material is improved (2006). They also reported that PLA/ kenaf fiber and PLA/ kenaf fiber/flexibilizers (which is a copolymer of lactic acid and aliphatic polyester) show good practical characteristics for housing materials of electronic products as compared to petroleum based plastic used in housing such as glass-fiber reinforced acrylonitrile-butadiene-styrene (ABS) resin. Khristova et al. reported that the soda-AQ pulp blends from kenaf and sunflower stalks results, in considerable improvement of the sunflower-pulp properties (1998).

- **Micro fibrillated materials:** Taniguchi and Okamura developed new type of micro fibrillated materials (MM) from natural fibers such as wood pulp fibers, cotton fibers, tunicin cellulose, chitosan, silk fibers and collagen by a super grinding method (1998). They made films using MM from natural fibers and the films obtained had 3–100$\mu$m thickness and were homogeneous, strong and translucent. Their tensile strength was much superior to those of commercial print grade papers.
- **Micro fiber bundles:** Yano and Nakahara used plant micro fiber bundles with a nanometer unit web like network to obtain a moulded product and found that the mouldings have a combination of environmentally friendly and high strength properties (2003).
- **Silkworm silk:** Silkworm silk is also a natural fiber, Shao and Vollrath reported that the mechanical properties of silkworm silks can approach those of spider dragline silk, when reeled under controlled conditions (2002).

- **Oil palm empty fruit bunch fiber:** A comparative study of polypropylene composites reinforced with oil palm empty fruit bunch fiber and oil palm derived cellulose was made by Khalid et al. (2008). The structure and mechanical properties of sheets prepared from cellulose was studied by Nishi et al. (1990).
- **Spider dragline silks:** Spider dragline silks are exceptionally strong and extensible (Kaplan et al., 1994). In fact their toughness equals that of commercial polyaramid (aromatic nylon) filaments, which is used to make materials ranging from radial tyres and bullet proof clothing to reinforced composites for aircraft panels. Heslot, Asakura and Kaplan reported that the exceptional toughness of dragline silk is achieved under benign conditions in contrast to current techno polymers based on petrochemicals (1998; 1994). The spider spins its totally recyclable fibers at ambient temperature, low pressures and with water as solvent. Tirrell et al. concluded that genetically engineered silk with specially tailored properties and spun using 'green' processes could replace the ubiquitous plastics, which are often detrimental to the environment in both production and disposal (1994).
- **Wood fibers and paper fibers (Alternative natural fibers):** Wood fibers are cellulose fibers made by mechanical and chemical methods whereas paper fibers are mechanically ground and chemically made organic material from trees and used in paper industry. In paper industry, both fibers are mixed in order to combine their properties.

Paper fibers contribute too many properties of composites like renewablity, biodegradability, low price, found in abundance and strength. The wood fibers sources are eucalyptus, pine, spruce, larch, beech, white beech, oak, aspen and linden trees.

Wood fiber consists mainly of cellulose, hemicellulose and lignin.

- Cellulose forms the frame of the cell wall and it is the main component of wood fiber, which is responsible for the properties of fibers and makes it possible to use them in paper making industry.
- Hemicellulose and lignin forms the surrounding intercellular substance and hemicellulose, which affects the ability of fibers to form bonding between each other.
- Lignin bonds fibers and gives stiffness to wood.

Wood can be used in various forms as reinforcement for example in the form of wood sheets, wood flours or fiber (paper). The mechanical properties of the wood fibers can be compared with those of synthetic fibers, for example glass fiber. Cellulose fibers and plastic composites could offer higher specific strength than glass fiber at a low cost.

The mechanical properties of the composites can be improved by using wood fibers. The properties of the composites are affected by the wood fiber form, dimensions, treatment of fibers and fillers. The strength of the composites depends on the length of fiber, the longer the fiber, the better is the strength whereas shorter fibers should disperse more easily and as a consequence increases properties of the fiber more than adhesion (Bolton, 1994).

The most important factor to affect the properties of the composites is the adhesion between fibers and matrix. The lignin forms better adhesion between hydrophobic and non-polar polymers such as polyolefins, if used in wood plastic composites. If adhesion is poor, wood fibers work as filler in the matrix and therefore, the tensile strength can further decrease. When the adhesion is good, the wood fiber work as reinforcement in the matrix. Adhesion can be improved by coupling agent for example, silanes.

Low melting points materials such as polyolefins, thermosets and polystyrenes are used as matrix in wood fiber composites. More reported that the composites made of polyolefins and wood fiber will absorb less moisture than commercial hardboards, where wood flour is used as filler (1997). The benefits of using polyolefins as a matrix material are that they are easily available and having low processing temperature so that wood fiber cannot degrade.

Sameni et al. used rubber matrix and biopolymer matrix for their studies (2002). Amnuay et al. used cellulose fibers from recycled newspaper as reinforcement for thermoplastic starch in order to improve its mechanical, thermal and water resistance properties (2011). They prepared composites from corn starch plasticized by glycerol as matrix, which was reinforced with microcellulose fibers; obtained from used newspaper. They reported that thermoplastic starch reinforced with recycled newspaper cellulose fibers could be fruitfully used as commodity plastic being strong, low cost, abundant and recyclable.

### 12.2.4 NATURAL POLYMER SOURCES

The challenge of green composites involves basically obtaining 'green' polymers that are used as matrix for the production of the composites. Polymer is said to be green, when it possesses environmentally favorable properties such as renewability and degradability. Biodegradation implies degradation of a polymer in natural environment that includes changes in the chemical structure, loss of mechanical and structural properties and changing into other compounds that are beneficial to the environment (Jamshidian et al., 2010; Pandey et al., 2007).

Polymers from natural sources such as starch, lignin, cellulose acetate, polylactic acid (PLA), polyhydroxy alkanoates (PHA), polyhydroxy butyrate (PHB) and some other synthetic sources such as aliphatic and aromatic polyesters, polyvinyl alcohol, modified polyolefins and so on, that are degradable are classified as biopolymers. The most important biopolymers from an overall market perspective are:

*POLYLACTIC ACID (PLA)*

It is a class of crystalline biodegradable thermoplastic polymer with relatively high melting point and excellent mechanical properties. All PLA resins are manufactured using renewable agricultural resources such as corn and sugar beets. Corn has the advantage of providing the required high purity lactic acid than other sources. The PLA is synthesized by condensation polymerization of D- or L-lactic acid or ring opening polymerization of the lactide (Fang and Hanna, 1999; Garlotta, 2002).

The PLA is commercially interesting because of its good strength properties, film transparency, biodegradability, biocompatibility and availability from renewable resources. Under specific environmental conditions, pure PLA can degrade to $CO_2$, $H_2O$ and $CH_4$ over a period of several months to 2 years, a distinct advantage compared to other petroleum plastics that need much longer periods. The final properties of PLA strictly depend on its molecular weight and crystallinity. The PLA has been studied as a biomaterial in medicine, but only recently, it has been used as a polymer matrix in composites.

Avella et al. present a brief review of the most suitable and commonly used biodegradable polymer matrices and natural fiber reinforcement in eco-composites and nano composites with special focus on PLA based material (2009).

In last few years, different natural fiber have been employed in order to modify the properties of PLA. The most studied natural fiber reinforcements for PLA were kenaf (Avella et al. 2009; Huda et al., 2009), flax (Bux and Mussig, 2008), hemp (Hu and Lin, 2007), bamboo (Tokoro et al., 2008), jute (Shikamoto et al. 2007) and wood fibers (Huda et al., 2006). Besides conventional natural fibers, reed fibers have been tested in appropriate PLA composites in order to improve the tensile modulus and strength (Huda et al., 2008).

## *POLYHYDROXY ALKANOATES (PHA)*

Poly-R-3-hydroxy butyrate (PHB) is the simplest family of PHA. The PHA's are synthesized biochemically by microbial fermentation and represent natural polyesters. Bacteria are still the only source of these polyesters. *Alcaligenes eutrophus* has been used to produce the commercial product, a polyhydroxy butyrate covalerate PHB (commercial name Biopol®). It is biotechnologically prepared polyester that constitute a carbon reserve in a wide variety of bacteria (Avella et al., 1996) and has attracted much attention as a biodegradable thermoplastic polyester (Doi, 1990). A large number of PHBV random copolymers can be produced from *A. eutrophus* depending upon the carbon substrate. Carbon sources include propionic acid, pentanoic acid, 4-hydroxybutyric acid, 1,4- butanediol and so on. Following the formation process, the dilute aqueous broth is extracted to obtain biopolymers that must be then isolated and purified. The PHB of very high purity can be produced by continuous fermentation in combination with advanced isolation procedures.

One of the main commercial developments in PHA technology has been the production of PHBV in the form of Biopol®. The first products from this polymer were shampoo bottles and cosmetics containers. Future applications for PHB based polymers could be in disposable products such as fast food utensils, garbage bags and diapers. Hocking and Marchessault reviewed the use of PHB or PHBV in a variety of products such as films, bottles and containers (1994).

## *STARCH*

Starch is produced in plants and some microorganisms. It is a mixture of linear amylase and branched amylopectin. The amount of amylase and amylopectin, the size and shape of starch granules depends upon its source. Amylase is the minor component of starch ranging from 20 to 30%. The amylopectin is responsible for crystalline properties of starches. The relative proportions of amylopectin and amylase in starch are determined by genetic and environmental control during biosynthesis. As a result, wide variations are found among plant raw materials. Starch is one of the cheapest biodegradable materials available in the market today. It is a versatile biopolymer with immense potential for use in the non-food industries.

The primary source of starch is maize and other sources are rice, wheat, potato and so on. Starch is an interesting alternative to thermoplastic material in situation where long term durability is not required and rapid degradation is an advantage. Shogren has reviewed the properties and applications of starch (1998). Starch can be made thermoplastic through destructurization process in specific extrusion conditions (Bastioli, 1998). In destructurization of starch, the semicrystalline starch granules are converted into a homogeneous amorphous polymer matrix. The mechanical properties of destructurized starch depend upon its degree. As destructurization increases, tensile strength and elongation also increases but the elastic modulus is reduced. This means that the material becomes increasingly flexible.

The properties of thermoplastic starch can be improved by adding plasticizers (e. g., water, urea, ammonia, diethylene glycol, citric acid etc.), lubricants (e. g., lipids, fatty acid, talc, silicon etc.) and fillers (e. g., proteins, water soluble polysaccarides and water soluble polymers). Thermoplastic starch products with different viscosity, water absorption properties and water solubility have been prepared by altering the moisture content, the amylase/amylopectin ratio of the raw material and the temperature and pressure in the extruder (Donovan, 1979; Mercier and Feillet, 1975). Thermoplastic starch is sensitive to humidity and is, therefore, unsuitable for most food packaging applications. The thermoplastic starch alone is mainly used in soluble compostable foams such as loose fillers, expanded trays, expanded layers and shape moulded parts as a replacement for polystyrene.

### *12.2.5 BIOCOMPOSITES*

A composite is a material made of two different phases that is a matrix phase and a disperse phase. The disperse phase may consist of synthetic material (e. g., fibers) or natural materials (e. g., natural fibers). The matrix phase can be a synthetic or a natural polymer; the matrix can also be classified as thermoplastic or thermoset. Thermoplastic are recyclable and biodegradable, where as thermoset are biodegradable only and is not recyclable. The combination of a plastic matrix and reinforcing fibers give rise to composites having the best properties of each component, since plastics are soft, flexible and light weight as compared to fibers, their combination provides a high strength to weight ratio for the resulting composite for various applications such as packaging, product casing, housing, automotive panels, furniture and so on.

**Natural Fiber Biocomposites:** Natural composites or green composites are emerging as a viable alternative to glass fiber reinforced composites especially in automotive and building product applications. It can also be effectively used as a material for structural, medical and electronic applications.

Natural composites reinforced with different natural fibers are as follows:

#### *CELLULOSE FIBER BASED BIOCOMPOSITES*

Cellulose materials are used in the polymer industry for a wide range of applications, including laminates, panel products and fillers, alloys and blends, composites and cellulose derivatives (Maldas and Kokta, 1993). The mechanical properties of cellulose-polymer composites can be improved by using graft copolymers of the matrix material and by the addition of a polar group (Felix and Gatenholm, 1991). Cellulose fibers can also reinforce thermoset polymers like polyester, epoxy, amino and phenolic resins (Flodin, 1986). Short cellulosic fiber-reinforced elastomer composites have gained practical and economic interest in the rubber industry (Setua, 1986).

The cellulose based composites are not fully biodegradable because of non-biodegradable synthetic matrix components. The processing and properties of biodegradable composites of bacteria-produced polyesters (Biopol®) reinforced with wood cellulose have been reported by Gatenholm et al. although cellulose fibers improved the strength and stiffness of the (PHB), the composites were very brittle (1992). The effect on the tensile modulus by the incorporation of cellulose fibers into three different thermoplastics like polypropylene, polystyrene and poly-R-3-hydroxy butyrate (PHB) has also been investigated, which revealed that the tensile modulus increased for each composite with increasing fiber content. The stiffing effect of cellulose fiber in PHB was in the same order as in polystyrene. Curvelo et al. proposed that the properties of thermoplastic starch composites can be improved using reinforcing fibers and fillers particularly cellulose fibers (2001).

Thermoplastic starch/cellulose fiber composites have been prepared by using fibers from different sources, such as flax and ramie fibers (Wollerdarfer and Bader, 1998), potato pulp fibers (Dufresne and Vignon 1998), bleached leaf wood fibers (Averous et al., 2001) and wood pulp fibers (Carvalho et al., 2002). Most of these authors have shown an improvement of the mechanical properties of the composites that was attributed to the chemical compatibility between the two polysaccharides, that is starch and vegetal fibers (Averous and Boquillon, 2004; Curvelo et al., 2001; Wallerdorfer and Bader, 1998). As a result, water resistance of the composites substantially increased (Dufrene et al., 2000) as a direct consequence of the addition of the less hydrophilic fibrous filler (Ma et al., 2005).

## *JUTE BASED BIOCOMPOSITES*

There are many reports about the use of jute as reinforcing fibers a thermoplastics (Karmaker and Hinrichseh, 1991; Karmaker and Youngquist, 1996; Rana et al., 1998) and thermoset (Bledzki and Gassan, 1996; Pal, 1984; Sahoo et al., 1999). Jute fibers are having high tensile modulus and low elongation at break. The specific modulus of jute is superior to glass fibers and on a modulus per cost basis, jute is far superior. The specific strength per unit cost of jute is almost equal to that of glass fiber.

Jute reinforced thermoplastic, thermosets and rubber based composites has been reviewed by Mohanty and Mishra (1995). It is essential to pretreat jute so that the moisture absorption would be reduced and wettability of the matrix polymer would be improved. Mitra et al. have reported the studies on jute reinforced composites, their limitations and some solutions through chemical modifications of fibers (1998). The effect of different additives on performance of biodegradable jute fabric-Biopol® composites has been reported by Khan et al. (1999). In order to study effects of additives, the jute fabrics were soaked with several additives solution of different concentrations, dicumyl peroxide was used as the initiator during the treatments.

The superior strength of alkali treated jute may be attributed to the fact that alkali treatment improves the adhesive characteristics of jute surface removing natural and artificial impurities thereby, producing a rough surface topography (Bisnada et al., 1991). In addition, alkali treatment leads to fiber fibrillation that is, breaking down of fabrics fiber bundle into smaller fibers. This increases the effective surface area available for contact with matrix polymer.

## *FLAX AND HEMP BASED BIOCOMPOSITES*

Flax and hemp are also used as reinforcement in composites. It was suggested that flax and sisal based composites are used for making vehicle interior parts (Haager, 1995). The reinforcement of polyisocyanate bonded particle boards with flax fibers led to products comparable to those of carbon and glass fiber reinforced particle boards (Barbu and Fritz, 1996).

Biocomposites containing natural fibers and biodegradable matrices are patented for applications as building materials (Herrman et al., 1994). These materials contain natural fibers for example, flax, hemp, ramie, sisal or jute and biodegradable matrix such as cellulose diacetate, or a starch derivative.

## *NATURAL POLYMER BIOCOMPOSITES*

### *THERMOPLASTIC AND THERMOSET BIOCOMPOSITES*

Thermoplastic composites are composites that make use of a thermoplastic polymer as a matrix. The properties of these polymers are toughness, chemically inert and recyclability. The advantage of thermoplastic is that they can be rapidly heated and cooled without any damaging effects to their microstructure. In thermoplastics, most of the work reported deals with polymers such as polyethylene, polypropylene, polystyrene and polyvinyl chloride. The natural fibers used to reinforce thermoplastics include wood, cotton, flax, hemp, jute, sisal, banana, pineapple and sugarcane fibers (Yao et al., 2008). Thermoset polymers are also used as a matrix material for most structural composite materials. The single biggest advantage of thermoset polymers is that they have a very low viscosity and therefore, they can be introduced into fibers at low pressures. These composite materials are chemically cured to a highly cross linked 3D network structure and are highly solvent resistant, tough and creep resistant. The common thermosetting materials are epoxy resins and unsaturated polyesters, phenolic resins, amino resins and polyurethane.

### *THERMOPLASTIC STARCH BASED COMPOSITES*

Carvalho et al. (2001) first reported the use of thermoplastic starch for the production of composites by melt intercalation in twin screw extruder. The composites were prepared with regular corn starch plasticized with glycerin and reinforced with hydrated kaolin. Biotech of Germany has conducted R and D along the lines of starch based thermoplastic materials. The company's three product lines are Bioplast granules, Bioflex film and Biopur foamed starch.

Novamont of Italy produces four classes of biodegradable materials Z, Y, V and A under the Mater- Bi trademark. All four classes of materials contain starch and differ in synthetic grades and has been developed to meet the requirements of specific applications. The physico-mechanical properties of Mater- Bi are similar to those of conventional plastics like plates, cutlery, cup lids and so on packaging like wrapping films, film for dry food packaging, board lamination and so on, stationary like pens, cartridges, pencil and so on, personal care and hygiene and others like toys, bags and so on. Different starch plastics with various trade names are now available in the market. In comparison to thermoplastic, biodegradable products based on starch still reveal many disadvantages such as hydrophilic character of starch polymers. In spite of many advantages, thermoplastic starch based materials are still at an early stage of development and the markets for such products are expected to grow in future as the properties are more improved, prices still declined and an infrastructure for composting becomes more established.

## *POLYLACTIC ACID (PLA) BASED COMPOSITES*

The PLA based materials are a new class of materials of interest developed in recent years due to the continuously increasing environmental awareness throughout the world. They can be considered as the 'green' evolution of the more traditional eco-composites, essentially consisting of synthetic polymers based composites reinforced with natural fibers or other micro or nanofiller.

In the past years, different natural fibers have been employed in order to modify the properties of PLA. Up to now, the most studied natural fiber reinforcements for PLA were kenaf (Avella et al., 2008; Huda et al., 2008), flax (Bax and Mussig, 2008), hemp (Hu and Lim, 2007), bamboo (Tokoro et al., 2008), jute (Shikamoto et al., 2007) and wood fibers (Huda et al., 2006). Huda et al. worked on kenaf fiber reinforced PLA laminated composites prepared by compression moulding using the film- stacking method (2008). It was found that standard PLA resins are suitable for the manufacture of kenaf fiber reinforced laminated biocomposites with useful engineering properties. The mechanical properties of PLA composites reinforced with cordenka rayon fibers and flax fibers (examples for completely biodegradable composites) were tested and compared. The promising impact properties of the presented PLA/cordenka composites show their potential as an alternative to traditional composites (Bax and Mussig, 2008).

The effects of the alkali treated natural fibers on the mechanical properties of PLA/hemp fibers were studied by Hu and Lim (2007). The result show that the composite with 40% volume fraction of alkali treated fiber possessed the best mechanical properties. Bamboo fiber reinforced PLA composites were prepared in order to improve the impact strength and heat resistance of PLA (Tokoro et al., 2008). The suitability of wood fibers as natural reinforcement in PLA based composites has also been demonstrated in comparison with wood fiber reinforced polypropylene composites (Huda et al., 2006). Silkworm silk fibers are recognized as reinforcements in PLA natural fiber composites for tissue engineering application due to the fact that the silk fiber surface bonds well with the polymer matrix. These fibers can be good material, as reinforcements for the development of polymeric scaffolds for tissue engineering applications (Cheung et al., 2008). The nano composites based on PLA are of special interest for medical purpose. Special attention has been paid to novel nanomaterials capable of facilitating the biorecognition of anticancer drugs. These novel nanocomposites imply some potential valuable applications as a kind of drug carriers in view of the respective good biocompatibility of PLA and large surface area of the nanoparticles (Song et al., 2008).

## *CELLULOSE BASED BIOCOMPOSITES*

Cellulose from agricultural products has been identified as a source of biopolymer that can replace synthetic polymer. Cellulose acetate is considered as potentially useful polymers in biodegradable applications (Buchanan et al., 1993; Gu et al., 1992; Rivard et al., 1992). Green nanocomposites have been successfully produced from cellulose acetate, triethyl citrate plasticizer and originally modified clay via melt compounding (Mishra et al., 2004).

Recently, a cellulose nanocomposite material has been investigated as a flexible humidity and temperature sensor (Mahadeva et al., 2011). Cellulose was obtained from cotton pulp via acid hydrolysis using a solution of lithium chloride and N,N-dimethyl acetamide. An active antimicrobial packaging material has been developed using methyl cellulose as the base material with montomorillionile as reinforcement (Tune and Duman, 2011). Researchers have evaluated the use of cellulose based nanocomposite with hydroxylapatite for medical applications (Zadegan et al., 2011; Zimmermann et al., 2011).

## *MISCELLANEOUS BIOCOMPOSITES*

Soy based polyurethane can be used as a matrix for the production of bio-based nanocomposites (Tate et al., 2010). Relatively water resistant biodegradable soy protein composite is resulted through blending of special bioabsorbable polyphosphate fillers, biodegradable soy protein isolate, plasticizer and adhesion promoter in a high shear mixer followed by compression moulding (Otaigbe, 1998). The degradable composite films composed of soy protein isolate and fatty acids as well as soy protein isolate and propylene glycolalginate have been prepared by Rhim et al. (1999). Luo and Netravali (1999) have reported the mechanical and thermal properties of bio/green composites obtained from pineapple leaf fibers (28% fiber content) and Biopol®, that is PHBV resin.

The blending of two or more polymers to achieve a polymer that is biodegradable has drawn such research interest. These polymers have been tested for their degradability and mechanical properties and thus, recommended for use as degradable polymers for composite applications. Such polymer blend starch/PLA blends, poly butylenes succinate/cellulose acetate blends, starch/modified polyester blends, polycarprolactone/polyvinyl alcohol blends and thermoplastic starch/polyesteramide blends have been reported (Averous et al., 2000; Ke and Sun, 2001; Kesel et al., 1997; Martin and Averous, 2001; Uesaka et al., 2000; Willett and Shogren, 2002). Recently, use of binary and ternary blends of PLA, polycaprolactone and thermoplastic starch (TPS) as composites has been reported by Sarazin et al. (2008).

### 12.2.6 PROPERTIES OF BIOCOMPOSITES

The properties of the composite are determined by:

- The properties of fiber
- The properties of resin
- The geometry and orientation of the fibers in the composite and,
- The surface interaction of fiber and resin.

#### *PROPERTIES OF FIBER*

In fiber reinforced composites, the strength of the composite is determined by the strength of the fiber and by the ability of the matrix to transmit stress to the fiber plant. Fibers consists of microfibrils, which are interconnected via lignin and hemicellulose fragments. The more parallel the microfibrils are arranged to the fiber axis, the higher is the fiber strength. The microfibril angle is one of the major factors in determining the mechanical properties of the fiber. In the spiral structure, the microfibril angle is the angle between the cellulose microfibrils and the longitudinal cell axis. The tensile strength and the Young's modulus decreased with the increase in the microfibril angle. Different fibers used in composites have different ways. The mechanical properties of natural fibers such as jute, hemp, flax and sisal are very good and may compete with glass fiber in specific strength and modulus. Natural fibers show higher elongation to break than glass or carbon fibers, which may enhance composite performance. Thermal conductivity of natural fibers is low and as a consequence, they make a good thermal barrier. The stiffest and strongest composites are based on unidirectional aligned natural fiber bundles impregnated with synthetic or natural resins.

In order to maximize the composite strength and stiffness and in order to transfer stress effectively between natural fibers and resin matrix, the following factors should be considered:

(a) Fibers should contain high cellulose content.
(b) The winding angle of cellulose in plant cell with respect to fiber axis should be small.
(c) Fibers should be disposed in a direction parallel to an applied uniaxial stress but should be cross laminated, if the stresses are bidirectional.
(d) The plant fiber should be surface treated to a good interfacial bond between fiber and matrix.

The resulting properties of the composites are influenced by the manufacturing stages of plant fiber composites because the exact nature of the interaction between plant fiber and polymer matrix is complex.

## *PROPERTIES OF RESIN*

(a) Resin should be thermosetting polymer to avoid plastic deformation at low stresses.
(b) Most natural resins are hydrophilic and are susceptible to changes in strength, chemistry or dimensions with water, so a composite will have an application period during which no such change occurs, followed by onset of degradation after the required lifetime.
(c) Composites with the same polymer in both fiber and matrix offer strong interfacial adhesion due to transcrytallinity in the matrix near fibers and also due to melting or dissolution of some of the fiber in the molten matrix near the fiber surface and as a consequence, the interfacial strength is increased.
(d) Resin should have good mechanical, adhesive and toughness properties.
(e) Resin should have good resistance to environmental degradation.
(f) Thermally stable, non-yielding composites require thermosetting rather than thermoplastic matrix.

## GEOMETRY AND ORIENTATION OF THE FIBERS IN THE COMPOSITE

(a) Orientation of the fibers is of advantage in providing maximum properties in the direction of orientation. Woven fibers are expected to provide the best strength properties but they are expensive then non-woven.

Jordan et al. (2003) states that a simple weave will have maximum mechanical properties in the two perpendicular directions of the fiber while in the diagonal and other directions these properties will decrease. The weaving of the fibers provides an interlocking that increases the strength better than can be achieved by fiber matrix adhesion.

(a) Fiber diameter is an important factor. More expensive smaller diameter fibers provide higher fiber surface areas, spreading the fiber/matrix interfacial loads.

(b) The geometry of the fibers in a composite is also important since fibers have their highest mechanical properties along their length, rather than across their widths. This leads to the highly anisotropic properties of composites. This means that it is very important, when considering the use of composites to understand at the design stage; both, the magnitude and the direction of the applied loads.

## SURFACE INTERACTION OF FIBER AND RESIN

The properties of composites mainly depend on the interface. The ultimate mechanical properties of fiber reinforced polymeric composites depend not only on the properties of the fibers and the matrix but also on the degree of interfacial adhesion between the fiber and the polymer matrix (Hong et al., 2008).

The following factors affect strength, toughness and stiffness of the natural fiber composites:

(a) The surface energies of the fiber and matrix phases

(b) The nature of fiber chemical pretreatment, for example, alkalization, acetylation and silane treatment.

(c) The addition of adhesion promoters to the resin, for example, maleic anhydride modified polypropylene.
(d) The degree of shrinkage of the matrix onto the fibers during manufacture of composites.

### 12.2.7 APPLICATIONS OF GREEN COMPOSITES

Eco-friendly biocomposites from natural fiber and bio-plastic are novel engineering materials of the twenty first century and would be of great importance to the material world, not only as a solution to growing environmental threat but also as a solution to the uncertainty of petroleum supply. The use of materials from renewing resources is attaining increased importance, and the world's leading industries and manufacturers are therefore interested in composites derived from natural fibers and polymers.

Green composites have several applications ranging from leisure goods to construction. Some of the areas in which the green composites find applications are:

#### AUTOMOBILES

Composites made from natural fibers are attractive because these are light in weight and have good mechanical properties comparable to widely used glass fiber reinforced plastics. In Europe, a large and still expanding market for natural fiber reinforced composites plastics is in automotive applications. It was estimated that more than 20,000 tones of natural fiber, mainly flax and hemp were used in automotive components (Karus et al., 2000). Green composite materials were first used extensively in production and its construction was the Eastern European Trabant. The body of this car was composed of panels of a natural fiber reinforced plastic composite called Doroplast, which were screwed to the galvanized steel frame of the sub structure.

Natural fibers have been used for thermal and acoustic insulation that is, low performance applications in interior automotive situations. The introduction of natural fiber composites for door panels in the Mercedes-Benz E-Class provided a step towards higher performance applications. In this particular application, a flax/ sisal mat was used as reinforcement in an epoxy matrix. A weight reduction of approximately 20% was claimed over the existing wood fiber material (Schuh, 1999). Wood plastic composites were also used for construction applications and natural fiber based synthetic polymers for automotive applications. Another example of application of natural composites in automotives is a resin made out of soy bean oil on reinforcement with glass fiber to be used in parts of newest tractors produced by John Deere.

### *AIRCRAFTS, SHIPS AND TRAINS*

The green composites are used in aircrafts and ships because of their light weights and also biodegradability. It is known that the fuel consumption will come down certainly if the weight of the vehicle is reduced. These types of green composites are also used in trains for the above reason. One of the first, true synthetic composites, potentially capable of being used in structural applications was 'Gordon-Aerolite' (Mc Mullen, 1984). This was a composite consisting of unidirectional aligned unbleached flax thread impregnated with phenolic resin. Since biocomposites are organic materials, they are combustible. So, one of the most important requirements for biocomposites is its use for paneling in railways or aircraft as it has a certain degree of flame resistance. The new aspects of eco-friendly polymer flame retardant systems have been reported (Zaikov and Lomakin, 1997).

### *PACKAGING*

Green composites have barrier properties, chemical resistance and surface appearance and these properties make it an excellent material for packaging application such as in beer and carbonated drinks bottles and paper board for fruit and dairy. The pallets, which have traditionally been made from low grade timber, are now being manufactured from wood fiber-plastic composites to fulfill the demands of hygiene, safety and longevity. Crates and boxes are other examples of applications for wood fiber-plastic composite in packaging.

### *CONSTRUCTION AND BUILDING PRODUCTS*

Green composites may find use in applications where timber, synthetic polymers or synthetic composites are currently used. In North America, wood fiber-plastic composites frequently occupy the position of timber products specially decking. Wood fiber-plastic composite have several advantages over wood. These include easy processing and treatment, no splintering, good appearance, improved resistance to biodegradation, no termite attack and low maintenance and so on. Good dimensional stability is a further attractive feature of wood fiber-plastic composites. Other building products such as railing, fencing, window and floor profiles, siding and shingles are manufactured from this material.

### *MOBILE PHONES AND COMPUTERS*

Green composites are used for mobile phone's body. Kenaf and PLA composites are used in mobile phones part in Japan to reduce the amount of $CO_2$ emissions during fabrications.

The NTT Docomo is one of the models of mobile phones in Japan, where green composites are used for such purpose. Components such as covers for mobile phones, casing for computers and monitors could be produced from biodegradable composite materials.

### *MISCELLANEOUS APPLICATIONS*

Green composites are used for indoor structural applications in housing. The composite used for the interior decoration is banana fiber and its composites. The walls and flooring can be covered with the boards, which will be attractive and will decrease the cost of construction. Fiber reinforced composites have unique properties in bio-medical applications, such as being transparent to X-rays. Titanium implants versus fiber reinforced composite implants was also studied. The stress range of the fiber reinforced composite implant was close to the stress level for optimal bone growth and the stress at the bone around the fiber reinforced implant was more even than that of the Ti implant.

### *12.2.8 REUSE, RECYCLING AND DEGRADATION OF COMPOSITES*

Recycling and reuse of materials is by no means a new concept, since over the last few decades and earlier waste newspaper, papers, cardboards and glass were recycled and reused. However, polymer based products gained popularity towards the end of seventies due to significant increase in costs of materials because of the unexpected rise in crude oil price. In recent years, there has been an increase in the use of composite products, particularly in the construction and automobile industries. These consume nearly half of all composites manufactured and therefore, the issue of composite recycling and use is becoming very important. In case of advanced composite materials, with an increasing number of additives and reinforcement materials, the only solution to reducing waste is to crush the compound and to recycle. Recycle means to recover a product at the end of its useful life, break it down into its constituent components and reincorporate it into new product that has an inherent value equal to the original product.

Recycling of plastics consists of four phases of activity:

(i) Collection,
(ii) Separation,
(iii) Processing and
(iv) Marketing.

In recycling thermoplastics, a mixed waste stream occurs due to problem in collection and sorting, associated with visual similarity and similar physical properties of commonly used polymers. Many polymers are not compatible and yield low properties, when blended and moulded directly. To overcome these problems, chemical compatibilizers are required.

Processing methods of recycled composites were reported to have a large effect on the quality of final product. The recycling of thermoplastic and their composites are much easier compared to the recycling of thermosets, so the widespread use of thermoplastic composites in different industries over the last decades is expected to have a more favorable environmental impact.

Recycled thermoplastic composites show a degradation of their mechanical performance and the extent, which depends on the recycling process and on the service conditions history. Furthermore, the possibility of recycling offers sound economical benefits because of the high price of the virgin material (Papanicolaou et al., 2008). The recycling processes do not cause very significant changes in flexural strength and thermal stability of the composites, particularly polypropylene based composites reinforced with kenaf fibers are less sensitive to reprocessing cycles with respect to rice hulls reinforced polypropylene based composites.

The properties of these composites remain unchanged after recycling processes and the recycled composites are suitable as construction materials for indoor applications (Srebrenkoska et al., 2008). Bourmaud and Baley have investigated the mechanical properties of sisal/polypropylene and hemp/polypropylene as a function of recycling and they found that both tensile modulus and strength were quite stable even after up to seven injection cycles, but the initial value was quite low due to the relatively poor mechanical properties of polypropylene (2007).

Duigou et al. observed the recyclability of flax/PLLA biocomposites elaborated with the injection moulding process, and compared their behavior with that of polypropylene composites (2008). It was found that repeated injection cycles had shown to influence many parameters such as reinforcement geometry, mechanical properties, molecular weight of PLLA, thermal behavior and rheological behavior.

Although biocomposites became more brittle with recycling, they retain a large part of their properties, at least until the third injection cycle. In an industrial situation, 100% of the recycled biocomposite is not be used but the recycled material is always mixed with virgin material. Thermoset matrix recycling is unfeasible because of the thoroughly cross linked nature and the inability to be remoulded. Nevertheless, technologies are now being developed that can reprocess scrap composites to recover some of the value in the material and then prevent disposing the solid in a landfill, which has been the only option earlier.

The four main processes in recycling of thermosetting composites are: (i) Grinding, (ii) Chemical degradation and fibers recovery, (iii) Pyrolysis and (iv) Incineration.

Recycling of thermosetting composites by grinding enables reuse of glass fibers, $CaCO_3$ and polymeric matrix without separation of the components. The composite is shredded, granulated to small fiber and used as filler in a new process of manufacture. These products have the same or even better mechanical properties as the virgin composite materials (Inoh et al., 1994; Petterson and Nilsson, 1994). In addition, grinded recyclate has a lower density and contributes to reduction in weight, when compared to conventional fibers.

(i) **Grinding:** In this process, the composite material is shredded to a convenient size using shredders designed for high torque and low speed. Hammer mills are used to reduce the size of recyclate further.

(ii) **Chemical degradation of composites:** It involves partial or selective degradation of polyester/styrene polymer network in the presence of water, ethanol, KOH and various amides (Winter et al., 1995). This process is inferior when compared with the quality of grinding recyclate. A neutralization step is also required, which generates large quantities of waste water and adds to the cost. Fluidized bed thermal processing technique has been developed to recover energy and fibers in a form suitable for recycling into high value products (Kennerley et al., 1996). These techniques are suitable for contaminated and mixed scrap material from end of life applications especially in the automotive industry.

(iii) **Pyrolysis:** It is a simple and well controlled process that recovers a good part of glass fibers and $CaCO_3$, which can be reused as filler and reinforcement. The process separates organics from inorganics, due to a significant difference between the temperatures of their thermal decompositions. Low temperature pyrolysis (<200°C) is applicable to thermoplastic composites and yields excellent recovery of glass fibers and inorganic fillers (Allred,

1996). High temperature pyrolysis (~750°C) leads to a considerable reduction in the strength of glass fibers, which prevents their reuse as a high quality reinforcement.

(iv) **Incineration:** Incineration of plastics and composites involves annihilation of the material, with inevitable air and land pollution, due to poor combustion and to gaseous/liquid/solid products of the process.

The advantages of an increased content of plastic waste for incineration are –

(a) Higher temperature of incineration due to volatiles, which reduce leachability of the fly ash and
(b) Shorter combustion zones and more intensely burning fire.

The disadvantages from increased plastic waste are:

(a) Increased formation of $NO_x$ with an increase in incineration temperatures, which automatically contradicts the above listed advantages,
(b) Increased CO emissions and
(c) Other excessive emissions such as $Cl_2$, dioxins and furans.

### *12.2.9 BIODEGRADATION OF COMPOSITES*

Biodegradable polymers have offered a possible solution to waste disposal problems associated with traditional petroleum derived plastics. Biodegradation is defined as a process, which takes place through the action of enzymes and/or chemical decomposition associated with living organisms like bacteria, fungi and so on. and their secretion products (Albertsson and Karlsson, 1994). Also abiotic reactions like photodegradation, oxidation and hydrolysis may alter the polymer before, during or instead of biodegradation because of environmental factors.

Almost all biosynthetic polymers are biodegradable within a reasonable time scale. There are four biodegradation environments for polymers and plastic products. These are (i) Soil, (ii) Aquatic, (iii) Landfill and (iv) Compost. Each environment contains different microorganisms and has different conditions for degradation (Wypych, 2003). In soil, fungi are mostly responsible for the degradation of organic polymers. The aquatic environment contains two types of bacteria, on the surface and in the sediment. The bacterial concentration in water decreases with increasing depth. Microorganisms biodegrade organic materials by the use of their enzymes; however, microorganisms do not synthesize polymer specific enzymes capable of degrading and consuming synthetic polymers of recent origin. Enzyme activity is inhibited by the hydrophobic nature of the plastic and high molecular weight. The amount of non-biodegradable plastic waste can be greatly reduced by proper development of biodegradable polymers and composites for short term products.

Biodegradation can takes place by two mechanisms:

(i) Hydro-biodegradation and
(ii) Oxo-biodegradation

Hydro-biodegradation is much more important in case of hydrophilic natural polymers such as cellulose, starch and polyesters, whereas the oxo-biodegradation predominates in the case of other natural polymers such as rubber and lignin. Synthetic flexible polymers do not hydrolyze under normal environmental conditions but biodegrade readily in the presence of a variety of thermophilic microorganisms in the surface layers of the polymer after transition metal catalyzed thermal peroxidation. The surface of the polymer after biological attack is physically weak and readily disintegrates under mild pressure (Bonhomme et al., 2003).

Polymers containing mainly covalent bonds in its main chain show little or no susceptibility to enzyme catalyzed degradation reactions, especially those with higher molecular weights. To overcome this problem, 'weak-links' are inserted in the backbones of such polymers by insertion of functional groups in the main chain, especially ester groups, which can be cleaved by chemical hydrolysis, and insertion of functional groups in, or on, the main chain that can undergo photochemical chain cleavage reactions, particularly of carbonyl groups. These can then be utilized and consumed by microorganisms through biodegradation processes.

Every material has its own characteristics property and therefore, composites are prepared, which can have different properties than its components or may have hybrid property of its component. Green composite is the requirement of the day and the nature is doing this job from time immemorial.

## KEYWORDS


- **Fibers**
- **Green Composites**
- **Life Cycle Assessment**
- **Oxo-biodegradation**
- **Petroleum feedstock**
- **Polymers**


## REFERENCES


1. Abe, K., & Ozaki, Y. (1998). *Soil Science and Plant Nutrition, 44,* 599–607.
2. Albertsson, A. C., & Karlsson, S. (1994). *Chemistry and technology of biodegradable polymers*. Glasgow: Blackie.
3. Allred, R. E. (1996). *SAMPE Journal, 32,* 46–51.
4. Amnuay, W., Katavut, P., Supranee, K., Pichan, S., & Claudio M. S. (2011). *Journal of Science and Technology, 33,* 461–167.
5. Asakura, T., & Kaplan, D. L. (1994). *Encyclopedia of Agricultural Science, 4,* 1–11.
6. Avella, M., Bogoeva-Gaceva, G., Buzarvska, A., Errico, M. E., Gentile, G., & Grozdanov, A. (2008). *Journal of Applied Polymer Science, 108,* 3542–3551.
7. Avella, M., Buzarovska, A., Errico, M. E., Gentile, G., & Grozdanov, A. (2009). *Materials, 2,* 911–925.
8. Avella, M., Immirzi, B., Malinconico, M., Martuscelli, E., & Volpe, M. G. (1996). *Polymer International, 39,* 191–204.
9. Averous, L., & Boquillon, N. (2004). *Carbohydrate Polymers, 56,* 111–122.
10. Averous, L., Fauconnier, N., & Moro, L. (2000). *Journal of Applied Polymer Science, 76,* 1117–1128.
11. Averous, L., Frigant, C., & Moro, L. (2001). *Polymers, 42,* 6565–6572.
12. Barbu, M., & Fritz, T. (1996). *Chemical Abstracts, 124,* 292690s.
13. Bastioli, C. (1998). *Polymer Degradation and Stability, 59,* 263–272.
14. Bax, B., & Mussig, J. (2008). *Composites Science and Technology, 68,* 1601–1607.
15. Bisanda, E. T. N., & Ansell, M. P. (1991). *Composites Science and Technology, 41,* 165–178.
16. Bledzki, A. K., & Gassan, J. (1996). *Angewandte Makromol Chemie, 236,* 129–138.
17. Bolton, A. (1994). *Journal of Materials Technology, 9,* 12–20.
18. Bonhomme, S., Cuer, A., Delort, A. M., Lemaire, J., Sancelme, M., & Scott, G. (2003). *Polymer Degradation and Stability, 81,* 441–452.
19. Bourmaud, A., & Baley, C. (2007). *Polymer Degradation and Stability, 92,* 1034–1045.

20. Buchanan, C. M., Gardner, R. M., & Komarek, R. J. (1993). *Journal of Applied Polymer Science, 47,* 1709–1719.
21. Carvalho, A. J. F. D., Curvelo, A. A. S., & Agnelli, J. A. M. (2002). *International Journal of Polymeric Materials, 51,* 647–660.
22. Carvalho, A. J. F., Curvelo, A. A. S., & Agnelli, J. A. M. A. (2001). *Carbohydrate Polymers, 45,* 189–194.
23. Cheung, H. Y., Lau, K. T., Tao, X. M., & Hui, D. A. (2008). *Composites Part B-Eng, 39,* 1026–1033.
24. Curvelo, A. A. S., Carvalho, A. J. F. & Agnelli, J. A. M. (2001). *Carbohydrate Polymers, 45,* 183–188.
25. Doi, Y. (1990). *Microbial Polyesters*. New York: VCH Publishers.
26. Donovan, J. W. (1979). *Biopolymers, 18,* 263–275.
27. Dufresne, A., Dupeyre, D., & Vignon, M. R. (2000). *Journal of Applied Polymer Science, 76,* 2080–2092.
28. Dufresne, A., & Vignon, M. R. (1998). *Macromolecules, 31,* 2693–2696.
29. Duigou, A. L., Pillin, I., Bourmaud, A., Davies, P., & Baley, C. (2008). *Composites, A: Applied Science and Manufacturing, 39,* 1471–1478.
30. Fang, Q., & Hanna, M. A. (1999). *Industrial Crops and Products, 10,* 47–53.
31. Felix, J. M., & Gatenholm, P. (1991). *Journal of Applied Polymer Science, 42,* 609–620.
32. Flodin, P., & Zadorecki, P. (1986). *Composite systems from natural and synthetic polymers*. Elsevier Science Publishers.
33. Garlotta, D. A. (2002). *Journal of Polymer and Enviornment, 9,* 63–84.
34. Gatenholm, P., Kubat, J., & Mathiasson, A. (1992). *Journal of Applied Polymer Science, 45,* 1667–1677.
35. Gu, J. D., Gada, M., Kharas, G., Eberiel, D., McCarthy, S. P., & Gross, R. A. (1992). *Journal of American Chemical Society and Polymer Prepr, 67,* 351–352.
36. Haager, S. (1995). *Energia, 47,* 50.
37. Han, J. S., Miyashita, E. S., & Spielvogel, S. J. (1999). *Kenaf Properties, Processing and Products*. In T. Sellers, N. A. Reichert (Eds.) (pp. 267–283). Mississippi: Mississippi State University.
38. Herrmann, A. S., Hanselka, H., & Niederstadt, G. (1995). Eur. Pat. Appl. EP 687, 711, 20 Dec.1995, DE Appl. 4,420, 817 16 June 1994.
39. Heslot, H. (1998). *Biochimie, 80,* 19–31.
40. Hocking, P. J., & Marchessault, R. H. (1994). Biopolyesters. In Griffin, G. J. L. (Ed.) *Chemistry and Technology of Biodegradable Polymers*, (pp. 48–96). Glasgow: Blackie.
41. Hong, C. K., Hwang, I., Kim, N., Park, D. H., Hwang, B. S., & Nah, C. (2008). *Journal of Industrial and Engineering Chemistry, 14,* 71–76.
42. Hu, R., & Lim, J. K. (2007). *Journal of Composite Materials, 41,* 1655–1669.
43. Huda, M. S., Drzal, L. T., Mohanty, A. K., & Misra, M. (2008). *Composite Interfaces, 15,* 169–191.
44. Huda, M. S., Drzal, L. T., Mohanty, A. K., & Misra, M. (2008). *Composites Science and Technology, 68,* 424–432.
45. Huda, M. S., Drzal, L. T., Mohanty, A. K., & Misra, M. (2006). *Journal of Applied Polymer Science, 102,* 4856–4869.

46. Inoh, T., Yokoi, T, Sekiyama, K. I., Kawamura, N., & Mishima, Y. (1994). *Journal of Thermoplastic Composite Materials, 7,* 42–55.
47. Jamshidian, M., Tehrany, E. A., Imran, M., Jacquot, M., & Desobry, S. (2010). *Comprehensive Reviews in Food Science and Food Safety, 9,* 552–571.
48. Jordan, N. D., Bassett, D. C., Olley, R. H., Hine, P. J., & Ward, I. M. (2003). *Polymer, 44,* 1133–1143.
49. Kaplan, D. L., Adams, W. W., Vincy, C., & Farmer, B. L. (1994). *Silk polymers: Materials science and biotechnology*. Washington: ACS Books.
50. Karmaker, A. C., & Hinrichsen, G. (1991). *Polymer-Plastics Technology and Engineering, 30,* 609–629.
51. Karmaker, A. C., & Youngquist, J. (1996). *Journal of Applied Polymer Science, 62,* 1147–1151.
52. Karus, M., Kaup, M., & Lohmeyer, D. (2000). *Study on markets and prices for natural fibres* (Germany and EU). Nova Institute (Viewed at: http://www.nova-institut.de Accessed on 26 January 2004).
53. Ke, T. Y., & Sun, X. Z. (2001). *Journal of Applied Polymer Science, 81,* 3069–3082.
54. Kennerley, J., Fenwick, N. J., Pickering, S. J., & Rudd, C. D. (1996). *Proceedings ANTEC,* 890–894.
55. Kesel, C. D., Wauven, C. V., & David, C. (1997). *Polymer Degradation and Stability, 55,* 107–113.
56. Khalid, M., Ratnam, C. T., Chuah, T. G., Salmiaton, A., & Thomas, S. Y. C. (2008). *Materials and Design, 29,* 173–178.
57. Khan, M. A., Ali, K. M. I., Hinrichsen, G., Kopp, C., & Kropke, S. (1999). *Polymer-Plastics Technology and Engineering, 38,* 99–112.
58. Khristova, P., Bentcheva, S., & Karar, I. (1998). *Bioresource Technology, 66,* 99–103.
59. Luo, S., & Netravali, A. N. (1999). *Polymer Composites, 20,* 367–378.
60. Ma, X., Yu, J., & Kennedy, J. F. (2005). *Carbohydrate Polymers, 62,* 19–24.
61. Mahadeva, S. K., Yun, S., & Kim, J. (2011). *Sensors and Actuators A: Physical, 165,* 194–199.
62. Maldas, D., & Kokta, B. V. (1993). *TRIP, 1,* 174–178.
63. Martin, O., & Averous, L. (2001). *Polymer, 42,* 6209–6219.
64. McMullen, P. (1984). *Composites, 15,* 222–229.
65. Mercier, C., & Feillet, P. (1975). *Cereal Chemistry, 52,* 283–297.
66. Misra, M., Park, H., Mohanty, A.K., & Drzal, L. T. (2004). *Presented at the global plastics environmental conference*. Michigan: Detroit.
67. Mitra, B. C., Basak, R. K., & Sarkar, M. (1998). *Journal of Applied Polymer Science, 67,* 1093–1100.
68. Mohanty, A. K., & Misra, M. (1995). *Polymer-Plastics Technology and Engineering, 34,* 729–792.
69. More, S. (1997). *Modern Plastics International,* 41–42.
70. Nishi, Y., Uryu, M., Yamanaka, S. et al. (1990). *Journal of Materials Science, 25,* 2997–3001.
71. Otaigbe, J. U. (1998). *Plast Engineering, 54,* 37–39.
72. Pal, P. K. (1984). *Plastics and Rubber Processing and Applications, 4,* 215–222.
73. Pandey, J. K., Chu, W. S., Lee, C. S., & Ahn, S. H. (2007). Presented at the International Symposium on Polymers and the Environment: *Emerging Technology and*

*Science, Bio Environmental Polymer Society* (BEPS) (pp. 17–20). Washington: Vancouver.
74. Papanicolaou, G. C., Karagiannis, D., Bofilios, D. A., Van Lochem, J. H., Henriksen, C., & Lund, H. H. (2008). *Polymer Composites,* 1026–1035.
75. Petterson, J., & Nilsson, P. J. (1994). *Journal of Thermoplastic Composite Materials, 7,* 56–63.
76. Rana, A. K., Mandal, A., Mitra, B. C., Jacobson, R., Rowell, R., & Banerjee, A. N. (1998). *Journal of Applied Polymer Science, 69,* 329–338.
77. Rhim, J. W., Wu, Y., Weller, C. L., & Schnepf, M. (1999). *Science and Ailments, 19,* 57–71.
78. Rivard, C. J., Adney, W. S., Himmel, M. E., Mitchell, D. J., Vinzant, T. B., Grohmann, K., Moens, L., & Chum, H. (1992). *Applied Biochemistry and Biotechnology, 34/35,* 725–736.
79. Sahoo, S., Pati, D., Misra, M., Tripathy, S. S., Nayak, S. K., & Mohanty, A. K. (1999). *International Symposium: Polymers*. Society of Polymer Science, India, (pp. 542–545), January.
80. Sameni, J. K., Ahmad, S. H., & Zakaria, S. (2002). *Plastics, Rubber and Composites (UK), 31,* 162–166.
81. Sarazin, P., Li, G., Orts, W. J., & Favis, B. D. (2008). *Polymer*, *49,* 599–609.
82. Schuh, T. G. (1999). *Natural Fibres Performance Forum,* Copenhagen, 27–28 May.
83. Setua, D. K. (1986). *International Polymer Science and Technology, Vol 33,* Plenum Press.
84. Shao, Z. Z., & Vollrath, F. (2002). *Nature, 418,* 741.
85. Shikamoto, N., Ohtani, A., Leong, Y. W., & Nakai, A. (2007). In 22nd Technical Conference of the American Society for Composites 2007: *Composites: Enabling a New Era in Civil Aviation*; Curran Associates, Inc., (pp. 151, 1–151, 10). New York: Red Hook.
86. Shin Serizawa, Kazuhiko, I., & Masatoshi, I. (2006). *Journal of Applied Polymer Science, 100,* 618–624.
87. Shogren, R. L. (1998). In Kaplan, D. L. (Ed.) *Biopolymers from Renewable Resources, Macromolecular Systems-Materials Approach,* (pp. 30–46). Berlin: Springer-Verlag.
88. Song, M., Pan, C., Chen, C., Li, J., Wang, X., & Gu, Z. (2008). *Applied Surface Science, 255,* 610–612.
89. Srebrenkoska, V., Gaceva, G. B., Avella, M., Errico, M. E., & Gentile, G. (2008). *Polymer International, 57,* 1252–1257.
90. Taniguchi, T., & Okamura, K. (1998). *Polymer International*, *47,* 291–294.
91. Tate, J. S., Akinola, A. T., & Kabakov, D. (2010). *Journal of Technology Studies, 1,* 25–32.
92. Tirrell, J. G., Fournier, M. J., Mason, T. L., & Tirrell, D. A. (1994). *Chemical & Engineering News, 72,* 40–51.
93. Tokoro, R., Vu, D. M., Okubo, K., Tanaka, T., Fujii, T., & Fujiura, T. (2008). *Journal of Materials Science, 43,* 775–787.
94. Tune, S., & Duman, O. (2011). *Food Science and Technology*, *44,* 465–472.
95. Uesaka, T., Nakane, K., Maeda, S., & Ogihara, N. (2000). *Polymer, 41,* 8449–8454.

96. Willett, J. L., & Shogren, R. L. (2002). *Polymer, 43,* 5935–5947.
97. Winter, H., Mostert, H. A. M., Smeets, P. J. H. M., & Paas, G. (1995). *Journal of Applied Polymer Science, 57,* 1409–1417.
98. Wollerdorfer, S. F., & Bader, H. (1998). *Industrial Crops and Products, 8,* 105–112.
99. Wypych, G. (2003). Biodegradation. *In Handbook of Material Weathering*, 3rd Edn. (Chapter 18, pp. 523–533). ChemTech Publishing.
100. Yano, H., & Nakahara, S. (2003). *Journal of Materials Science, 39,* 1635–1638.
101. Yao, F., Wu, Q., Lei, Y., & Xu, Y. (2008). *Industrial Crops and Products, 28,* 63–72.
102. Zadegan, S., Hosainalipour, M., Rezaie, H. R., Ghassai, H., & Shokrgozar, M. A. (2011). *Materials Science and Engineering, 31,* 954–961.
103. Zaikov, G. E., & Lomakin, S. M. (1997). *Polymer-Plastics Technology and Engineering, 36,* 527–546.
104. Zimmermann, K. A., LeBlanc, J. M., Sheets, K. T., Fox, R. W., & Gatenholm, P. (2011). *Materials Science and Engineering, 31,* 43–49.

CHAPTER 13

# GREEN MANUFACTURING PROCESSES

JITENDRA VARDIA, DIPTI SONI, and RAKSHIT AMETA

## CONTENTS

## 13.1 INTRODUCTION

Green manufacturing involves production processes, which use inputs with relatively low environmental impacts, which are highly efficient and generates little or no waste or takes care of pollution. Green manufacturing involves source reduction (also known as waste or pollution minimization or prevention), recycling and green product design.

Source reduction is broadly defined to include any actions reducing the initially generated waste. Recycling includes using or reusing waste as ingredients, as a process or as an effective substitute for a commercial product, or returning the waste to the original process, which generated it as a substitute for raw material feedstock. Green product design involves creating products whose design, composition, and uses minimize the environmental impact throughout their life cycle. Green manufacturing goals are to conserve natural resources for future generations. The benefit of green manufacturing is to create a great reputation to the public, saves cost, and promotes research and design.

## 13.2 SOME MAJOR GREEN MANUFACTURING PROCESSES

The options for green manufacturing can be divided into four major areas:

### *13.2.1 PRODUCTION PROCESS CHANGES*

Major process changes fall into the following categories:

(i) Changing dependence on human intervention: Production dependence on active human intervention has a significant failure rate. This may lead to various problems ranging from off specification products to major accidents. A strategy that can reduce the dependence of production process on active human intervention is having machines to take over parts of what humans use to do. Automated process control, robots for welding purpose and numerically controlled cutting tools all may reduce waste.

(ii) Use of continuous process instead of a batch process: Continuous process causes less environmental impact than batch process. This is due to the reduction of residuals in the production machinery and thus, to reduce need for cleaning and better opportunity for the process control allowing the improved resource and energy efficiency and decreasing off-specification products. There are opportunities for environmentally improved technology in batch processes. For chemical batch process for instance, the main waste prevention methods are:
    (a) Eliminate or minimize unwanted byproducts possibly by changing reactant, processes or equipment.
    (b) Recycle the solvent used in the reaction and extractions.
    (c) Recycle excess reactants.

(iii) Changing the nature of steps in the production process: Physical, chemical or biological process can affect its environmental impact. Such changes may involve switching from one chemical process to another or from a chemical to a physical or biological process or vice versa. In general, using a selective production route such as through inorganic catalyst and enzymes will be environmentally beneficial by reducing inputs and their associated waste, for example, the banning of chlorofluorocarbon led to other ways of producing flexible polyurethane forms. Another example of an environmentally beneficial change in the physical nature of a process is using electrodynamics in spraying. A major problem of spraying process is that a significant amount of material misses its target. In such cases, waste may be generally reduced by giving the target and the spread material opposite electrical charges.

(iv) Eliminating steps in the production processes: In this process, we prevent waste because each step typically creates waste. In a chemical industry, there is a trend to eliminate the neutralization steps, which generates the waste salts as by product. This is mainly achieved by using a more selective type of synthesis.

(v) Changing the cleaning processes: Cleaning is the source of considerable environmental impacts from the production processes.

These impacts can partially be reduced by changing inputs in the cleaning process, for example, using water based cleaners rather than solvents. In other processes, reduced cleanliness is achieved by minimizing carry over from one process step to the next. The switch from batch to continuous process will also reduce the need for cleaning.

### 13.2.2 CHANGES OF INPUTS IN THE PRODUCTION PROCESS

Changes in inputs are important tools in the green manufacturing. Both major and minor product ingredients, which contribute to the production without being incorporated in the end product, may be worth changing, for example, where changing a minor input in production may substantially reduce its environmental impact, for example, in cars and airplanes. The introduction of powder based and high solid paints substantially reduce the emission of volatile organic compounds. Substituting water based coating by solvent based coating may have less environmental impact.

### 13.2.3 INTERNAL REUSE OF WASTE

The potential for internal reuse is often substantial with many possibilities for the reuse of water, energy, some chemicals, and metals. Washing, heating, and cooling in a counter current process will facilitate the internal reuse of energy and water in a closed-loop process. Water recycling which replaces single pass system is usually economically attractive with both water and chemicals, which are potentially being recycled. In some production processes, there may be possibilities for cascade-type reuse in which water is used in one process step; where quality requirements are less stringent.

### 13.2.4 BETTER HOUSEKEEPING

Good housekeeping refers to generally simple, routinized and non resource intensive measures that keep a facility in good working and environmental order. It includes segregating wastes, minimizing chemical and waste inventories, installing overflow alarms and automatic shutoff valves, eliminating leaks and drips, and putting collecting devices at places, where spills may occur. Frequent inspection aimed at identifying environmental concerns and potential malfunctioning of the production processes, instituting better controls on operating conditions (flow rate, temperature, pressure etc.), regular fine tuning of machinery, and optimizing maintenance schedules. These types of actions often offer quick, easy and inexpensive ways to reduce chemicals and wastes.

The green manufacturing processes have a wide scope in all industries, but in recent years, some of the main industries, which got special attention, are pharmaceutical, polymer, petroleum, fine chemicals industries, and so on.

### 13.2.5 PHARMACEUTICAL INDUSTRY

The pharmaceutical industry is well known for its intensive use of many petrochemicals as starting materials, synthetic route with conventional techniques, high energy requirements for industrial processes, high use of organic solvents for separation and purification of high volume waste. Pharmaceutical industry produces more waste per kg of product than other chemical industries (petrochemical, bulk and fine chemical, polymer etc). The pharmaceutical industry uses 6–8 steps organic synthetic routes and generates 25–100kg of waste for each kg of product (Fortun and Confalone, 2007; Fortunak, 2009). The big pharmaceutical company uses in its manufacturing large amount of solvents and its known water liquid waste contents are about 85–90% organic solvents (Bruggink, 2003; Slater, 2007). In the last decade, pharmaceutical manufacturers embraced green chemistry ideas to promote their environmental credential and increase the efficiency of their manufacturing processes. Slater et al. (2012) designed a green engineering program for the purification and recovery of isopropanol (IPA) in waste streams of celecoxib process.

Pharmaceutical processes use large amount of organic solvents and liquid waste is 85% non aqueous in nature. The reduction in the use of organic solvents is an important issue in most of the pharmaceutical industries. Some new organic synthetic routes with minimum of zero solvent are in the research state (Sheldon, 2005). The solvents which are more acceptable for organic synthetic processes, having low toxicity are normally acetone, ethanol, 2-propanol, ethyl acetate, isopropyl acetate, methylethylketone, 1-butanol, and so on. Solvents that are used for their ability to dissolve other chemicals despite of their toxicity are cyclohexane, n-hepane, toluene, methylhexane, acetonitrile, THF, DMSO, acetic acid, and ethylene glycol. The pharmaceutical industry has initiated many studies on the replacement of such toxic solvents with solvent that are benign to human health and the environment (Jimenez-Gonzalez et al., 2001, Jimenez-Gonzalez and Constable, 2002).

Enzymes can accelerate a reaction, lower the use of energy, use alternative starting materials, and reduce the use of solvents and production of waste. Enzymes are biomaterials that can biodegrade under environmental conditions. The enzymes are used for biocatalysis of the basic steps in the synthesis, reducing the use of solvents by 90% and the starting material by 50%. The company will reduce its industrial waste by 200.000 metric tons, compared to the old method.

Doramectin was synthesized as the antiparasitic drug. By changing the biocatalysis synthesis, the efficiency of the reaction is increased by 40% and the byproducts are reduced. Biocatalysis improved the synthetic routes for the industrial production operation of these drugs; Oselravimit and Pelitrexol (Harrington et. al., 2004; Hu et. al., 2006). Biocatalysis has been primarily used in the production of chiral chemical intermediates required for the production of medicines. The "green" biocatalytic synthesis of the active substances like atorvastatin has been reported (Ma et. al., 2010). An improved efficient method was developed to make ibuprofen using only three steps instead of six steps used in earlier method. In this case, all starting materials are converted to products, reclaimed as by product or completely recycled in the process. Thus, the generation of wastes has been practically eliminated (Cann and Connelly, 2000).

### 13.2.6 POLYMER INDUSTRIES

In polymer industry, we are developing useful biopolymers from naturally occurring renewable resources as an alternative to the petrochemical-based sources, which is in common use presently, utilizing lignocellulosic and other biomasses, especially residues from food and agricultural processes, one way is to develop new polymer products from industrial wastes or recyclate streams. Hilonga et al. reported a two-step rapid route of synthesizing inexpensive mesoporous silica, which was used in manufacturing of green tyre (2012). Zhuo et al. (2011) developed a versatile CNT synthesis process, where they have used waste solid polymers like polyethylene, polypropylene, polystyrene and so on. Bertaud et al. (2011) have used natural phenolic polymers of tannins and lignin as substitutes of petro-based chemicals, which were used in wood panels in order to reduce formaldehyde emissions and to develop green adhesives. Chang and Abu-Zahra (2011) developed a robust and cost-effective method to maximize the use of recycled poly (vinyl chloride) (PVC) back into its virgin compounds. In this method, 70% of recycled or regrind PVC was successfully implemented in the extrusion process for manufacturing high-quality foam PVC profiles, which may be used in building industry.

#### *BIOPLASTICS*

Some green processes have been developed for the synthesis of polyhydroxyalkanoates (PHAs), a kind of non-petrochemical bioplastics. Bioplastic synthesized by living organisms are genrally biodegradable. PHAs continue to attract increasing industrial interest as renewable, biodegradable, biocompatible, and extremely versatile thermoplastics (Steinbüchel and Lütke-Eversloh, 2003; Suriyamongkol et al., 2007). PHAs are the only water-proof thermoplastic materials available that are fully biodegraded both; in aerobic and anaerobic environments. Two classes of PHAs are distinguished according to their monomer composition: short-chain length (SCL) PHAs and medium-chain length (MCL) PHAs. SCL-PHAs are polymers of 3-hydroxyacid monomers with a chain length of three to five carbon atoms, such as poly(3-hydroxybutyrate) (PHB, the most common PHA); whereas MCL-PHAs contain 3-hydroxyacid monomers with six to sixteen carbon atoms. All of them are optically active *R*- compounds. This versatility is partly due to the wide substrate range of the PHA-synthesizing enzymes, and gives PHAs an extended spectrum of associated properties, which is a clear advantage *vis-à-vis* to other bioplastics. Around 200 different monomer constituents were found in the polymers analyzed so far.

The versatile copolymer P (HB-co-HV) was initially manufactured as shampoo bottles and other cosmetic containers (Hocking and Marchessault, 1994). Later on, pens, cups, and packaging elements (e.g., films) made with PHAs also appeared in the market. PHAs are biocompatible and for this reason, they have also attracted attention as raw material to be used in medical devices. Being composed by R-(—) monomers, PHAs are source of chiral compounds with a high demand from the pharmaceutical industries (Chen and Wu, 2005). However, the manufacture of PHAs is carried out at small facilities and, as a consequence, it lacks the economic benefit of a large scale production (Chanprateep, 2010).

### 13.2.7 OTHER INDUSTRIES

Tu et al. (2012) developed a green method for manufacturing $CuFe_2O_4$ from industrial Cu sludge. It was successfully applied for combustion of volatile organic compounds (VOCs) derived from isopropyl alcohol. Zhang et al. (2012) converted formic acid into glycerine with a yield of 31.0%. Their work may be explored for the production of formic acid from renewable biomass. Triammonium citrate was manufactured by using neutralization of citric acid solution with liquid ammonia directly by Yang et al. (2012). Their method has advantages of 100% raw materials utilization ratio and totally zero emission during mass production over conventional route. Liu et al. (2012) demonstrated the versatile use of $CO_2$ in organic synthesis as the alternative carbonyl source of phosgene for the manufacturing of some compounds such as cyclic carbonates, oxa-zolidinones, ureas, isocyanates, and polymers. Pirzada et al. (2012) used a green chemical approach to synthesize silica xerogels from sodium silicate through a cost effective way. It was synthesized using the waste material (hexafluorosilicic acid, $H_2SiF_6$) of phosphate fertilizer industry and sodium silicate ($Na_2O{\cdot}SiO_2$). Cationic surfactant hexadecyl trimethyl ammonium bromide (CTAB) was used as the structure template.

Shanmuganathan et al. (2011) observed that photocurable mixture of a multifunctional acrylate, a tetrafunctional thiol, and a photoinitiator can be processed into continuous fibers by *in situ* photopolymerization during electrospinning under ambient conditions. The prepared fibers were mechanically robust and have excellent chemical and thermal stability. A novel eco-sustainable catalytic pathway was developed by Falletta et al. (2011) for the synthesis of 3-hydroxypropionic acid from allyl alcohol. This method highlights the good potential of gold-based and bimetallic catalysts in the aerobic oxidation of allyl alcohol. Lovell et al. (2010) investigated green lubricant combinations, which were prepared by homogeneously mixing of nano-, submicrometer- and micrometer-boric acid powder additives with canola oil in a vortex generator.

Pallavkar et al. (2010) studied the use of microwave energy to accomplish high temperature destruction of p-xylene in a packed bed reactor, which was performed using SiC (silicon carbide) foam, while Cambronero et al. (2009) manufactured Al-Mg-Si alloy foam using calcium carbonate as foaming agent. The prepared foam showed a low degree of aluminium draining, no wall cell cracks, and a good fine cell size distribution. The cradle-to-cradle (C2C) manufacturing is being increasingly used by chemical industries as a profitable and environmental friendly manufacturing process (Scott, 2009).

Menzler et al. (2010) manufactured anode-supported solid oxide fuel cells (SOFC) by different wet chemical powder processes and subsequent sintering at high temperatures. The cell was characterized for its slurry viscosity, green tape thickness, relative density, substrate strength, electrical conductivity, and shrinkage. Li et al., (2009) studied an environmentally-benign and cost-effective production method of nanoscale zero-valent iron (nZVI) using a precision milling system.

A new green path for epichlorohydrin used in the production of glycerol, plastics, epoxy glues, resins, and elastomers has been developed. They have also developed a process to convert glycerol into monopropylene glycol (MPG) that could lead to new outlets for glycerol made from biodiesel.

Various solid acids like zeolites, ion-exchange resins, and mixed metal oxides are known catalysts in the esterification of dodecanoic acid with 2-ethylhexanol, 1-propanol, and methanol out of which sulphated zirconia is considered as most promising material (Kiss et al., 2006). Chromium compound is an important chemical for many industries, which normally shows quite low utilization efficiency of resources and energy. Discharge of chromium-containing toxic solid wastes and gas results in serious pollution problems. A green manufacturing process of chromium compounds has been developed with the design objective of eliminating pollution at the source. This green process achieves higher resource utilization efficiency and zero emissions of chromium-containing waste residue by alteration of process chemistry, change of reactor, operation, regeneration and recycle of reaction media, and comprehensive use of resources.

The vat dye manufacturing process is responsible for a large amount of contamination of both soil and groundwater. Recently, dyeing methods have been improved with better engineering, better control, and the ability to hot wash, which has enabled dyers to process textiles in the right manner considering the environmental impacts. Solid acids, especially, which are based on micelle-templated silicas and other mesoporous high surface area support materials play an important role in the greening of fine chemicals manufacturing processes. The bamboo is considered as a renewable and bio-degradable resource in textiles, as it is the growing demand for more comfortable, healthier and environment friendly products in textile industry (Rekha and Sudam, 2009).

Many volatile organic compounds are the source of paints and their smell is harmful to the health and environment. Some biobased paints are known, which in addition to lower odor, has better scrub resistance and better opacity.

Recently, novel biocatalytic pharmaceutical processes have been developed to replace chemical routes, which contain poorer process efficiency and higher manufacturing costs (Chen et al., 2007; Tao and Xu, 2009). Green chemistry challenges the innovators to take care to design and utilize matter and energy in a way that increases performance and value, while protecting human health and the environment (Manley et al., 2008). In present scenario, the key issues of switching to renewable resources, avoiding hazardous and polluting processes, manufacturing and using safe and environmentally compatible products, one needs to develop sustainable and green chemical product supply chains. Organic chemicals and materials need to operate under agreed and strict criteria and need to start with widely available, totally renewable and low cost carbon (the only source is biomass) and the conversion of biomass into useful products will be carried out in biorefineries (Clark, 2007).

It is very important to turn concept or experiments on laboratory scales into manufacture to industrial scales. Efforts are being made to use green chemical routes to develop green manufacturing processes, which is most desired but less taken care off.

## KEYWORDS

- **Biocatalytic pharmaceutical processes**
- **Bioplastics**
- **Green manufacturing processes**
- **Pharmaceutical industry**
- **Polymer industry**

## REFERENCES


1. Bertaud, F., Tapin-Lingua, S., Pizzi, A., Navarrete, P., & Petit-Conil, M. (2011). *ATIP - Association Technique de l'Industrie Papetière, 65,* 6–12.
2. Bruggink, A., Straathhof, A. J., & Vander, W. L. A. (2003*). Advances in Biochemical Engineering and Biotechnology, 80,* 69–113.
3. Cambronero, L. E. G., Ruiz-Roman, J. M., Corpas, F. A., & Ruiz Prieto, J. M. (2009). *Journal of Materials Processing Technology, 209,* 1803–1809.
4. Cann, M. C., & Connelly, M. E. (2000). *Real-world cases in green chemistry*. Washington, DC: American Chemical Society.
5. Chang, H., & Abu-Zahra, N. (2011). *Journal of Vinyl and Additive Technology, 17,* 17–20.
6. Chanprateep, S. (2010). *Journal of Bioscience and Bioengineering, 621–632*, 1389–1723.

7. Chen, Z., Liu, J., & Tao, J. (2007). *Progress in Chemistry, 19,* 1919–1927.
8. Chen, G. G. Q., & Wu, Q. (2005). *Applied Microbiology and Biotechnology, 67,* 592–599.
9. Clark, J. H. (2007). *Journal of Chemical Technology and Biotechnology, 82,* 603–609.
10. Falletta, E., Della Pina, C., Rossi, M., He, Q., Kiely, C. J., & Hutchings, G. (2011). *Journal of Faraday Discussion, 152,* 367–379.
11. Fortun, J. M., & Confalone, P. N. (2007). *Current Opinion in Drug Discovery and Development, 10,* 651–653.
12. Fortunak, J. M. (2009). *Future Medicinal Chemistry, 1,* 571–575.
13. Harrington, P. J., Brown, J. D., Foderaro, T., & Hughes, R. C. (2004). *Organic Process Research and Development, 8,* 86–91.
14. Hilonga, A., Kim, J. K., Sarawade, P. B., Quang, D. V., Shao, G. N., Elineema, G., & Kim, H. T. (2012). Korean Journal of Chemical Engineering (in press).
15. Hocking, P. J., & Marchessault, R. H. (1994). *Biopolyesters.* In G. J. L. Griffin (Ed.), *Chemistry and technology of biodegradable polymers.* Glasgow: Blackie Academic and Professional.
16. Hu, S., Kelly, S., Lee, S., Tao, J., & Flahive, E. (2006). *Organic Letters, 8,* 1653–1655.
17. Jimenez-Gonzalez, C., & Constable, D. J. C. (2002). *Clean Technologies and Environmental Policy, 4,* 44–53.
18. Jimenez-Gonzalez, C., Curzous, A. D., & Constable, D. J. C. (2001). *Clean Technologies and Environmental Policy, 3,* 35–41.
19. Kiss, A. A., Dimian, A. C., & Rothenberg, G. (2006). *Advanced Synthesis and Catalysis, 348,* 75–81.
20. Li, S., Yan, W., & Zhang, W. X. (2009). *Green Chemistry, 11,* 1618–1626.
21. Liu, A. H., Li, Y. N., & He, L. N. (2012). *Pure and Applied Chemistry, 84,* 581–602.
22. Lovell, M. R., Kabir, M. A., Menezes, P. L., & Higgs III, C. F. (2010). *Philosophical Transactions of the Royal Society A: Mathematical Physical and Engineering Sciences, 368,* 4851–4868.
23. Ma, S. K., Gruber, J., & Davis, C. (2010). *Green Chemistry, 12,* 81–86.
24. Manley, J. B., Anastas, P. T., & Cue Jr., B. W. (2008). *Journal of Cleaner Production, 16,* 743–750.
25. Menzler, N. H., Schafbauer, W., & Buchkremer, H. P. (2010). *Materials Science Forum, 638-642,* 1098–1105.
26. Pallavkar, S., Kim, T. H., Lin, J., Hopper, J., Ho, T., Jo, H. J., & Lee, J. H. (2010). *Industrial & Engineering Chemistry Research, 49,* 8461–8469.
27. Pirzada, T., Demirdoogen, R. E., & Shah, S. S. (2012). *Journal Chemical Society of Pakistan, 34,* 177–183.
28. Rekha, R., & Sudam, A. (2009). *Man Made Textiles in India,* **52**, 397–401.
29. Scott, A. (2009). *Chemical Week, 171*(26), 24.
30. Shanmuganathan, K., Sankhagowit, R. K., Iyer, P., & Ellison, C. J. (2011). *Journal of Materials Chemistry, 23,* 4726–4732.
31. Sheldon, R. A. (2005). *Green Chemistry, 7,* 267–278.
32. Slater, C. S., & Savefski, M. (2007). *Journal of Environmental Science and Healt. Part A: Environmental Science and Engineering & Toxic and Hazardous Substance Control, 42,* 1595–1605.

33. Slater, C. S., Savelski, M., Hounsell, G., Pilipauskas, D., & Urbanski, F. (2012). *Clean Technologies and Environmental Policy, 14,* 687–698.
34. Steinbüchel, A., & Lütke-Eversloh, T. (2003). *Biochemical Engineering, 16,* 81–96.
35. Suriyamongkol, P., Weselake, R., Narine, S., Moloney, M., Shah, S. (2007). *Biotechnology Advances, 25,* 148–175.
36. Tao, J., & Xu, J. H. (2009). *Current Opinion in Chemical Biology, 13,* 43–50.
37. Tu, Y. J., Chang, C. K., & You, C. F. (2012). *Journal of Hazardous Materials, 229–230,* 258–264.
38. Yang, C., Zhao, W., Wang, H., & Zhang, X. L. (2012). *Advanced Materials Research, 518–523,* 3908–3911.
39. Zhang, Y. L., Zhang, M., Shen, Z., Zhou, J. F., & Zhou, X. F. (2012). *Journal of Chemical Technology and Biotechnology*. (In Press)
40. Zhuo, C., & Hall, B., Levendis, Y., Richter, H. (2011). *Materials Research Society Symposia Proceedings, 1317,* 177–183.

## CHAPTER 14

# PRESENT SCENARIO AND FUTURE TRENDS

SURESH C. AMETA

## CONTENTS

## 14.1 INTRODUCTION

Most of the green starting materials either come from biomass or biomass derived materials, but it is not possible to fulfill all the requirements of useful chemicals in the society, like food materials, polymers, textiles, healthcare products, cosmetics, energy sources, detergents, pesticides, paints, and so on, may not be prepared by only biomass or biomass derivatives. It is, therefore, necessary to find out newer green alternatives for starting materials and this may be the future prospects in this direction.

All the products used may not be biodegradable or may be recycled and thus create a problem of their disposal. Wood, iron, textile, and so on are being rapidly replaced by some or the other kind of polymers and therefore, this is called a polymer era. This may be appreciated on one hand but on the other hand it also generates large amount of waste. Therefore, there is an urgent need to search for biodegradable polymers. The same is true for detergents, which are almost replaced by washing soap. It is predicted that more and more products will come in future, which are either biodegradable or degrade into less harmful products.

Most of the chemical reactions are slow enough and require catalyst for the enhancement of their reaction rate. Catalysts available at present are either metal or metal based, which are toxic in nature. At this stage, enzymes are more preferred to drive a reaction but these are quite specific or have their own limitations, so, searching newer green catalysts is a demand of the day, which are specific in catalyzing reactions and are less toxic as compared to existing conventional catalysts. Next few decades will witness some more green catalysts.

Volatile organic solvents create a lot of pollution by their vapors. Thus, there is an urgent requirement of some high boiling solvents, which will dissolve majority of organic and inorganic reactants. This requirement is fulfilled by ionic liquids. One of the beauties of ionic liquid is that it can be designed as per our requirement. These are truly designated as designer solvents. Future will see some more interesting designed ionic liquids.

Some chemical reactions can now be carried out in supercritical carbon dioxide ($scCO_2$) and supercritical water ($scH_2O$); however, the conditions are quite cumbersome. It is hope that these supercritical fluids find a proper place in the field of organic synthesis in years to come.

Water is a universal as well as a green solvent. Many other green solvents like cyclopentyl methyl ether (CPME), 2-methyltetrahydrofuran, polyethylene glycol, 1, 3-dioxolane, and so on have find their place as green solvents, but the search is still incomplete. Efforts are being made to carry out many more reactions in water and other green solvents. Sidewise newer green solvents may be searched out, which fulfill the demand of any chemical reaction.

Photocatalysis is an emerging technology for waste water treatment. It is an advance oxidation process, which completely mineralizes most of the organic pollutants. It provides an electron-hole pair, which can be used for both; oxidation and reduction. Newer photocatalyst can be synthesized with different energy levels of conduction and valence bands or available semiconductor can be modified by sensitization, metallization or doping to have better photocatalytic activities.

Photo-Fenton reagent is an eco-friendly reagent as ferrous and ferric ions are recycled. This reagent is more active than Fenton reagent. It also degrades number of pollutants. There is a lot more scope in developing photo-Fenton like reagents, where iron is replaced by some other metal ion with variable oxidation states.

Ultrasound has found many applications but it has been used to carry out chemical reactions of synthetic importance only a few decades back. It creates very high temperature and pressure to drive the chemical reactions. Its combination with some other advanced oxidation processes may add a new chapter in the field of combating against ever increasing water pollution. We can also see some more sonochemical reactions of importance in organic synthesis in years to come.

Microwave assisted organic synthesis (MAOS) has opened an avenue in the field of organic synthesis. As this type of reaction is less time consuming, provides more yield with greater purity and needs less solvent or it is almost solvent free. Majority of reactions have been carried out under microwave irradiation and the list is becoming quite long. MAOS has been successfully used in laboratory conditions and in some cases, even on industrial scale and time is not far off, when MAOS may substitute majority of organic synthesis.

Nature has provided us biocomposites and there are good examples of green composites like networking of pipeline of water supply in plants, spider web, and so on. Green composites have combined properties of two materials of different properties like hardness, softness, flexibility, and so on, but most important is that these composites should be biodegradable. Some more new green composites will find their place in the coming decades.

Green chemical processes have many established examples on laboratory scale but only limited examples are available on industrial scale. It is utmost necessary to bring these laboratory exercises and prove their worth on industrial level, the so called green manufacturing processes. While designing a green manufacturing process, one has to take care of environment along with its economic viability so as to make this process acceptable by the society.

## KEYWORDS

- **Biocomposites**
- **Biodegradable polymers**
- **Green solvents**
- **Green chemical processes**
- **Sonochemical reactions**

# INDEX

## D

## N

## O

## P

## Q



## R



## S

## T

## U



## V



## W

## Y



## Z

Printed in the United States
by Baker & Taylor Publisher Services